a Novel Capacitor-less DRAM Cell

FLOATING BODY CELL

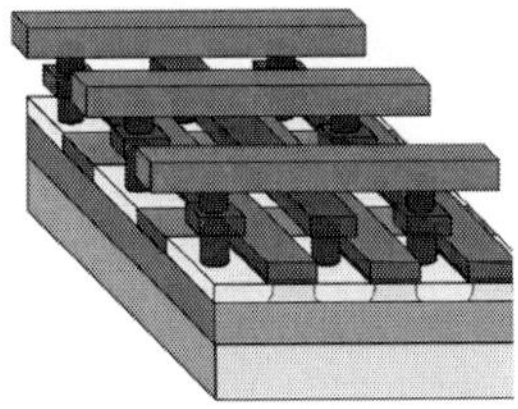

Takashi Ohsawa
Takeshi Hamamoto

a Novel Capacitor-less DRAM Cell

FLOATING BODY CELL

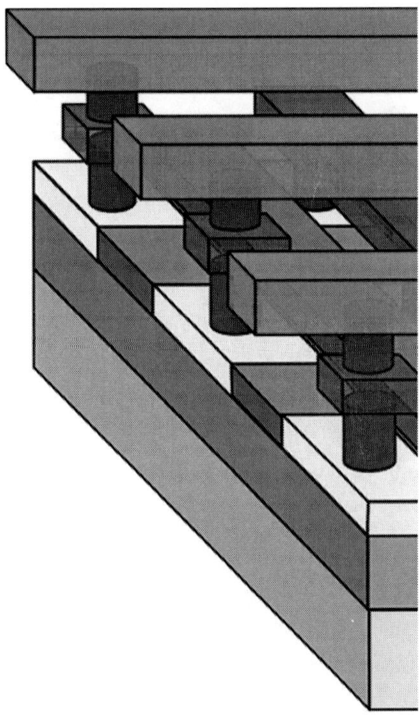

PAN STANFORD PUBLISHING

Published by

Pan Stanford Publishing Pte. Ltd.
Penthouse Level, Suntec Tower 3
8 Temasek Boulevard
Singapore 038988

Email: editorial@panstanford.com
Web: www.panstanford.com

British Library Cataloguing-in-Publication Data
A catalogue record for this book is available from the British Library.

Floating Body Cell: A Novel Capacitor-less DRAM Cell

ISBN 978-981-4303-07-1 (Hardcover)
ISBN 978-981-4303-08-8 (eBook)

Printed in the USA

Contents

Preface

Floating body cell (FBC) is one of the simplest semiconductor memory cells ever proposed, which consists of just a single metal–oxide–semiconductor field-effect transistor (MOSFET) with its body floating. The flash memory cell also consists of a single MOSFET having a floating gate between the gate and the channel. And the both cells rely on a very similar data writing mechanism, i.e., they change the threshold voltage of a MOSFET as a means of memorizing data; the flash cell does it through electrical charges stored in the floating gate, and FBC does it through those stored in the floating body. However, the mere difference in data storage node in a MOSFET makes the flash a non-volatile data storage memory and FBC a volatile random access memory (RAM).

Though the idea of storing data in a floating body of a MOSFET is rather old, the FBC research and development have become especially active since the year 2000 when a useful selective write method by controlling the gate and the drain voltages was invented. FBC consists of a single-transistor cell structure on silicon-on-insulator (SOI) without a capacitor which is used for storing data in the conventional one-transistor and one-capacitor (1T-1C) dynamic random access memory (DRAM). The cell size of FBC is about a half of the conventional DRAM cell. The process to fabricate FBC is almost compatible with the standard CMOS one. These features are attractive especially for embedded memories, for a memory with density larger than the DRAM can be implemented into logic LSIs as freely as many other standard libraries just like the static random access memory (SRAM) without suffering from the difference in process between making the logic transistors and the FBC cells. Furthermore, it can even break through the scaling problems the present-day DRAM is struggling to surmount, because the scaling of FBC with no explicit capacitor to store data is much easier than the DRAM cell.

Many efforts have been made and are still under way for FBC to be phased into real memory products of embedded memory macros and of stand-alone memories alike in a circumstance where the conventional 1T-1C DRAM is suffering from being scaled further. In some cases, the SOI substrate which FBC is necessary to be built on became a bottleneck from a chip cost point of view. In other cases, the data retention time which is shorter than that of the 1T-1C DRAM cell became a showstopper of the project for implementing the small and easy-to-make cell into a stand-alone memory. However, recent proposals of new FBC structures permit the cells to be built on bulk silicon substrate. A new report encourages further efforts to optimize the device structures for the retention time to meet the stand-alone DRAM specifications. Furthermore, the autonomous refresh, an FBC-specific refresh operation, has been shown to contribute to improving the data retention characteristics, if successfully applied.

This monograph describes a broad spectrum of FBC technologies from the basic principles to its memory design practices, providing a comprehensive summary of the FBC process, devices, circuits, and memory designs. In Chapter 1, the 1T-1C DRAM cell technologies are briefly reviewed with their difficulties in scaling remarked, followed by a short history of capacitor-less DRAM cells. In Chapter 2, the concept and the operational principles of FBC is explained with TCAD simulation results which verify the principles. The write selectivity is explained in detail in relation to bit line disturb. A comprehensive summary of write and read methods for FBC is given. In Chapter 3, the mechanism of generating the signal of FBC is analyzed and the signal-to-noise ratios (SNRs) are defined with their criteria for achieving a high-density memory estimated theoretically. In Chapter 4, a 128 Mb memory using FBC is introduced with various data shown for verifying the prediction by the theories. The experimental data on retention time are also presented including its degradation by the bit line disturb. In Chapter 5, the scaling of FBC is discussed. A scaling law in which the SNR is kept constant is shown to hold by TCAD Monte Carlo simulations. In Chapter 6, cell array architectures and sense amplifier designs are explained by comparing the two array architectures, the twin cell and the single cell schemes. A useful dummy cell system to help enhance the chip yield is explained with simulated and experimental data. An attractive array architecture specific to FBC is presented which mitigates inter bit line coupling noise without performance degradation. In Chapter 7, two design practices are discussed and compared; one is a twin-cell-based high-speed embedded memory and the other is a single-cell-based high-density memory. The unique autonomous refresh which can make all cells in an array refreshed at once is explained in detail. As a special example of the autonomous refresh, it is theoretically predicted that FBC can be an SRAM cell if the gate direct tunneling current is used to extract holes from the body with the impact ionization current combined to realize two stably stationary points in retaining data. In Chapter 8, past activities in FBC technology research and development are summarized comprehensively, followed by categorizing them into several high-potential variations. Finally, future directions of FBC technologies are discussed.

Senior undergraduate and graduate students in electrical engineering and microelectronics who are learning SOI technologies will grasp many practical aspects realized in electrical characteristics of a floating body MOSFET through understanding the FBC's operational principles and features described in this book. It also provides memory design practices in device and circuits, which will be useful for engineers and scientists involved in research and development of materials, process, device, and circuits for emerging memories. The device or circuit engineers who are striving to make FBC usable in a real memory product will understand this particular cell's design points along with its potentials and constraints. We hope that this book will help contribute to a possible breakthrough to make the emerging memory cell be a reality.

Many colleagues we worked with have contributed to the contents of this book. We would like to take this opportunity to thank them all for their contributions to the fruitful results which have grown the FBC technology as is today and is still in progress. We are quite indebted to the following individuals for process, designs, simulations, experiments, measurements, tests, and supports: Nobutoshi Aoki, Yoshiaki Asao, Toshiyuki Enda, Shuso Fujii, Katsuyuki Fujita, Ryo Fukuda, Yoshiaki Fukuzumi, Hironobu Furuhashi, Tohru Furuyama, Kosuke Hatsuda, Tomoki Higashi, Fumio Horiguchi, Tamio Ikehashi, Kazumi Inoh, Hidemi Ishiuchi, Yoshihisa Iwata, Yasuyuki Kajitani, Takeshi Kajiyama, Shigeru Kawanaka, Yusuke Kohyama, Naoki Kusunoki, Fumiyoshi Matsuoka, Yoshihiro Minami, Mutsuo Morikado, Hiroomi Nakajima, Akihiro Nitayama, Yukihito Oowaki, Tomoaki Shino, Kazumasa Sunouchi, Hiroyoshi Tanimoto, Testunori Wada, Shigeyoshi Watanabe, Yohji Watanabe, Takashi Yamada, and Makoto Yoshimi.

Takashi Ohsawa
Takeshi Hamamoto
Yokohama, Japan
February 2011

Chapter 1

Introduction

Technologies of the one-transistor and one-capacitor (1T-1C) dynamic random access memory (DRAM) bit cell are briefly summarized along with serious problems they are facing in scaling further, followed by a history of single-transistor DRAM cells which have been proposed for gaining performance and potential scalability over the 1T-1C DRAM cell.

1.1 OPERATIONAL PRINCIPLE OF ONE-TRANSISTOR AND ONE-CAPACITOR DYNAMIC RANDOM ACCESS MEMORY

The dynamic random access memory (DRAM) which is currently being produced has the bit cell which consists of a capacitor for storing data and a metal–oxide–semiconductor field-effect transistor (MOSFET) for selecting the bit whose equivalent circuit is shown in Fig. 1.1.[1]

It is commonly called 1T-1C DRAM cell. The 1T-1C DRAM cell is much smaller in size than six-transistor cell of the static random access memory (SRAM), another random access memory which is used in high speed applications. The small number of elements which constitute the bit cell in DRAM is one of the main reasons why the 1T-1C DRAM is used in main memories of mainframes, workstations, and personal computers for more than 30 years[2] and recently in working memories of mobile phones and digital electrical home appliances, even though it is not necessarily easy to use because of cumbersome refresh operation which is necessary for retaining data. And its importance as a key device which supports recent information-oriented society remains unchanged today, achieving 2 Gbit products using 50 nm minimum feature size. These days, furthermore, DRAM has found a way out to embedded applications in various logic LSIs, providing large memories with high performance and low power as well.[3] One of the important applications is seen in the on-chip DRAM cache

Floating Body Cell: A Novel Capacitor-less DRAM Cell
By Takashi Ohsawa and Takeshi Hamamoto

www.panstanford.com

memories for microprocessors which was traditionally an exclusive territory of SRAM.[4]

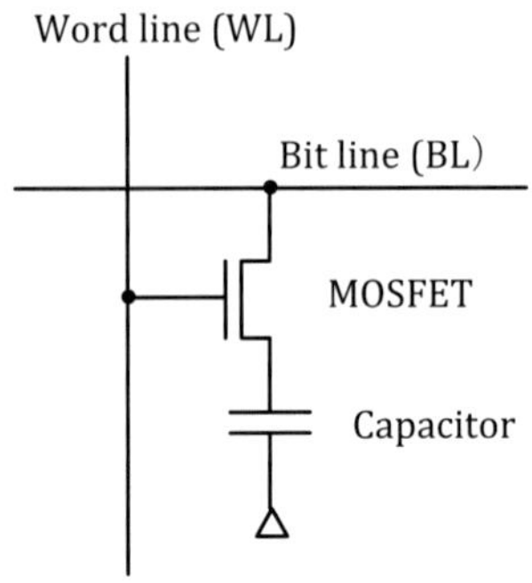

Figure 1.1 Equivalent circuit of the 1T-1C DRAM cell.

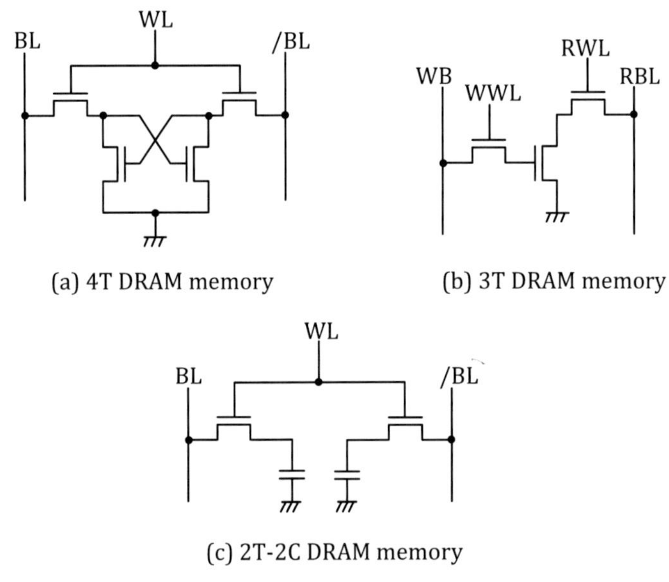

Figure 1.2 Transition of DRAM cells (excluding the 1T-1C cell).

Here, we give a glimpse into a history of DRAM cells. The four-transistor cell in flip-flop configuration was invented at almost the same time as the 1T-1C DRAM cell's proposal, the equivalent circuit of which being shown in Fig. 1.2a.[5] The cell develops large signal because it has a function of amplifying signal in itself. And all cells in an array can be refreshed simultaneously by raising word line (WL) with a pair of bit lines (BL and /BL) kept at a high voltage, increasing drastically the refresh efficiency with low retention power and busy rate. The three-transistor cell shown in Fig. 1.2b was then presented at a conference.[6] This is the first commercialized cell in the *i1103*

1 kbit DRAM by *Intel* in 1970. Though it has an ability to draw current in itself,[a] refresh operation is required in which a sense amplifier circuit (S/A) reads the signal stored in a cell and restores the amplified data back to the cell before leakage current degrades the data too much to be detectable by S/A because it has no ability to refresh itself as the four-transistor cell has. The 1T-1C cell has been standard from 4 kbit generation. And it is possible to write complementary data into a pair of 1T-1C cells (two-transistor and two-capacitor) to enhance the signal amount, as is shown in Fig. 1.2c.

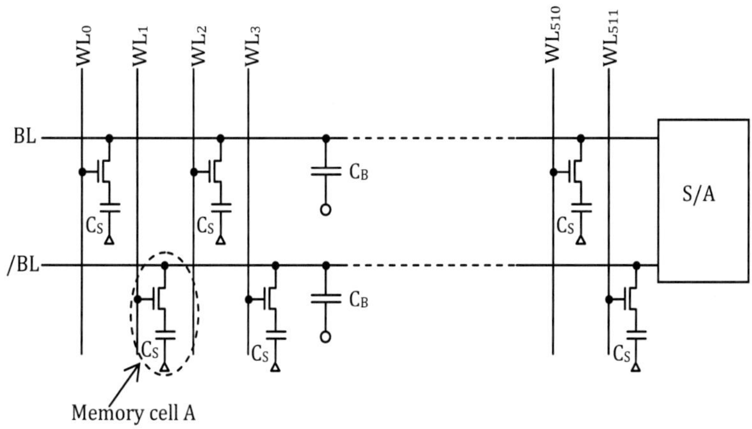

Figure 1.3 DRAM cell array and sense amplifier (S/A).

Now let us consider the data write and read operations of the 1T-1C DRAM cell. Figure 1.3 shows schematically a column in a two-dimensional cell array which consists of 1T-1C DRAM cells wired by a pair of bit lines BL and /BL and word lines $WL_0 \sim WL_{511}$. A S/A is connected to the pair of bit lines. Although this schematic depicts only 512 cells connected a S/A, 4096 pairs of bit lines and S/As, for example, are arranged to shape a 512 × 4096 = 2 Mbit cell array except for some redundant cells which are designed to replace failed cells. A stand-alone memory chip or an embedded memory macro is composed of multiple cell arrays. The capacitance of the cell capacitor and the bit line are denoted by C_S and C_B, respectively. All bit lines including BL and /BL are precharged to $0.5 \times V_{AA}$ prior to write or read operation by a circuit which is not shown, where V_{AA} is an array voltage which is the maximum positive voltage applied to BLs. All word lines $WL_0 \sim WL_{511}$ are set at 0 V prior to read operation, making all MOSFETs in cells cut off. First, the operation to write data to the cell A is explained. WL_1 is raised from 0 V to a voltage larger than $V_{AA} + V_{th}$, making the MOSFET in the cell A turn on, where V_{th} stands for the threshold voltage of the MOSFET. Then, V_{AA} or 0 V is applied to /BL

[a]The ability is called "gain."

depending on "1" write or "0" write, respectively, where "1" is defined as the state with high voltage stored in the capacitor of the cell and "0" is defined as the state with 0 V stored. After writing data, WL_1 is returned back to 0V, followed by BL's return to $0.5 \times V_{AA}$. This completes the write operation and 0 V WL cuts off the MOSFET, holding the data within the storage node (the capacitor node where V_{AA} or 0 V is written) isolated from /BL. Then, the operation to read the data from the cell A is explained. After both BL and /BL are left floating (the voltages must be kept at the precharged level $0.5 \times V_{AA}$), WL_1 is raised to $V_{AA} + V_{th}$. This operation begins to redistribute electrical charge between /BL and the cell's storage node. If we assume that data "1" was written to the storage node, the charge stored in the storage node was $C_S \times V_{AA}$. And the charge stored in the bit line /BL was $C_B \times 0.5 \times V_{AA}$. Since the charge is redistributed to the node with capacitance $C_S + C_B$, according to the law of conservation of charge, we have

$$C_s \times V_{AA} + C_B \times 0.5 \times V_{AA} = (C_S + C_B) \times V_1, \tag{1.1}$$

i.e.,

$$V_1 = \frac{C_S + 0.5 \times C_B}{C_S + C_B} V_{AA}, \tag{1.2}$$

where V_1 is an equilibrium voltage of /BL and the cell storage node for data "1" read. Since the voltage of BL, another bit line which consists of the pair connected to a S/A, is left floating at $0.5 \times V_{AA}$, the voltage difference ΔV_{B1} between BL and /BL for data "1" read is expressed as

$$\begin{aligned}\Delta V_{B1} &= V_1 - 0.5 \times V_{AA} \\ &= \frac{C_S + 0.5 \times C_B}{C_S + C_B} V_{AA} - 0.5 \times V_{AA} = \frac{0.5 \times C_S}{C_S + C_B} V_{AA}.\end{aligned} \tag{1.3}$$

Similarly, if we assume that data "0" is written to the storage node of the cell A, the voltage V_0 which appears at the bit line /BL after raising the word line WL_1 is calculated from the law of conservation of charge

$$C_B \times 0.5 \times V_{AA} = (C_S + C_B) \times V_0, \tag{1.4}$$

as

$$V_0 = \frac{0.5 \times C_B}{C_S + C_B} V_{AA}. \tag{1.5}$$

The voltage difference ΔV_{B0} between BL and /BL for data "0" read is, therefore, expressed as

$$\Delta V_{B0} = 0.5 \times V_{AA} - V_0$$
$$= 0.5 \times V_{AA} - \frac{0.5 \times C_B}{C_S + C_B} V_{AA} = \frac{0.5 \times C_S}{C_S + C_B} V_{AA}. \tag{1.6}$$

As calculated, the signal which appears between BL and /BL is independent of the data polarity and is written to be

$$\Delta V_B = \Delta V_{B1} = \Delta V_{B0} = \frac{0.5 \times C_S}{C_S + C_B} V_{AA}, \tag{1.7}$$

which can be rewritten to

$$\Delta V_B = \frac{0.5}{1 + \lambda} V_{AA}, \tag{1.8}$$

where λ is the ratio of bit line total capacitance divided by the cell capacitance defined by

$$\lambda = \frac{C_B}{C_S}. \tag{1.9}$$

Equation 1.8 indicates that the read signal increases as λ decreases and/or V_{AA} increases. The dependence of ΔV_B on λ indicates that the signal increases as the cell capacitance increases with the bit line capacitance kept constant.

The following features can be easily understood based upon the operational principle of write and read operations for the 1T-1C DRAM cell which has been explained above:

1. The cell is a destructive read-out cell, i.e., the cell data is destroyed during read. Therefore, the read data needs to be restored to the cell in each read cycle.
2. Since the reference voltage to distinguish data "1" and "0" is provided automatically by the BL in the pair which is not connected to the selected cell, a dummy cell or reference cell is not necessary.
3. The WL must be raised at a high voltage V_{PP} which is larger than $V_{AA} + V_{th}$ when writing the data. Since this V_{th} should be taken as the value which is modified by body effect due to the source voltage of V_{AA} when writing "1," the value V_{PP} is much higher than V_{AA}, leading to gate dielectric reliability concern.
4. Decreasing V_{AA} leads to signal decrease.
5. If the number of cells per bit line increases, the chip size favorably decreases, because the number of S/As decreases. However, the signal decreases because C_B increases.
6. Decreasing C_S leads to signal decrease.

1.2 CHALLENGES IN SCALING 1T-1C DRAM CELL

It is getting more and more difficult for the 1T-1C DRAM cell to be scaled.[7] In fact, at the moment, no clear-cut method has been found to scale it down to less than 40 nm minimum feature size. When scaling the 1T-1C DRAM cell, it is necessary to keep the cell capacitance C_S constant. This scaling guideline directly results from Eq. 1.8. Although the bit line capacitance per cell decreases as generation proceeds, there is a tendency that the number of cells per a bit line increases to make the cell efficiency larger or to make the chip size smaller. Therefore, C_B is kept almost constant of generation on average. V_{AA} is gradually decreasing with generation. However, recently it tends to be saturated and almost constant of generation. As a consequence, Eqs. 1.8 and 1.9 tell that C_S must be kept constant for the read-out signal to be kept constant. It is one of the biggest challenges in scaling the 1T-1C DRAM cell that a capacitor with the same capacitance as a previous generation must continue be put into a smaller cell footprint in successive generations.

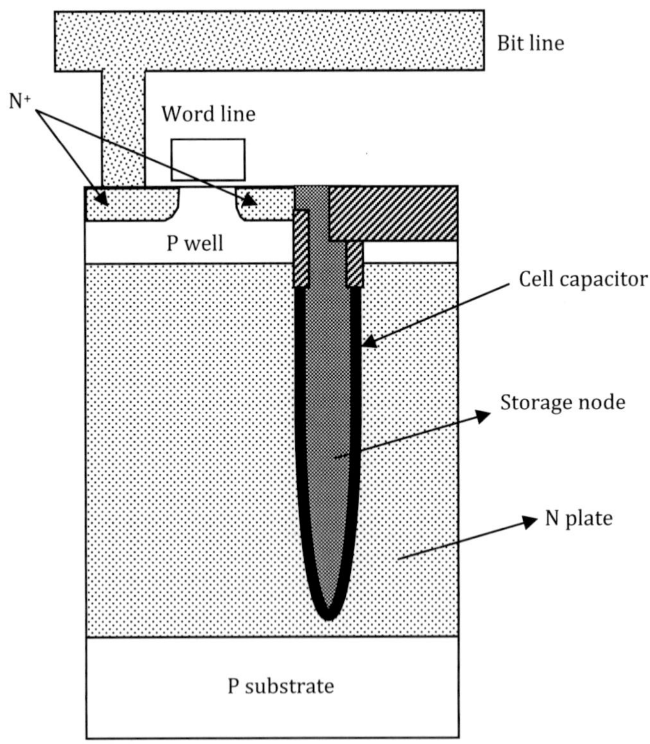

Figure 1.4 Deep trench capacitor cell.

Up to now, engineers have solved the problem by increasing the capacitor area in three-dimensional structures such as a deep trench capacitor or a stacked capacitor. Figures 1.4 and 1.5 show schematically the deep trench capacitor cell and the stacked capacitor cell, respectively. They have been introduced into products from the 4 Mbit generation ca. 1987. The trench capacitor cell enhances its area by digging a capacitor hole[8] and the stacked capacitor cell does it by building up capacitor nodes.[9] It is worth noting that the structure depicted in Fig. 1.5 is one called capacitor-over-bit line where the storage node is put over the bit line and that the storage node contact must be staggered so that it comes between adjacent bit lines.[10,11] These revolutionarily structural changes of the 1T-1C DRAM cell help contribute the evolution of the 1T-1C DRAM cell generations and both cells are used even today. However, their method of increasing the area has some limitations because the trench must be dug deeper and deeper and the stack must be build higher and higher as generation proceeds. It is said that dielectric material with higher dielectric constant must be introduced in the near future to the capacitor, increasing the concern that the DRAM development speed will be slowed down due to the lead time for the development of exotic materials including those for the electrode.

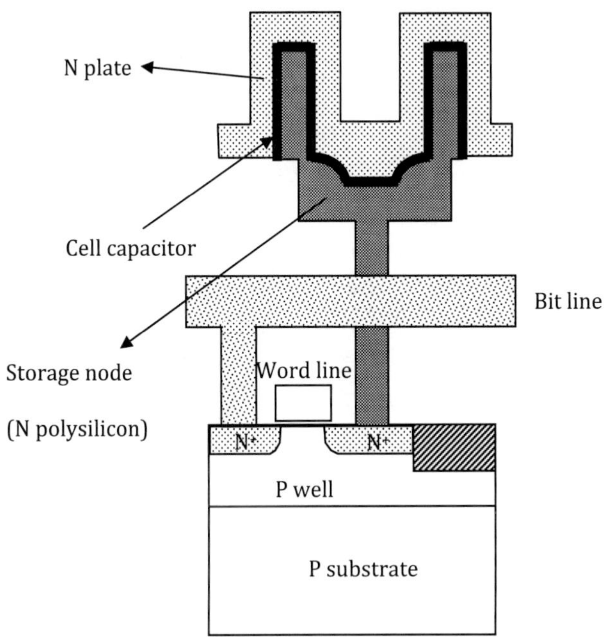

Figure 1.5 Stacked capacitor cell.

Another problem which deteriorates the scalability of the 1T-1C DRAM cell is the noise due to capacitive coupling between adjacent bit lines which decreases the read-out signal of the 1T-1C DRAM cell as cell size scales. There is no doubt that the scaling trend of the 1T-1C DRAM cell has been achieved by shrinking the minimum feature size F which defines the smallest pattern size of the cell's components usually represented by the bit line half pitch (blhp). However, the shrinking of F alone hasn't been enough for the cell to follow the trend completely.

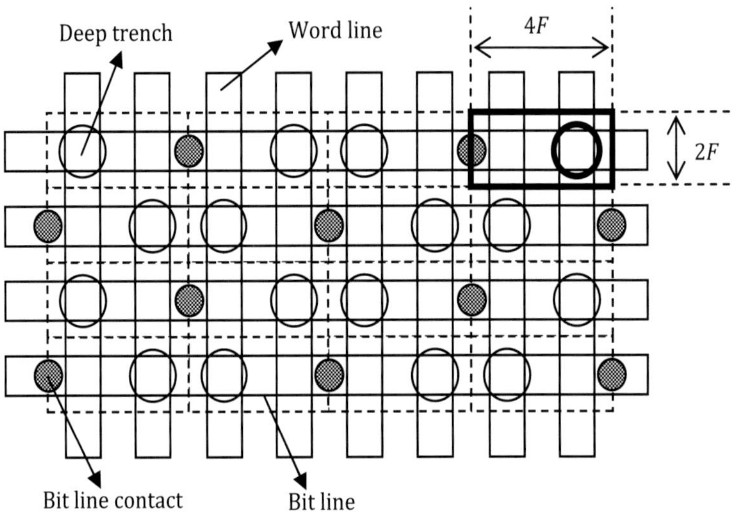

Figure 1.6 Array arrangement of $8F^2$ 1T-1C deep trench DRAM cell.

Therefore, the 1T-1C DRAM cell has changed its configuration so that the cell size shrinks from $8F^2$ to $6F^2$. The necessity of realizing the $4F^2$ size cell is presumed in the near future. The shrinkage of the cell area factor[b] from 8, 6 to 4 combined with the shrinkage of F itself have achieved the DRAM cell size trend and will follow the trend. Figures 1.6–1.8 show the cell arrays using the $8F^2$, $6F^2$, and $4F^2$ deep trench cell, respectively. It is understood that the cell array becomes denser as the cell area factor decreases from 8, 6 to 4 even with the same minimum feature size F. The area permitted for the cell capacitor decreases as the cell area factor shrinks, making the capacitor put under the transistor and the bit line contact for the ultimate $4F^2$ cell. The process to make the cell, therefore, becomes more difficult as the cell area factor shrinks. One important difference of the $4F^2$ cell and the $6F^2$ cell from the $8F^2$ cell is the way by which data appears on bit lines. Data appears on every other bit lines in $6F_2$ the $4F^2$ cell array, while they appear on all bit lines in the and $6F^2$ the $4F^2$ cell arrays. Therefore, S/As must be arranged as in Fig. 1.9 for

[b]When the cell size is expressed in the form of αF^2, α is called the cell area factor.

the $8F^2$ cell and as in Fig. 1.10 for the $6F^2$ and $4F^2$ cells. The former is called the folded bit line architecture and the latter is called the open bit line architecture. The pitch of bit line wirings gets small as cell size shrinks. However, the

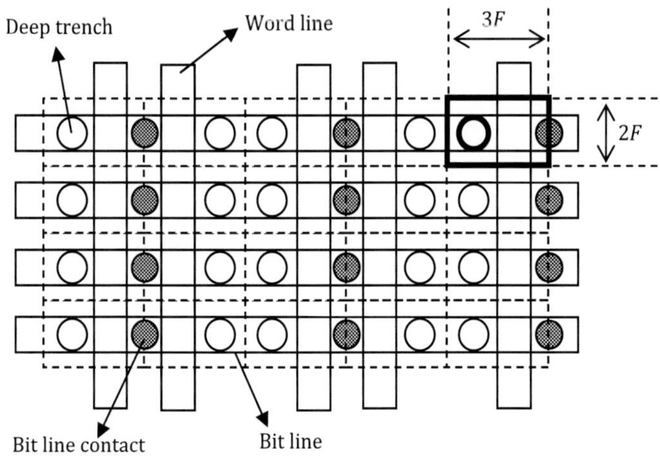

Figure 1.7 Array arrangement of $6F^2$ 1T-1C deep trench DRAM cell.

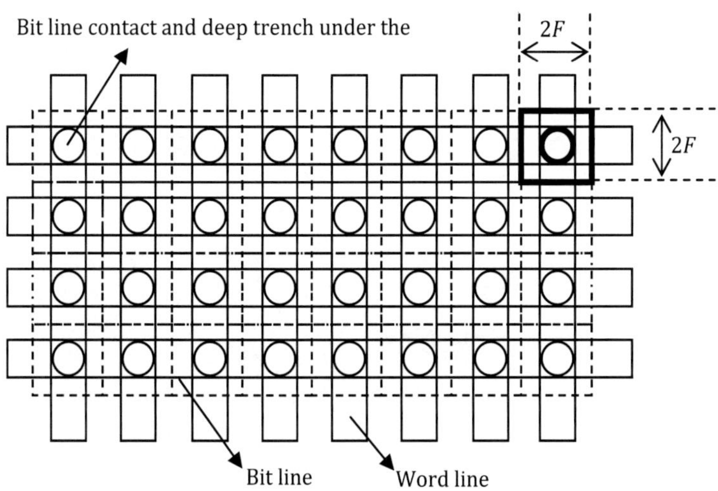

Figure 1.8 Array arrangement of $4F^2$ 1T-1C deep trench DRAM cell.

height of bit lines does not become small in the same scaling factor as the space of bit lines to avoid the increase of wiring resistance, making the ratio of component of bit line capacitance coupled to adjacent bit lines divided by the total bit line capacitance increases as cell size shrinks. This situation leads to the increase of the noise due to capacitance coupling between adjacent bit lines. This noise decreases the read-out signal from the value calculated by Eq. 1.8.

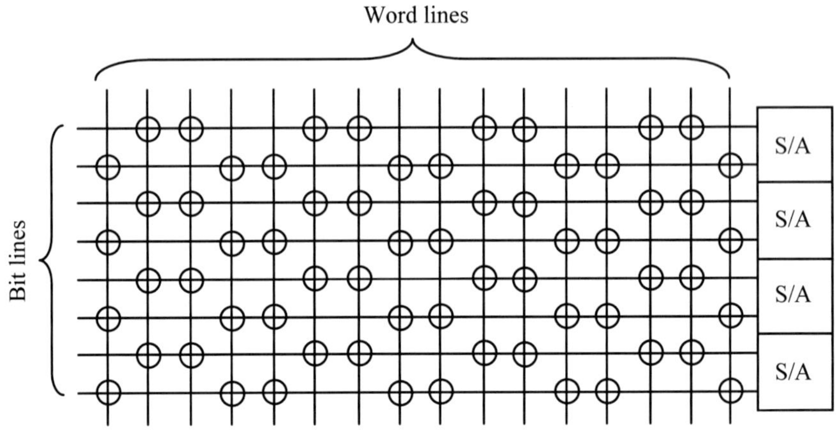

Figure 1.9 Folded bit line architecture.

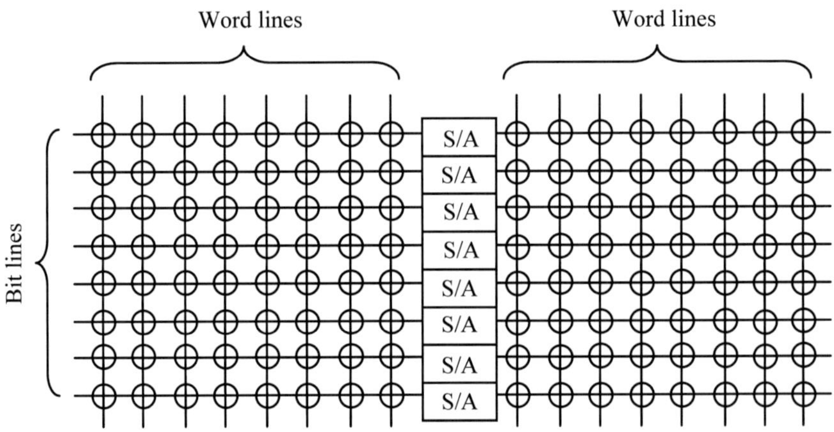

Figure 1.10 Open bit line architecture.

This problem has been solved in many products by introducing bit line twist techniques.[12,13] There are many variations in the bit line twist techniques. Figure 1.11 shows only two of them. However, the bit line twist techniques are not effective for the open bit line architecture. Therefore, the problem of the noise due to capacitance coupling between adjacent bit lines will capture the attention when the cell factor shrinks to 6 or smaller. If we must solve the problem without twisting bit lines, there will be no other way than increasing C_S. This brings about another big challenge in scaling the 1T-1C DRAM cell.

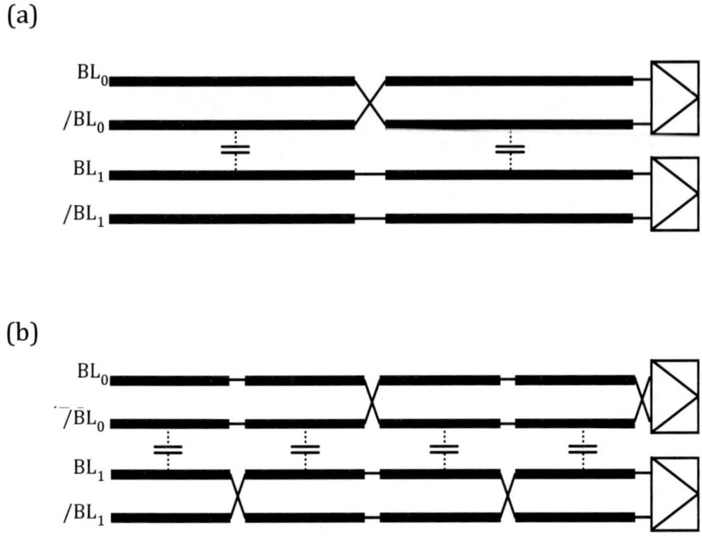

Figure 1.11 Two examples of bit line twist techniques.

1.3 HISTORY OF SINGLE-TRANSISTOR DRAM CELLS

There were a variety of cells proposed with the intention to replace the 1T-1C DRAM cell which has been used since the first half of the 1970s. Especially, many gain cells have been proposed which have abilities to draw current within the cells on the contrary to the 1T-1C DRAM cell which determines the signal by sharing the charge between the capacitors. Although there was a gain cell proposed for a DRAM cell which consists of three transistors before the domination of the 1T-1C DRAM cell,[6] there have been several single-transistor gain cells proposed for the purpose of replacing the 1T-1C DRAM cell. Since some of them have very similar concept as that of the floating body cell (FBC), here the history of single-transistor cells will be looked back on.

Figure 1.12 is a single-transistor cell named as taper isolated dynamic RAM cell which was presented by P. K. Chatterjee *et al.* in 1979.[14] This has a structure that separates p-type body from p-type substrate by n-type band in an n-type MOSFET, storing binary data by accumulating holes in the floating body. The n-type region which separates the body from the substrate acts also as a buried channel of the n-type MOSFET. When holes are accumulated in the body, the threshold voltage of the MOSFET decreases due to body effect and vice versa. The data "1" defined by the state where more holes are

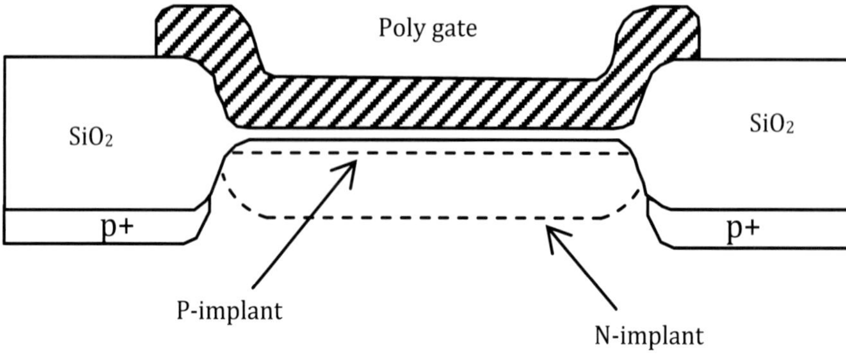

Figure 1.12 Concept of a single-transistor cell proposed by P. K. Chatterjee *et al.*[14]

accumulated in the floating body can be distinguished from the data "0" with fewer holes accumulated in the body by comparing the drain current difference between the MOSFETs with different data states at the same conditions of the drain and gate voltages. Writing data to the cell is performed by giving a short positive or negative voltage pulse to the gate, making the parasitic p-n-p bipolar transistor consisting of the body (p), the separating band (n), and the substrate (p) turned on. When the voltage pulse is triggered, the bipolar transistor is not turned on if the source and the drain voltages are fixed at a high voltage, inhibiting the writing operations. This realizes the write selectivity. This was an epoch-making single-transistor DRAM cell with small cell size with very few number of control wires. However, it has never been phased into products, probably because there remained high potential in scaling the 1T-1-C DRAM cell with little necessity of alternative technology for DRAM cells in those days.

Figure 1.13 shows a single-transistor memory cell built in silicon-on-sapphire which was proposed by N. Sasaki in 1981.[15] This cell stores data by changing the electrical charge amount accumulated in a floating body of a p-type MOSFET. The data "1" is defined as the state where plus charges are stored in the body and the data "0" as the state where the body voltage is 0 V. To write the data "1," –10 V is applied to the gate, extracting electrons from the body by making use of the charge pumping phenomenon. The charge pumping is a phenomenon where channel holes are recombined with the body electrons which are trapped in the interface states formed between the silicon–SiO_2 interface of a p-type MOSFET when applying a negative voltage pulse to the gate, flowing the current into the body when the body voltage is kept at a constant voltage with the p-n junction biased at a reverse bias.

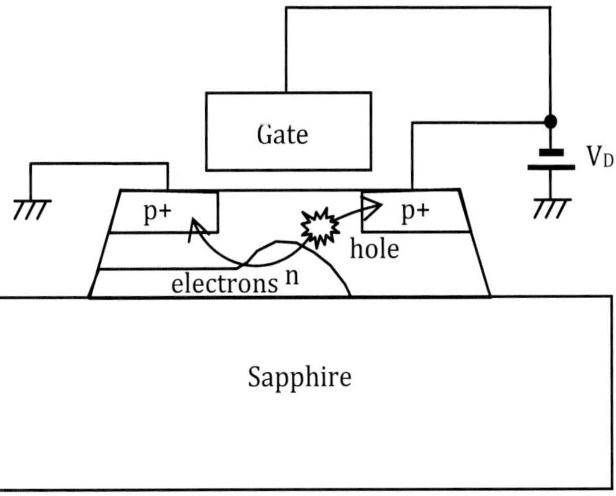

Figure 1.13 Concept of a single-transistor cell proposed by N. Sasaki.[15]

Or it is a phenomenon where the channel holes are recombined with the electrons approaching to the silicon surface when a positive voltage is applied to the gate of a p-type MOSFET, similarly flowing current into the body when the body is at a constant voltage.[16] The former is illustrated in Fig. 1.14 and the latter in Fig. 1.15. In the literature, data "1" is written by applying a negative voltage pulse to the gate of a p-type MOSFET cell. Data "0," on the other hand, is written by making use of drain avalanche multiplication which introduces the majority carries electrons into the body with a large minus voltage applied to the drain terminal as is shown in Fig. 1.13. However, all cells do not have a sufficient number of interface states nor the probability of channel carries remain at the transition from the inversion to the accumulation states to recombine with the body carriers is large enough when the cell size is scaled down, the method of writing data by making use of the charge pumping phenomenon is found not to be useful for scaled cells at a later time.

Figure 1.16 indicates a single-transistor memory cell which was presented by M. R. Tack *et al.* in 1990.[17] They proposed a memory cell of an n-type MOSFET formed in silicon-on-insulator (SOI) with its body floating. The data are stored in the body in the form of electrical charge accumulated there. The binary data correspond to the two states: one has more holes accumulated in the body with a lowered threshold voltage and the other has fewer holes accumulated there with as heightened threshold voltage, both due to body effect of the MOSFET. However, in those days there has not

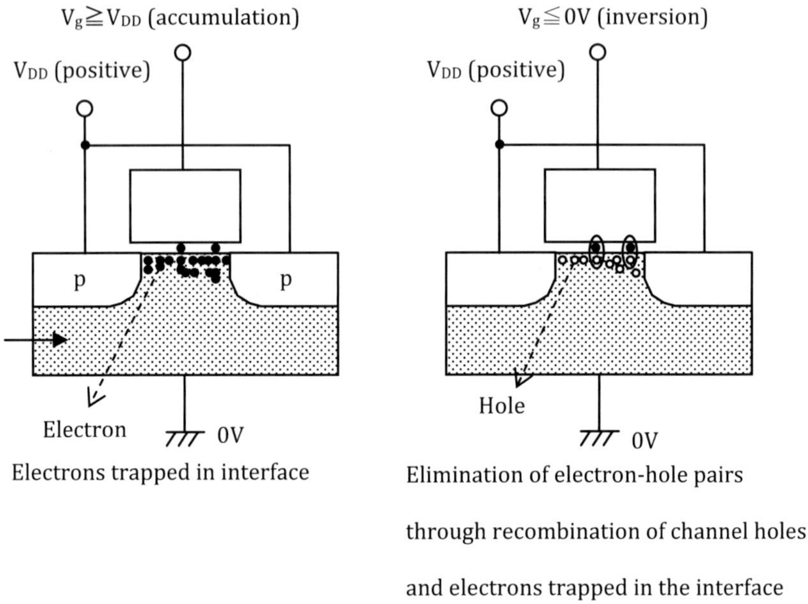

Figure 1.14 Charge pumping phenomenon due to trapped electrons in a p-type MOSFET.

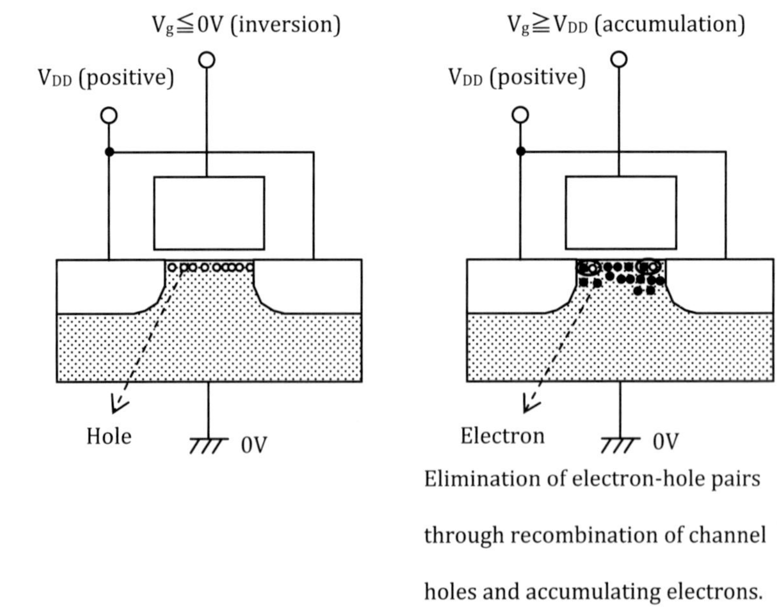

Figure 1.15 Charge pumping phenomenon due to remnant channel holes in a p-type MOSFET.

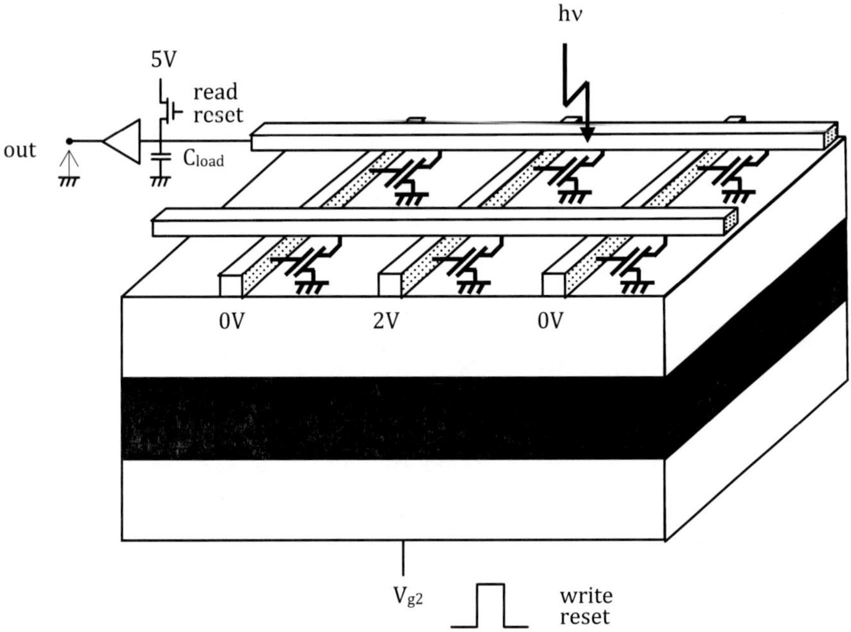

Figure 1.16 Concept of a single-transistor cell proposed by M. R. Tack *et al.*[17]

been discovered the way to extract holes from a cell by controlling the voltages of the gate, source, and drain terminals. Their method to write is achieved by injecting holes into selected cells by impact ionization after extracting holes from all cells' bodies in an array by applying a large positive voltage pulse (purging data) on the substrate under the thick buried oxide of the SOI. The technology was, therefore, unfinished as memory applications. Their interim application is intended for a photo detector where the initially purged "0" cells can be changed to the state "1" when they are hit by photons.

Figure 1.17 is a capacitor-less DRAM cell which was proposed by H. Wann and C. Hu in 1993.[18] Strictly speaking, this concept involves two MOSFETs in a cell, one n-type and the other p-type. However, since the two MOSFETs are combined very compactly to make a cell with their gates commonly formed, the cell is also categorized as a single-transistor cell. Though this is also a cell of a p-type MOSFET with its body floating for data storage, one of unique features of the cell is that the data charge is switched into and out through an n-type MOSFET combined with the p-type MOSFET. However, it has not been used in memory products, either, because the cell size is large and it requires many control wires such as WL, write bit line, read bit line, and purge line (buried WL).

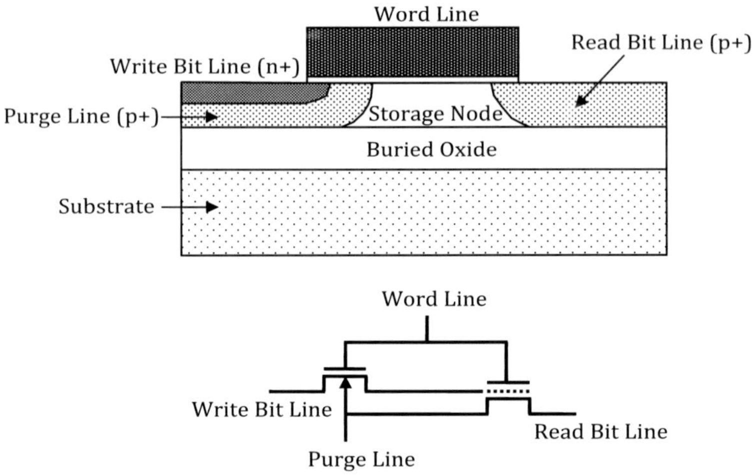

Figure 1.17 Concept of a single-transistor cell proposed by H. Wann and C. Hu.[18]

All single-transistor cells proposed before the year 2000, some of which were explained in this chapter as examples, have failed to be used in products and the effort to make capacitor-less DRAM cell usable in products seemed to fade away until two parties started independently to develop a truly single-transistor capacitor-less DRAM cell to which data can be written selectively ca. 2000.[19,20] Since then many companies and research organizations have begun to investigate the cell for making it productive and analyzing the physics behind its operational mechanism. This monograph is written exclusively and as comprehensively as possible about this novel particular capacitor-less DRAM cell which is named floating body cell.

References

1. R. H. Dennard, "Field-effect transistor memory," *U.S. Patent 3,387,286*, June 4, 1968.
2. K. Itoh, *VLSI Memory Chip Design*, Springer, Berlin, 2001.
3. H. Ishiuchi, T. Yoshida, H. Takato, K. Tomioka, K. Matsuo, H. Momose, S. Sawada, K. Yamazaki and K. Maeguchi, "Embedded DRAM technologies," *IEDM Tech. Dig.*, pp. 33–36, Dec. 1997.
4. R. E. Matick and S. E. Schuster, "Logic-base eDRAM: origins and rationale for use," *IBM J. Res. Dev.*, vol. 49, no. 1, pp. 145–165, Jan. 2005.
5. D. P. Spampinato, O. Park and L. M. Terman, "Pulsed power four device memory cell," *U.S. Patent 3,541,530*, 1970.
6. W. M. Regitz and J. Karp, "A three-transistor-cell, 1024-bit, 500 ns MOS RAM," *ISSCC Dig. Tech. Papers*, pp. 42–43, Feb. 1970.

7. J. A. Mandelman, R. H. Dennard, G. B. Bronner, J. K. DeBrosse, R. Divakaruni, Y. Li and C. J. Radens, "Challenges and future directions for the scaling of dynamic random-access memory (DRAM)," *IBM J. Res. Dev.*, vol. 46, no. 2/3, pp. 187–212, 2002.

8. H. Sunami, T. Kure, N. Hashimoto, K. Itoh, T. Toyabe and S. Asai, "A corrugated capacitor cell (CCC)," *IEEE Trans. Electron Devices*, vol. ED-31, pp. 746–753, 1984.

9. M. Koyanagi, H. Sunami, N. Hashimoto and M. Ashikawa, "Novel high-density, stacked capacitor MOS RAM," *IEDM Tech. Dig.*, pp. 348–351, Dec. 1978.

10. T. Ema, S. Kawanago, T. Nishi, S. Yoshida, H. Nishibe, T. Yabu, Y. Komada, T. Nakano and M. Taguchi, "3-dimensional stacked capacitor cell for 16M and 64M DRAMs," *IEDM Tech. Dig.*, pp. 592–595, Dec. 1988.

11. S. Kimura, Y. Kawamoto, T. Kure, N. Hasegawa, J. Etoh, M. Aoki, E. Takeda, H. Sunami and K. Itoh, "A new stacked capacitor DRAM cell characterized by a storage capacitor on a bit-line structure," *IEDM Tech. Dig.*, pp. 596–599, Dec. 1988.

12. M. Aoki, Y. Nakagome, M. Horiguchi, H. Tanaka, S. Ikenaga, J. Etoh, Y. Kawamoto, S. Kimura, E. Takeda, H. Sunami and K. Itoh, "A 60-ns 16-Mbit CMOS DRAM with a transposed data-line structure," *IEEE J. Solid-State Circuits*, vol. 23, no. 5, pp. 1113–1119, Oct. 1988.

13. H. Hidaka, K. Fujishima, Y. Matsuda, M. Asakura and T. Yoshihara, "Twisted bit-line architectures for multi-megabit DRAM's," *IEEE J. Solid-State Circuits*, vol. 24, no. 1, pp. 21–27, Feb. 1989.

14. P. K. Chartterjee, G. W. Taylor and M. Malwah, "Circuit optimization of the Taper Isolated Dynamic RAM cell for VLSI memories," *ISSCC Dig. Tech. Papers*, pp. 22–23, Feb., 1979.

15. N. Sasaki, "Charge pumping in SOS-MOS transistors," *IEEE Trans. Electron Devices*, vol. ED-28, no. 1, pp. 48–52, Jan. 1981.

16. J. S. Brugler and P. G. A. Jespers, "Charge pumping in MOS devices," *IEEE Trans. Electron Devices*, vol. ED-16, no. 3, pp. 297–302, March 1969.

17. M. R. Tack, M. Gao, C. L. Claeys and G. J. Declerck, "The multistable charge-controlled memory effect in SOI MOS transistors at low temperatures," *IEEE Trans. Electron Devices*, vol. ED-37, no. 5, pp. 1373–1382, May 1990.

18. H. Wann and C. Hu, "A capacitorless DRAM cell on SOI substrate," *IEDM Tech. Dig.*, pp. 635–638, Dec. 1993.

19. S. Okhonin, M. Nagoga, J. M. Sallese and P. Fazan, "SOI capacitor-less 1T-DRAM concept," *IEEE Int. SOI Conf.*, pp. 153–154, Oct. 2001.

20. T. Ohsawa, K. Fujita, T. Higashi, Y. Iwata, T. Kajiyama, Y. Asao and K. Sunouchi, "Memory design using one-transistor gain cell on SOI," *ISSCC Dig. Tech. Papers*, pp. 152–153, Feb. 2002.

Chapter 2

Concept of Floating Body Cell and Its Operational Principle

The operational principle and the features of floating body cell (FBC) are described in comparison with those of the one-transistor and one-capacitor dynamic random access memory cell. The method of selective write realizes one of the key characteristics, bringing about a trade-off between the selectivity and the bit line disturb which is necessary to be considered in designing the memory using FBC. The difference between FBCs on partially depleted silicon-on-insulator and on fully depleted silicon-on-insulator is summarized. Finally, various write and read methods are explained and compared.

2.1 COMPARISON BETWEEN FLOATING BODY CELL AND ONE-TRANSISTOR AND ONE-CAPACITOR DYNAMIC RANDOM ACCESS MEMORY CELL

In this section, the concept of a new single-transistor capacitor-less dynamic random access memory (DRAM) cell called floating body cell (FBC) will be explained in comparison with the conventional one-transistor and one-capacitor (1T-1C) DRAM cell.[1,2] As is shown in Fig. 2.1, the conventional 1T-1C DRAM cell stores data by charging the capacitor from a bit line (BL) through the transistor which works as a switch. FBC, on the other hand, stores data by charging the floating body of an n-type MOSFET. A p-type MOSFET is, of course, able to be used as FBC. Although the floating body of a MOSFET can be realized most commonly by building the MOSFET in silicon-on-insulator (SOI) substrate, there are many other options to make the body floating. Holes, the majority carriers of p-type semiconductor, are stored in the floating body in the case of the n-type MOSFET FBC. The data state where more holes are stored in the body is defined as the state "1," while that with fewer holes stored is the state "0." The body potential of the state "1"

Floating Body Cell: A Novel Capacitor-less DRAM Cell
By Takashi Ohsawa and Takeshi Hamamoto

www.panstanford.com

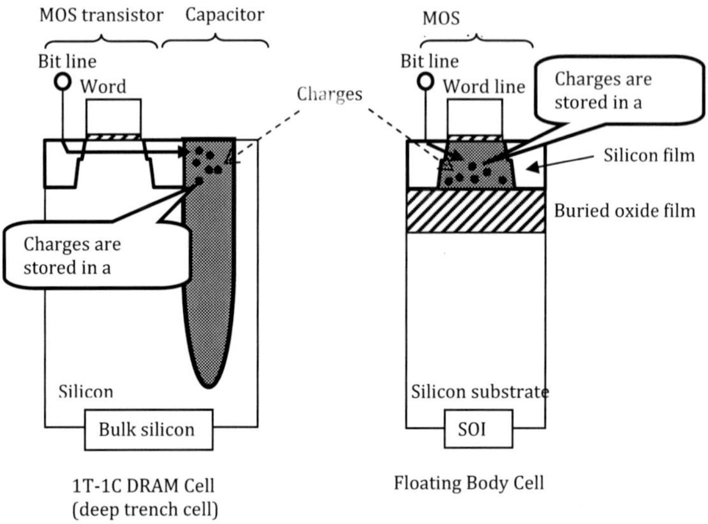

Figure 2.1 Comparison between the conventional deep trench 1T-1C DRAM cell and the floating body cell (FBC).

is higher than that of the state "0." The threshold voltage of the cell transistor of the state "1" is lower than that of the state "0" due to the body effect of MOSFET. Therefore, the drain current of the "1" cell I_1 becomes larger than that of the "0" cell I_0 at the same drain and gate voltages, making the data state "1" distinguished from the data "0" state at read operation in a current-sensing circuit by comparing them with a certain current $I_{1/2}$ which is generated as a value halfway between I_0 and I_1 to be supplied to the circuit.

Let us describe the read signal more quantitatively. According to the body effect formula, the dependence of the n-type partially depleted SOI (PD-SOI) MOSFET threshold voltage V_{th} on the body voltage (V_b) at read is expressed[3] as

$$V_{th} = V_{fb} + 2\phi_B \frac{\sqrt{2\varepsilon_{Si} q N_A}}{C_{ox}} \sqrt{2\phi_B - V_b}\,, \tag{2.1}$$

where V_{fb} is the flat band voltage, ϕ_B is the difference between Fermi level and intrinsic Fermi level which is written by

$$\phi_B = \frac{kT}{q} \ln\left(\frac{N_A}{n_i}\right), \tag{2.2}$$

where k is the Boltzmann constant, q is the elementary charge, T is temperature, N_A is the concentration of impurity atoms (acceptor) in the body, n_i is the intrinsic carrier concentration. In Eq. 2.1, ε_{Si} is the permittivity of silicon and C_{ox} is the capacitor of the gate dielectrics per unit area which is expressed by

$$C_{ox} = \frac{\varepsilon_{ox}}{t_{ox}}, \tag{2.3}$$

where t_{ox} is the thickness of the gate dielectric material and ε_{ox} is its permittivity. According to Eq. 2.1, the threshold voltage V_{th} becomes smaller when the body voltage of the MOSFET at read V_b is larger. If the body voltage of the "1" data FBC and that of the "0" data FBC both at read are expressed by V_{b1} and V_{b0}, respectively, the threshold voltage difference $\Delta V_{th} = V_{th0} - V_{th1}$ which is to be observed at read is calculated by Eq. 2.1 as

$$\Delta V_{th} = \frac{\sqrt{2\varepsilon_{Si} q N_A}}{C_{ox}} \left(\sqrt{2\phi_B - V_{b0}} - \sqrt{2\phi_B - V_{b1}} \right). \tag{2.4}$$

According to Eq. 2.4, to enlarge the body voltage difference at read $\Delta V_b = V_{b1} - V_{b0}$ certainly increases the read signal $\Delta V_{th} = V_{th0} - V_{th1}$. In addition, to make N_A and t_{ox} larger contributes to enhancing the read signal ΔV_{th}.

The 1T-1C DRAM cell consists of a capacitor which is a data storage device and a MOSFET which is a switching device to connect and disconnect the storage device and the associated BL. FBC is considered to consist of a single device in which a switching device is merged with a data storage device in the sense that the data storage capacitor is the floating body of the switching MOSFET. And data is read in the form of a conductivity change of the switching device which is affected by the data storage state. As is understood by the above description comparing the FBC concept with the 1T-1C DRAM cell's one, its structure is nothing other than some of the single-transistor memory cells proposed so far. The features which differentiate it from the past cells are related to the method of write operations which will be explained in section 2.3. By the way, FBC is a gain cell which has an ability in itself to draw current from BL or source line (SL).[2,4] Since the signal of the current is basically kept constant when the cell is shrunk, it has an advantage over the 1T-1C cell which loses the signal of charge when shrinking the capacitor. Furthermore, it is basically non-destructive read-out cell which means that the cell data are not destroyed during read, making it unnecessary for FBC to be restored after read, which contributes to achieve memories with fast random cycle time. Strictly speaking, however, it is not a perfect non-destructive read-out cell, because the data state "1" is slightly degraded during read due to charge pumping phenomenon which was explained in section 1.3[a] and the p-n junction forward bias between the body and the source during read. Therefore, the cell is categorized as a quasi non-destructive read-out cell,[5,6] still providing several performance merits over the destructive read-out cell. These two features of FBC, a gain cell and a quasi non-destructive read-out cell, are those which are not enjoyed by the 1T-1C DRAM cell.

---

[a] The data degradation by charge pumping will be explained in detail in section 6.1.

2.2 CELL ARRAY OF FBC AND ITS CROSS-SECTIONAL VIEW

Figure 2.2 shows a plan view of a cell array arrangement of FBC and its cross-sectional view cut along a BL. As is shown in this figure, a two-dimensional cell array is formed by replacing stripes of silicon films of SOI by shallow trench isolation (STI) lines and by arranging n-type MOSFETs continuously along the residual silicon stripes. The sources and drains of the MOSFETs in line are shared by adjacent cells and they are connected to the associated BL and SLs alternately. The gates form word lines (WLs). Each cell's body is made electrically isolated, for it is surrounded by gate dielectrics and buried oxide (BOX) of SOI in the vertical direction, STI in the WL direction, and p-n junctions in the BL direction (if reverse biased). The cell size is estimated to be about $3F \times 2F = 6F^2$, where F represents a minimum feature size or a BL half pitch defined in Fig. 2.2. This cell size is about the same as that of state-of-the-art 1T-1C DRAM cell. However, the DRAM cell needs to be constructed

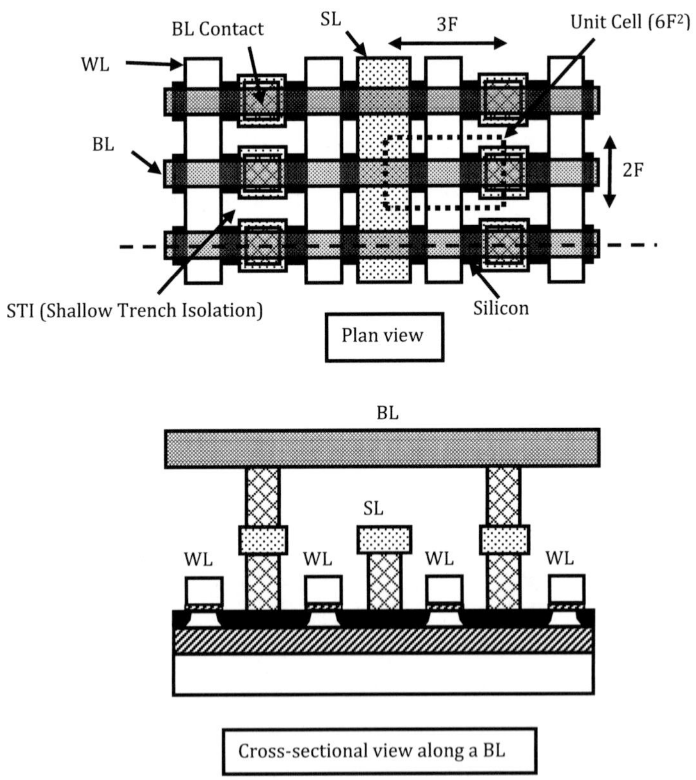

Figure 2.2 Plan view of an FBC cell array and its cross-sectional view along a BL.

by using a DRAM specific process technology of BL contact which is self-aligned to poly-silicon gates instead of the standard aligned contact technique to achieve $6F^2$ cell size. The process of making FBC is simple and compatible with the standard CMOS one, offering distinctive advantage over the DRAM cell, because FBC can be made by the standard process which opens contact holes for BL and SL in a simple technique of mask alignment. If FBC is designed under the assumption of using the self-aligned technology for BL and SL contacts, the cell size can be expected to be shrunk down to $4F^2$ size. Furthermore, the lithographic technology for patterning active region (or STI region) of the 1T-1C DRAM cell is very difficult, because the active area is in the form of island surrounded by STI region.

The process of making FBC, on the other hand, is very lithography-friendly one, because the active regions and the STI regions form very simple line and space patterns. This contributes to reduce the process cost for FBC compared to that of the conventional 1T-1C DRAM whose cost has recently increased by the introduction of super-resolution methods such as phase shift masks and optical proximity correction techniques.

One more metal layer would be necessary in wiring an FBC array in addition to those to wire the 1T-1C DRAM cell array, because SLs must be made of a low resistive metal layer. Although the metal layer can be omitted by wiring SLs with n-type diffused silicon stripes, it seems unrealistic since large current must flow in SLs in write and read (especially in refresh operation) as will be explained later. The array noise due to IR drop would lead to malfunction in cell write and read operations with n-type diffusion layer wiring for SLs.

2.3 BASIC WRITE AND READ OPERATIONS OF FBC

Next, typical write and read methods of FBC will be explained. Figure 2.3 shows schematically the typical write and read principles.[2,4] To write "1," the cell transistor is biased at saturation region by raising WL and BL at positive voltages V_{WLHW} and V_{BLH}, respectively, with respect to the SL voltage V_{SL} which is assumed to be fixed at GND (0.V). The impact ionization generates many pairs of electrons and holes at the drain side of the channel and electrons are swept into the drain terminal, while holes are to be accumulated in the potential well in the p-type floating body. To write "0," BL voltage is lowered to a negative voltage V_{BLL} with WL raised at the same voltage as "1" write operation V_{WLHW}. The p-n junction between the body and the drain is to be forward biased to eject accumulated holes into the drain terminal. To read, the cell transistor is biased at linear region by raising WL high V_{WLHR} ($V_{WLHR} = V_{WLHW}$ in Fig. 2.3) with BL biased at a slightly positive voltage V_{BLR}, for example 0.2 V in Fig. 2.3. The reason why the cell is biased at linear region during read is that the impact ionization may destroy the data "0" state during read if it is biased at saturation region.

Figure 2.4a shows TCAD simulation results of a transient analysis for body voltage response to the input of WL and BL voltages based on the write and read operations in Fig. 2.3. The analysis starts from an initial equilibrium state of "0" write. The "0" data is once instantaneously held by lowering WL to V_{WLL}= −1.5 V and setting BL at 0 V at 2 ns. Though data holding time is usually very long, the data is held just instantaneously in this simulation to save time without analyzing data retention characteristics. After that, data "1" is written between 10 ns and 20 ns and the data is read at 28 ns. Similarly, data "0" is written between 30 ns and 40 ns and the data is read at 48 ns. The data is instantaneously held between the data write and data read for both "1" and "0." There are two reasons why WL is biased at V_{WLL} = −1.5 V during data hold. One reason is that the body voltage is

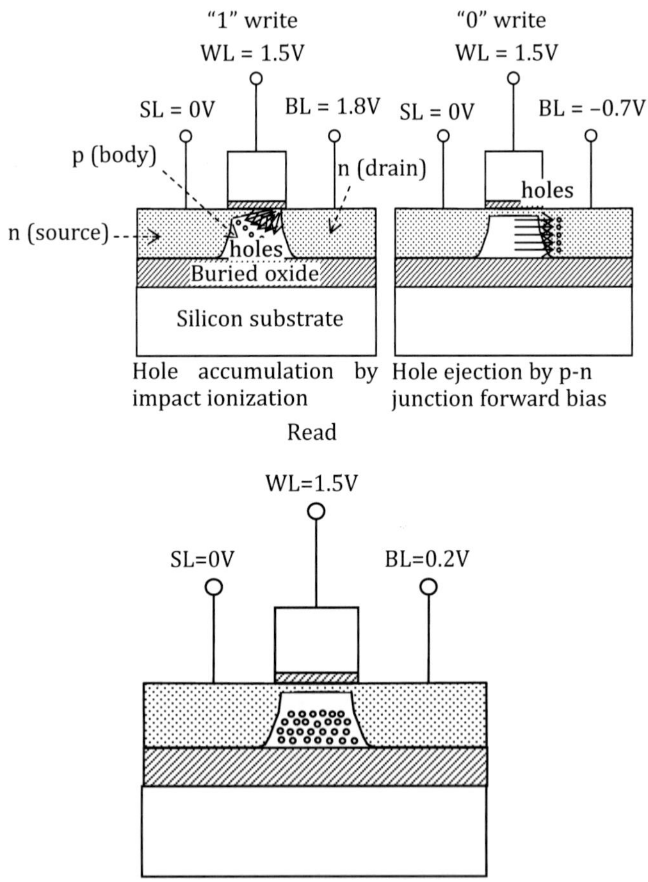

Drain current differs depending on the number of holes accumulated in the body due to body effect generations and device structures.

Figure 2.3 Write and read operational principles of FBC. The set of voltages shown in this figure is just an example. The optimum voltages may differ depending on cell generations and device structures.

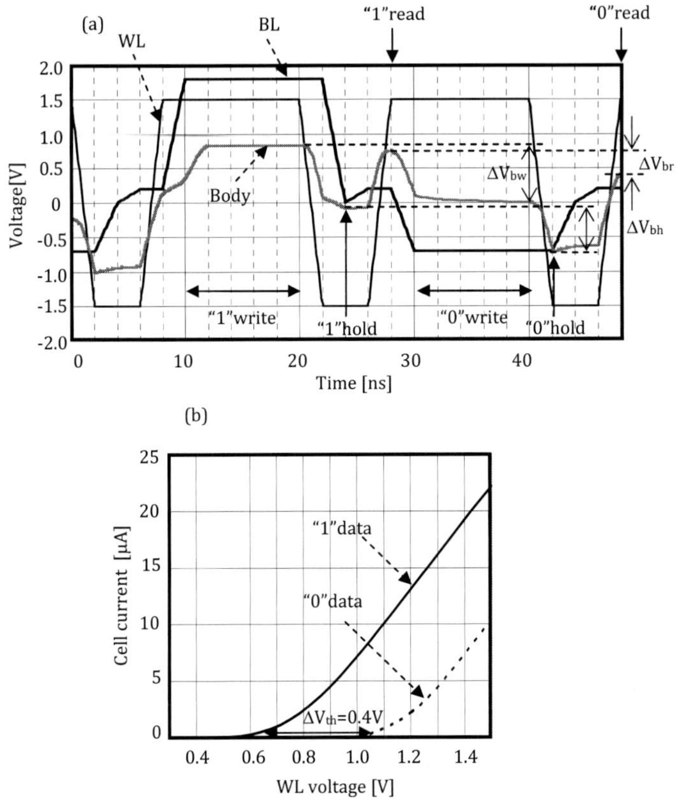

Figure 2.4 Results of TCAD transient analysis of write and read operations of FBC.

pushed down to a negative voltage due to the gate-body capacitive coupling during data hold, making the p-n junction between the body and the source and the drain reverse biased for long data retention time with low-level leakage current. The other reason is related to the "0" write selectivity. There are many cells connected to a BL. To write "0" to only a cell which is selected by a WL (raised to a positive voltage), the body voltages of many other non-selected "1" cells must be low enough not to be strongly forward biased by the negative BL voltage to write "0" to the selected "1" cells. Therefore, unselected cell which are in the data holding state must be pushed down to a negative voltage which is low enough not to be forward biased by the BL. The negative WL voltage at hold will guarantees the "0" write selectivity. This optimization of WL hold voltage level is one of most important design points of the FBC cell array and will be considered in detail in section 2.4 in connection with BL disturb issues. Figure 2.4b shows the cell read currents for the data "1" and "0" which are plotted as the relations between the cell transistor drain currents (I_{ds}) and the WL voltage (V_{gs}). The read current for "1" cell is taken from I_{ds} 28 between 26 ns and ns, while the read current for "0" cell is taken from I_{ds} between 46 ns and 48 ns. The difference represents the signal of FBC and the threshold voltage difference $\Delta V_{th} = 0.4$ V is observed. And the body voltage difference between the data "1" and "0" at the end of write is seen to be reduced at data hold and further reduced at read. These phenomena cannot be avoided, albeit mitigated. They are to be explained by the differences of gate-body and the drain-body capacitive couplings and the threshold voltage difference all between the data "1" cell and the data "0" cell. It is important to adopt optimum device structures and operational waveforms for minimizing this kind of signal reductions from write to read through hold, which will be explained in detail in section 3.1.

Here, the physical mechanisms of the write operations are discussed quantitatively. As is shown in Fig. 2.5, writing the data "1" is performed by accumulating holes in the potential well of a cell's p-type body which are generated by substrate current I_{sub} through impact ionization with the cell at saturation region. When the number of holes increases, the potential of the body becomes positive with respect to the source potential of the cell. In this situation, some holes start to flow out of the body to the source terminal, generating a forward current I_{fwd} at the p-n junction between the p-type body and the n-type source. Therefore, the body potential increase by the impact ionization is slowed down by the p-n junction forward current to finally reach an equilibrium state which is determined by the condition $I_{sub} = I_{fwd}$. This terminates the "1" write operation and the body voltage increase stops at a certain value V_{b1}. The time T_{w1} which is necessary to write the data "1" with the body potential from V_{b0} to V_{b1} is expressed

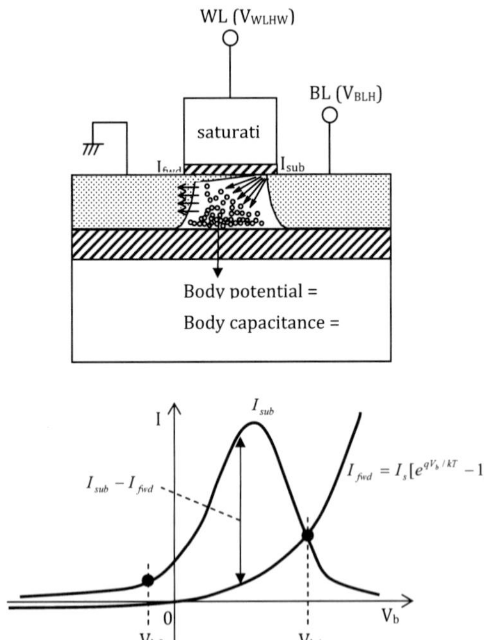

Figure 2.5 "1" write mechanism of FBC.

$$T_{w1} = C_b \int_{V_{b0}}^{V_{b1}} \frac{1}{I_{sub} - I_{fwd}} dV_b, \quad (2.5)$$

by where C_b is the capacitance of the body. As you can see in Fig. 2.5, since I_{sub} - I_{fwd} keeps relatively large value from V_{b0} to V_{b1}, "1" write operation terminates in a short period. In fact, the data "1" is reached within 2 ns in the TCAD simulation shown in Fig. 2.4. The situation is very different for

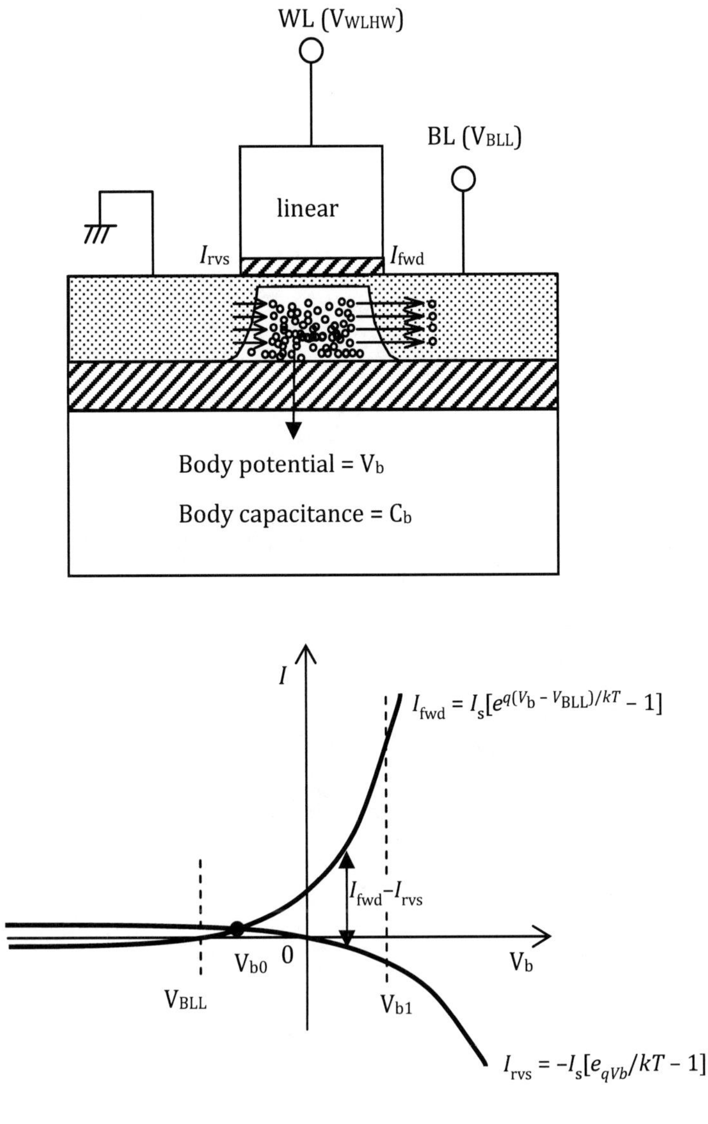

Figure 2.6 "0" write mechanism of FBC.

writing the data "0." As is shown in Fig. 2.6, writing the data "0" to FBC is performed by extracting holes from the body by forward biasing the p-n junction between the p-type body and the n-type drain terminal. The opposite p-n junction between the p-type body and the n-type source is reverse biased. The forward current $I_{\rm fwd}$ which flows between the body and the drain is written as

$$I_{\rm rvs} = -I_{\rm s}\left[e^{qV_{\rm b}/kT} - 1\right], \tag{2.6}$$

where $I_{\rm s}$ is the saturation current for the p-n diode and $V_{\rm BLL}$ stands for the BL low voltage which is applied to the drain terminal when writing the data "0." The reverse current $I_{\rm rvs}$ which flows between the body and the source terminal is written as

$$I_{\rm rvs} = -I_{\rm s}\left(e^{qV_{\rm b}/kT} - 1\right). \tag{2.7}$$

The equilibrium point $V_{\rm b0}$for "0" write is determined by

$$I_{\rm s}\left(e^{q(V_0 - V_{\rm BLL})/kt} - 1\right) = -I_{\rm s}\left(e^{qV_{\rm b}/kt} - 1\right) \tag{2.8}$$

giving the solution as

$$V_{\rm b0} = \frac{kT}{q}\ln\left[\frac{2}{e^{-qV_{\rm BLL}/kT} + 1}\right] = V_{\rm BLL} + \frac{kT}{q}\ln 2$$

$$= V_{\rm BLL} + 0.018\ {\rm V}\ @{\rm RT} \tag{2.9}$$

The time $T_{\rm w0}$ for the body voltage to reach this equilibrium value $V_{\rm b0}$ in writing the data "0" is expressed as

$$\begin{aligned} T_{\rm w0} &= -C_{\rm b}\int_{V_{\rm b1}}^{V_{\rm b0}} \frac{1}{I_{\rm fwd} - I_{\rm rvs}}\, dV_{\rm b} \\ &= -\frac{C_{\rm b}}{I_{\rm s}}\int_{V_{\rm b1}}^{V_{\rm b0}} \frac{dV_{\rm b}}{(e^{-qV_{\rm BLL}/kT} + 1)e^{qV_{\rm b}/kT} - 2} \\ &= \frac{C_{\rm b}}{2I_{\rm s}}\frac{kT}{q}\left[\frac{V_{\rm b0} - V_{\rm b1}}{kT/q}\ln\left\{\frac{(e^{-qV_{\rm BLL}/kT} + 1)e^{qV_{\rm b1}/kT} - 2}{(e^{-qV_{\rm BLL}/kT} + 1)e^{qV_{\rm b0}/kT} - 2}\right\}\right], \end{aligned} \tag{2.10}$$

where $V_{\rm b1}$ is an initial potential of the body before writing the data "0." A typical value of $T_{\rm w0}$ for an FBC with the gate length $L_{\rm g}$ = 150 nm can be calculated by using the parameters $I_{\rm s}$ = 0.16 fA, $C_{\rm b}$ = 0.2 fF, and $V_{\rm b1}$ = 0.7 V and is shown in Fig. 2.7. The calculation results indicate that the time for the body to reach the equilibrium point is more than 100 ms and that the cell cannot reach the equilibrium point within the write time which is commonly permitted in semiconductor memory specifications. If we assume that the body potential becomes $V_{\rm b}$ in lapse time t for writing the data "0,"

$$t = -C_{\mathrm{b}} \int_0^{V_{\mathrm{b}}} \frac{1}{I_{\mathrm{fwd}}} dV_{\mathrm{b}}, \tag{2.11}$$

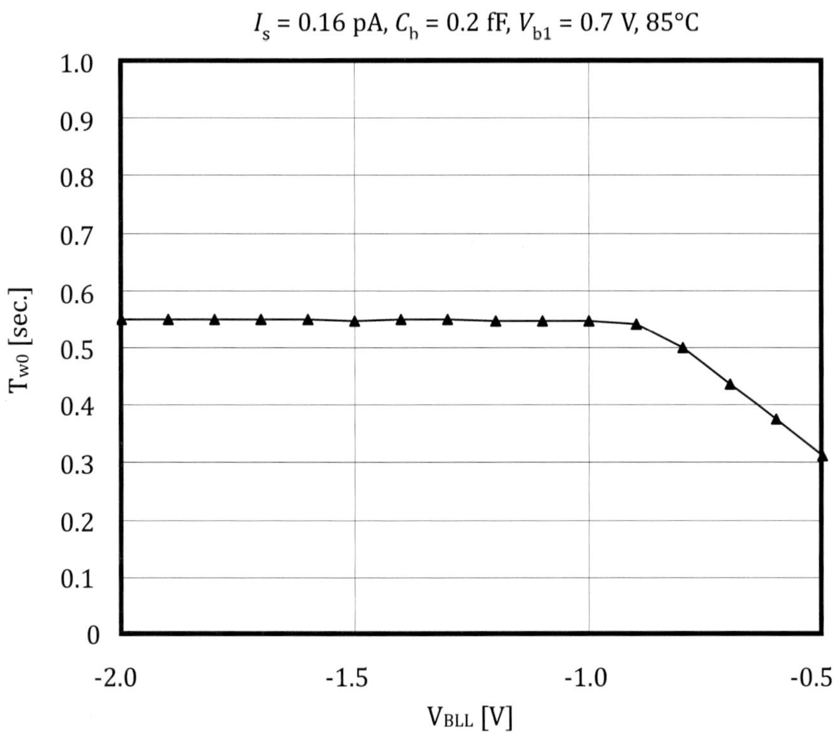

Figure 2.7 The time T_{w0} for the body to reach the "0" write equilibrium point calculated for 150 nm gate length generation.

$$I_{\mathrm{fwd}} = I_{\mathrm{s}} \left[e^{\frac{V_{\mathrm{b}} - V_{\mathrm{BLL}}}{v_{\mathrm{th}}}} - 1 \right], \tag{2.12}$$

where v_{th} is the thermal voltage defined by kT/q and the reverse current I_{rvs} is assumed to be negligibly small compared to the forward current I_{fwd}. The equations can be solved as

$$V_{\mathrm{b}} = V_{\mathrm{BLL}} + v_{\mathrm{th}} \times \ln \left\{ \frac{1}{1 - e^{-t/t_0} \left[1 - e^{V_{\mathrm{BLL}}/v_{\mathrm{th}}} \right]} \right\}, \tag{2.13}$$

where t_0 is a time constant which is defined by

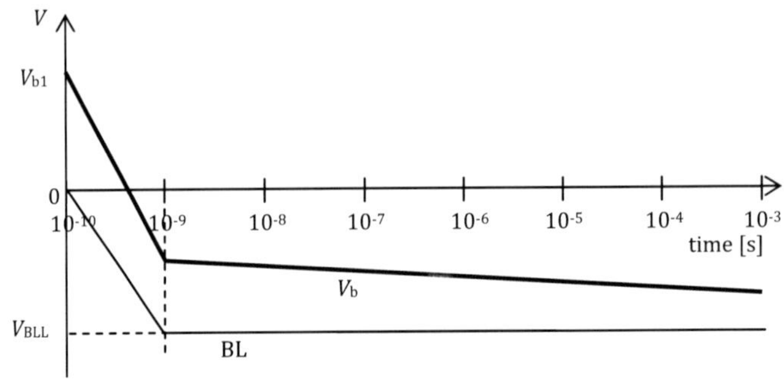

Figure 2.8 Schematic illustration showing the body potential variation during "0" write operation.

$$t_0 = \frac{C_b v_{th}}{I_s}. \tag{2.14}$$

This time constant t_0 is estimated to be about 30 ms for the above parameters for the 150 nm gate length generation. If we assume that t is much smaller than t_0, Eq. 2.13 is approximated to be

$$V_b = V_{BLL} + v_{th}\left[\ln(t_0) - \ln(t)\right]. \tag{2.15}$$

The body potential variation during "0" write operation is in the form of logarithmic function of time and is very slow as is shown in Fig. 2.8. This phenomenon can be observed in the TCAD transient simulation in Fig. 2.4a. Although the body potential goes down to a certain level suddenly at 30 ns, its level decreases very slowly from 30 ns to 40 ns. Therefore, it is impossible to define the "0" write level by an equilibrium point on the contrary to the "1" write operation where an equilibrium point can be shortly reached. Instead, the "0" write level must be defined by the write time during which BL level is lowered to V_{BLL} for semiconductor memories where fast write is requested.

Now, let us discuss about the temperature dependence of the threshold voltages of the data "1" cell and the data "0" cell, i.e., V_{th1} and V_{th0}. Figure 2.9 shows experimental results of the dependence of V_{th1}, V_{th0}, and their difference $\Delta V_{th} = V_{th0} - V_{th1}$ no temperature for the FCB with gate length $L_g = 150$ nm, channel width $W = 150$ nm, gate oxide thickness $t_{ox} = 6$ nm, silicon film thickness $t_{Si} = 55$nm, and BOX thickness $t_{Box} = 25$ nm. The BL voltages during "1" write and "0" write are $V_{BLH} = 2.4$ V and $V_{BLL} = 1.8$ V, respectively. The threshold voltage is defined by the gate voltage at which the drain current $I_{ds} = 53$ nA with the drain voltage at 0.2 V.[b] We measured them at three temperatures −40°C, 25°C, and 85°C. As is shown in this graph, both V_{th1} and V_{th0}, therefore also the signal ΔV_{th}, have almost no dependence on

[b] The actual method of measuring threshold voltages is due to one which is described in section 4.2. The data in Fig. 2.9 were taken with the read duration time RDT = 5 ns.

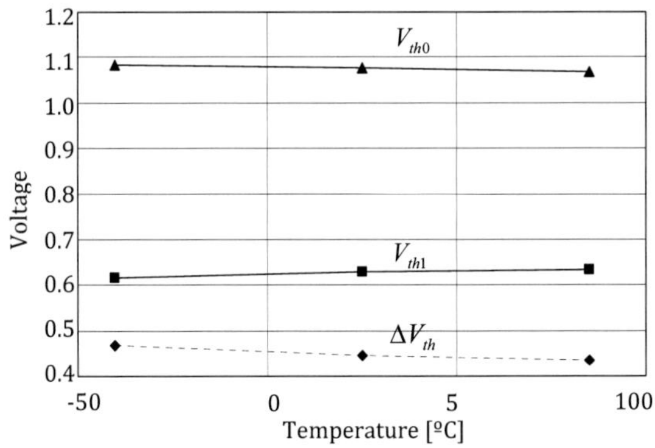

Figure 2.9 Temperature dependence of the threshold voltages for the data "0" and "1" cells.

temperature over a very wide range. Since the intrinsic threshold voltage of MOSFET which is defined by the threshold voltage of MOSFET with its body tied to a constant voltage decreases as temperature increases,[7] these observations indicate that some factors have an effect of increasing the threshold voltage as temperature increases which may balance with the intrinsic threshold voltage tendency of decreasing with temperature. The impact ionization current decreases as temperature increases.[8] And the p-n diode forward current increases as temperature increases.[9] Therefore, the equilibrium point V_{b1} in Fig. 2.5 may shift to the left as temperature increases, making the threshold voltage larger through body effect. This effect, when combined with the intrinsic threshold voltage tendency, can make the threshold voltage V_{th1}observed to be insensitive to temperature. Actually, V_{th1} in Fig. 2.9 is increasing slightly with temperature, making us presume that V_{b1} shift in Fig. 2.4 by temperature overcompensates the temperature variation of the intrinsic threshold voltage but almost in balance. It is very interesting and favorable as well from a point of view of the temperature stability of the FBC operation that the intrinsic threshold voltage variation can be almost compensated by the temperature variation of the body voltage of "1" data cell through body effect. A similar explanation is possible for the temperature dependence of V_{th0}. The body potential V_{b0} of "0" written cell may become lower with temperature, because the p-n forward diode current increases with temperature, which again can compensate the temperature variation of the intrinsic threshold voltage. Figure 2.9 indicates that the V_{b0} variation actually compensates the intrinsic threshold voltage variation. This is also a nice feature from an FBC operation point of view. As the results of the two phenomena, the signal of FBC or the threshold voltage difference ΔV_{th} between the "1" and "0" is almost constant of temperature.

On a final note in this section, many other operational methods have been proposed for writing data to and reading data from FBC than explained in this section, though the latter is the original and is most commonly studied and applied. The other methods will be explained and benchmarked in section 2.6.

2.4 WRITE SELECTIVITY AND BL DISTURB OF FBC

In the previous section, typical write operations are explained along with their physical mechanisms. Since a cell to write "1" is biased at saturation region for generating holes to be accumulated in the body with the associated BL and WL raised high, the "1" write is only possible to the cell placed at the cross point where the BL and the WL intersect with other cells inhibited from being written "1." However, the selectivity of writing "0" is not necessarily as trivial as that of writing "1"; therefore, it will be explained in this section. This feature, the "0" write selectivity, is a key point for FBC to be a candidate as a memory cell. In fact, FBC became the candidate by being equipped with the "0" write selectivity, making a big leap from the cell proposed by M. R. Tack et al.[10] which was not supposed to be used in memories but in photo detectors because the selective write was yet to be discovered.

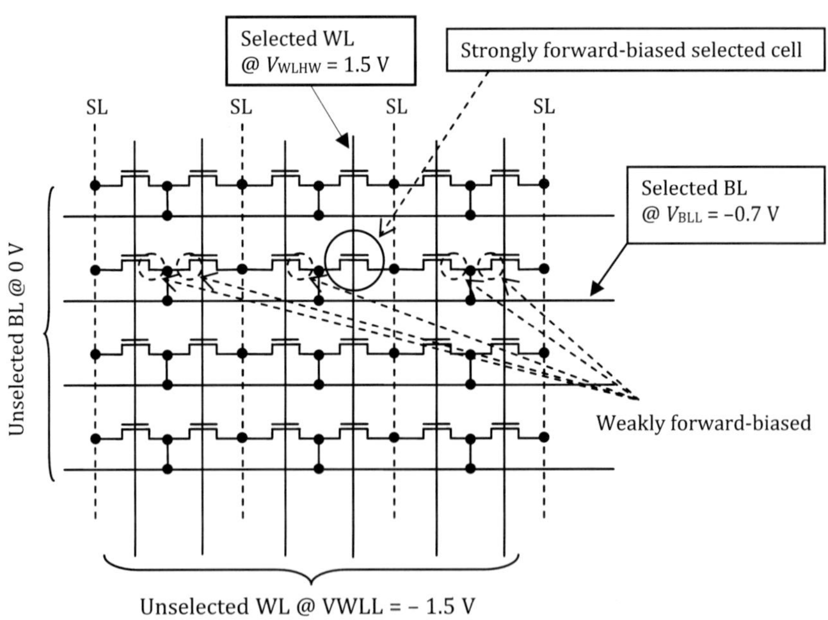

Figure 2.10 Bias conditions for writing "0" to a cell in an array.

Figure 2.10 exemplifies bias conditions by which a cell in an array is selected to be written the data "0." The WL hold level is assumed to be V_{WLL}

= −1.5 V and the BL hold level is 0 V which is the SL voltage V_{SL}. To select a cell (circled), the associated WL is raised to V_{WLHW} = 1.5 V to turn on the cell transistor, followed by lowering the associated BL to V_{BLL} = −0.7 V to make the p-n junction forward biased between the body and the drain of the selected cell. If holes are extracted only from the selected cell by these bias applications with data state of all other cells in the array unchanged, the "0" write selectivity is guaranteed. Looking back to Fig. 2.4a, we find that the body potential is raised to about 0.8 V during writing "1" but that it goes down to a low voltage which is slightly negative with respect to SL during holding "1" which is instantaneously shown at 24 ns. The latter is mainly due to the capacitive coupling between the neutral region of the body and the gate terminal whose voltage decreases from V_{WLHW} = 1.5 V to V_{WLL} = −1.5 V with a minor effect by the capacitive coupling between the neutral region of the body and the drain terminal which decreases form V_{BLH} = 1.8 V to V_{SL} = 0 V. From 26 ns to 28 ns, WL is raised to V_{WLHW} = 1.5 V to write "0" and the body potential is also coupled up to about 0.75 V again through the capacitive coupling between the body and the gate. And by lowering BL to V_{BLL} = −0.7 V, the p-n junction between the body and the drain of the selected cell is strongly forward biased to extract holes from the body very quickly, lowering the body potential accordingly. If the body potential remained unchanged in this event, the p-n junction was to be biased at 1.45 V, which would be very strongly forward. The concern is other cells on the same BL which hold data "1" if any. These unselected "1" cells on the BL hold their body potentials at about −0.05 V according to Fig. 2.4a. Therefore, the p-n junction between the body and the drain for the unselected "1" cells is biased about 0.65 V. Though this bias is certainly forward, it is much weaker compared with 1.45 V for the selected cell. This forward voltage difference between the selected cell and unselected cells makes the "0" write selectivity. In other words, holes are ejected from the body of the selected cell through the strongly forward biased p-n junction, while holes remain in the bodies of unselected cells with their p-n junctions weakly forward biased during a very short period of write time in the order of a few nanoseconds. It must be noted that this "0" write selectivity is not perfect, i.e., the holes are gradually leaking from the bodies of unselected "1" cells if their drain terminals are subject to V_{BLL} for a long time, because the p-n junctions between the bodies and the drains of unselected cells are forward biased weakly during "0" write to other cells connecting the same BL. This situation defines the BL "0" disturb where the data state of an unselected "1" cell degrades through weak leakage of holes by long and/or many "0" writings to other cells on the same BL. It is worth noting that the lower V_{WLL} becomes, the less severe the BL "0" disturb becomes, in other words, the better "0" write selectivity becomes under the same V_{BLL} condition. This is because the body potential of hold "1" cells gets lower by the capacitive coupling between the gate and the body with the body potential of selected "1" cell unchanged, making the difference in p-n junction forward biases for a selected and an unselected cells becomes larger.

Strictly speaking, the selectivity of writing "1" is not perfect. The BL "1" disturb cannot be avoided which degrades the data state of an unselected "0" cell through weak injection of holes by long and/or many "1" writings to other cells on the same BL. This phenomenon will be explained by using Fig. 2.11. When data "1" is written to a cell which is selected by raising WL and BL to V_{WLHW} = 1.5 V and V_{BLH} = 1.8 V, respectively, the gate voltage with respect to the drain voltage for an unselected cell on the same BL as the selected cell is V_{gd} = −3.3 V. This strong negative gate voltage generates gate-induced

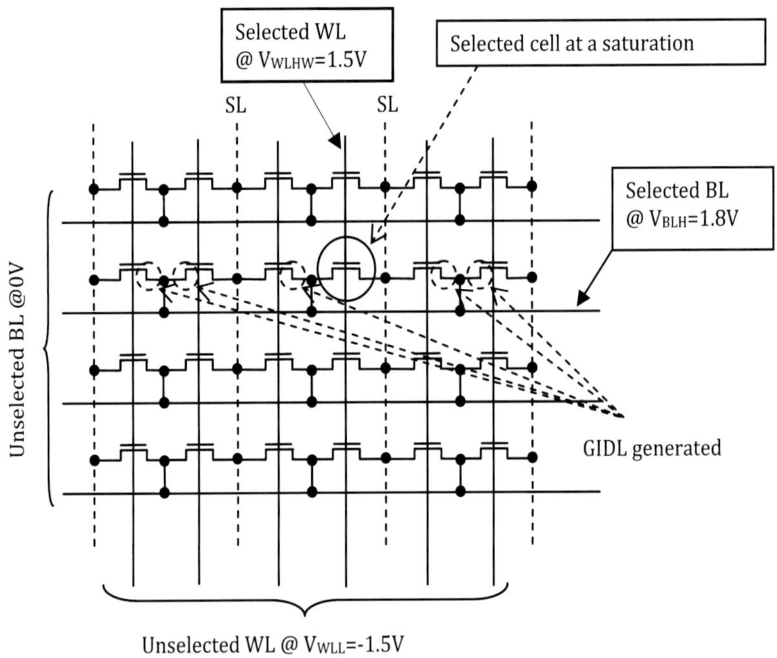

Figure 2.11 Bias conditions for writing "1" to a cell in an array.

drain leakage (GIDL) current due to the band-to-band tunneling which injects holes into the body of the unselected cell.[11,12] The unselected cell, when it is holding the data "0," may degrade to be failed, if its associated BL is raised to V_{BLH} for a long time and/or frequently. This is the BL "1" disturb. It is understood that the higher V_{WLL} becomes, the less severe the BL "1" disturb becomes, in other words, the better "1" write selectivity becomes under the same V_{BLH} condition. Therefore, there is a trade-off between the BL "0" disturb and the BL "1" disturb with respect to the WL hold voltage V_{WLL}. The

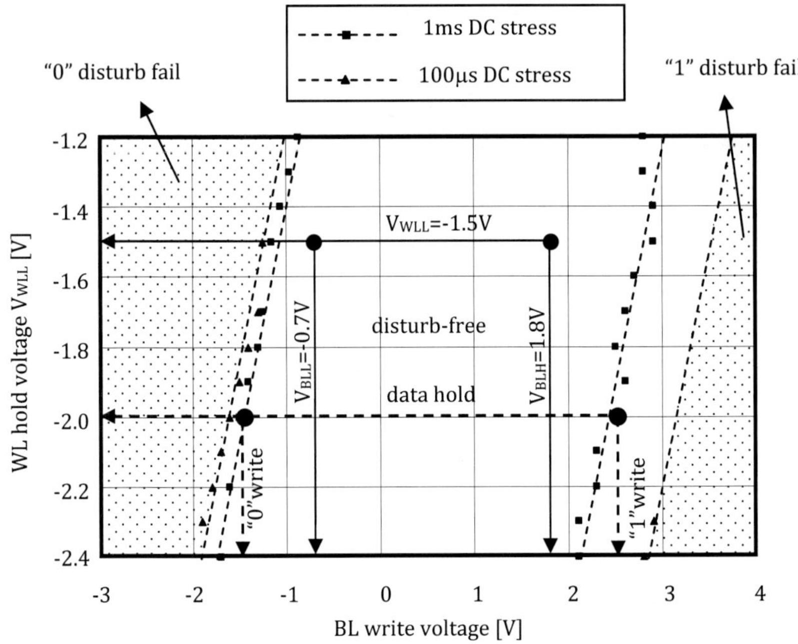

Figure 2.12 Margin map of BL disturb failure measured for FBC of 0.17 mm^2 size.

higher V_{WLL}, the milder the BL "1" disturb and the severer the BL "0" disturb. In other words, the lower V_{WLL}, the severer the BL "1" disturb and the milder the BL "0" disturb. Consequently, determining the WL hold voltage V_{WLL} is one of the most critical considerations in designing memory arrays using FBC. Figure 2.12 indicates operational margin map of an FBC of the size 0.17 μm^2 with channel length L_g = 150 nm, gate dielectric thickness t_{ox} = 6 nm, silicon thickness t_{Si} = 55 nm and BOX thickness t_{BOX} = 25 nm where 5% read current degradation or more is defined as fail.[13] The horizontal axis is BL voltage at "1" write V_{BLH} or BL voltage at "0" write V_{BLL}. The vertical axis is WL hold voltage V_{WLL}. There are two kinds of data points, i.e., those for 100 μs DC BL disturb stress and 1 ms DC BL disturb stress. The group of data in the right hand side define the border line between pass and fail for the BL "1" disturb, while the group of data in the left hand side define those for the BL "0" disturb. This graph clearly shows the trade-off between the BL "0" disturb and the BL "1" disturb with respect to the WL hold voltage V_{WLL}. The set of operating voltages used in Fig. 2.4, V_{BLH} = 1.8 V, V_{BLL} = −0.7 V, and V_{WLL} = −1.5 V, is shown to be within the pass area for both stress conditions. Even when the absolute values of the BL write voltages V_{BLH} and V_{BLL} are raised to 2.5 V and −1.5 V, respectively, this FBC can be made guaranteed to pass by lowering V_{WLL} to −2.0 V even for 1ms stress. Since the data discussed here were ones measured at room temperature and taken only for a single cell, they do not directly lead to the conclusion which guarantees memory operations. However, the data are very useful to understand the BL disturbs qualitatively and an averaged cell's quantities of BL disturb characteristics. The temperature dependence and the statistical characteristics of the BL disturbs will be discussed in section 4.4.

2.5 FBC ON PD-SOI AND FULLY DEPLETED SILICON-ON-INSULATOR

It is straightforward for FBC which stores majority carriers in the floating body to be built in PD-SOI, because a MOSFET in PD-SOI contains a charge neutral region in the body which has an ability to accumulate majority carriers. However, it is also possible to build FBC in fully depleted silicon-on-insulator (FD-SOI).[14] The operation of FD-SOI FBC is actually just realized as a PD-SOI MOSFET made by accumulating majority carriers in a potential well in the thin body at the interface to the BOX with the substrate biased at a proper voltage.[15] However, it is worth being examined and compared with PD-SOI FBC since the operation of FD-SOI FBC has some interesting features which PD-SOI FBC does not have.

We performed TCAD simulation to observe the current I_{ds} which flows between the drain and the source of an SOI MOSFET when the drain voltage is swept according to Fig. 2.13 with the gate and the source fixed at the constant voltages.

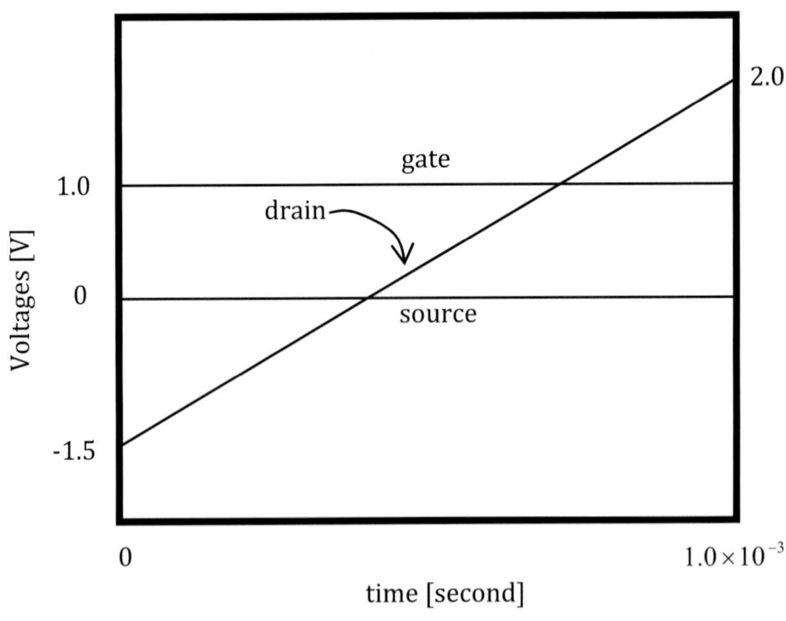

Figure 2.13 Sweep conditions for voltages which are applied to SOI MOSFETs.

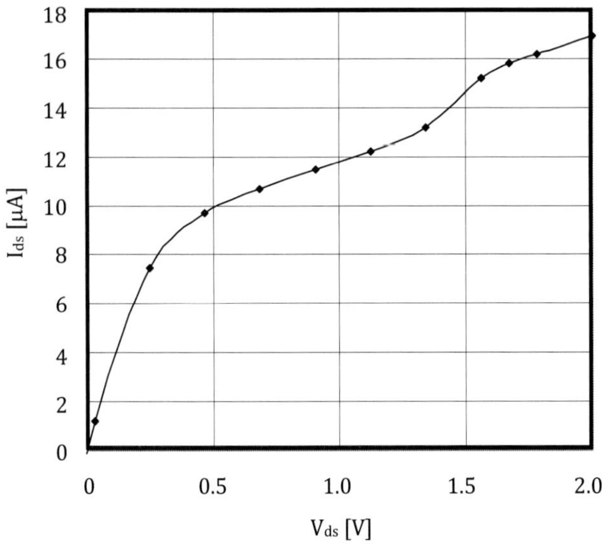

Figure 2.14 TCAD simulation of I_{ds} versus V_{ds} when voltages are applied according to Fig. 2. 13 for $L_g = 70$ nm, $t_{Si} = 50$ nm, $N_A = 5.0 \times 10^{18}$ cm^{-3}, and thick BOX.

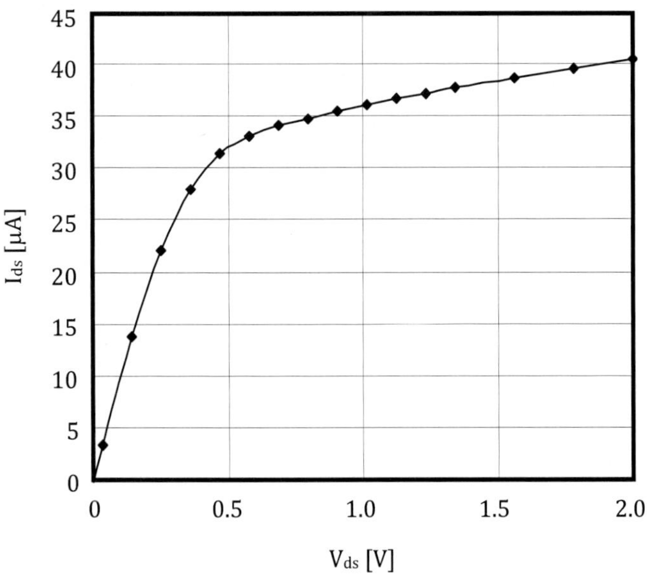

Figure 2.15 TCAD simulation of I_{ds} versus V_{ds} when voltages are applied according to Fig. 2. 13 for $L_g = 70$ nm, $t_{Si} = 50$ nm, $N_A = 5.0 \times 10^{17}$ cm^{-3}, and thick BOX.

Figure 2.14 is the result showing the I_{ds} as a function of drain voltage with respect to the source V_{ds} for a PD-SOI n-type MOSFET with $L_g = 70$ nm,

t_{Si} = 50 nm, N_A = 5.0 × 10^{18} cm^{-3}, and with its BOX thick enough. As is shown in this graph, I_{ds} for a PD-SOI MOSFET exhibits a "kink," indicating generation and accumulation of holes to increase the body potential and to decrease the threshold voltage of the MOSFET due to the body effect when it operates in saturation region. Since the mechanism of data storage for FBC is making use of the "kink" effect, we can judge the possibility for a MOSFET to be used as FBC by checking the existence of the "kink" in this simulation test. Next, we decreased the body impurity (boron) concentration by one order of magnitude to N_A = 5.0 × 10^{17} cm^{-3}. The result is shown in Fig. 2.15 which exhibits no "kink" phenomenon, showing that the boron concentration is too low for the MOSFET with 50 nm thick silicon to remain partially depleted and that holes cannot be accumulated in the body for the large gate voltage.

An FD-SOI MOSFET, thus, cannot be used as FBC as it is. Even if we decrease the BOX thickness to 4 nm, it exhibits no "kink" when its substrate is 0 V as is shown in Fig. 2.16. However, when we bias the substrate at −1 V as is shown in Fig. 2.17, "kink" resumes. Therefore, an FD-SOI n-type MOSFET may be used as FBC with its substrate biased at a minus voltage for holes to be accumulated in a potential well generated by a negative voltage application at the substrate.

The threshold voltage difference between the "1" cell and the "0" cell of PD-SOI FBC was expressed by Eq. 2.4. The threshold voltage difference of FD-SOI

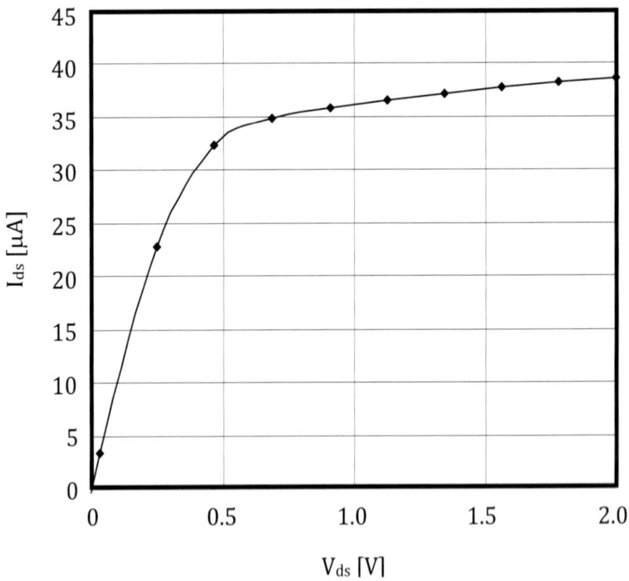

Figure 2.16 TCAD simulation of I_{ds} versus V_{ds} when voltages are applied according to Fig. 2. 13 for L_g = 70 nm, t_{Si} = 50 nm, N_A = 5.0 × 10^{17} cm^{-3}, and t_{BOX} = 4 nm. V_{sub} = 0 V.

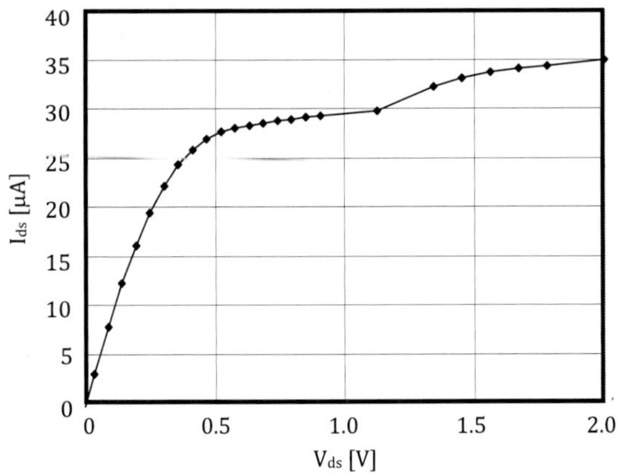

Figure 2.17 TCAD simulation of I_{ds} versus V_{ds} when voltages are applied according to Fig. 2.13 for L_g = 70 nm, t_{Si} = 50 nm, N_A = 5.0 × 10^{17} cm^{-3}, and t_{BOX} = 4 nm. V_{sub} = −1 V.

FBC is express as follows. By V_b we denote the voltage of holes accumulated in a potential well of FD-SOI FBC at read with a negative voltage applied to the substrate. The threshold voltage of the n-type FD-SOI MOSFET is written as[16]

$$V_{th} = V_{fb} + 2\phi_B + \frac{C_{Si}}{C_{ox}}(2\phi_B - V_b) - \frac{Q_{dep}}{2C_{ox}}, \tag{2.16}$$

where C_{Si} is the capacitance of silicon film per unit area which is expressed as follows by using the permittivity of silicon ε_{Si}

$$C_{Si} = \frac{\varepsilon_{Si}}{t_{Si}}, \tag{2.17}$$

and Q_{dep} denotes the charge of acceptor (boron) ions contained in the silicon film with thickness t_{Si} per unit area which is written by

$$Q_{dep} = -qt_{Si}N_A. \tag{2.18}$$

Therefore, the threshold voltage difference of FD-SOI is expressed as[14,17]

$$\Delta V_{th} = \frac{C_{Si}}{C_{ox}}(V_{b1} - V_{b0}) = \frac{C_{Si}}{C_{ox}}\Delta V_b, \tag{2.19}$$

where V_{b1} and V_{b0} are the body voltages of the “1” data cell and the “0” data cell, respectively, and $\Delta V_b = V_{b1} - V_{b0}$. Figure 2.18 shows the relations between the threshold voltage of an n-type PD-SOI MOSFET and those of an n-type FD-SOI MOSFETs with various silicon film thicknesses. While the threshold voltage of the PD-SOI MOSFET depends on the body potential V_b as ∝

$\sqrt{2\phi_B - V_b}$, the threshold voltage of the FD SOI MOSFET depends on V_b as $\propto (2\phi_B - V_b)$. The FD-SOI MOSFET with different silicon thickness has its own body effect straight line with different slope which is represented by a tangent of the PD-SOI body effect curve at the body potential where the maximum depletion layer width under the channel at the acceptor concentration and the body potential is just equal to the silicon thickness, where the maximum depletion layer width is written by

$$X_{dmax}(V_b) = \sqrt{\frac{2\varepsilon_{Si}(2\phi_B - V_b)}{qN_A}}. \tag{2.20}$$

This is understood by considering that the silicon surface potential of the FD-SOI MOSFET with the charge neutral region potential V_b at the interface

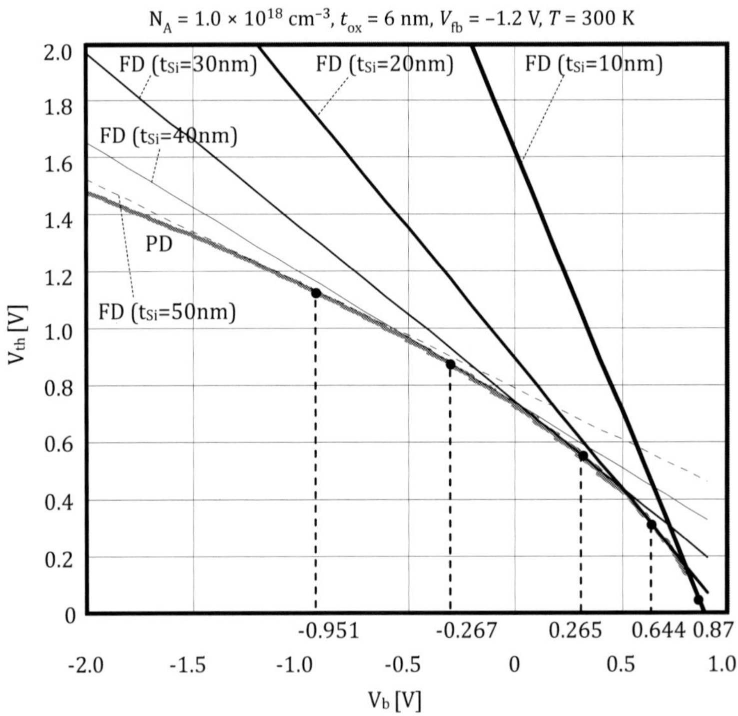

Figure 2.18 The comparison of body effects between PD-SOI and FD-SOI MOSFETs.

to BOX is the same as that of the PD-SOI MOSFET with the charge neutral region of the potential V_b which is separated from the surface by the depletion region of the width $X_{dmax}(V_b)$ as shown in Fig. 2.19. The parameter which characterizes the threshold voltage of a PD SOI MOSFET is the acceptor concentration N_A which determines the maximum depletion layer width X_{dmax}, while that of an FD-SOI MOSFET is the silicon film thickness t_{Si}.

If we assume that X_{dmax} (V_{b0}) for a PD-SOI MOSFET is equal to t_{Si} of an FD-SOI MOSFET, i.e.,

$$\sqrt{\frac{2\varepsilon_{\mathrm{Si}}(2\phi_{\mathrm{B}}-V_{\mathrm{b0}})}{qN_{\mathrm{A}}}}=t_{\mathrm{Si}}, \tag{2.21}$$

the read signal of the PD-SOI FBC Eq. 2.4 is expressed by

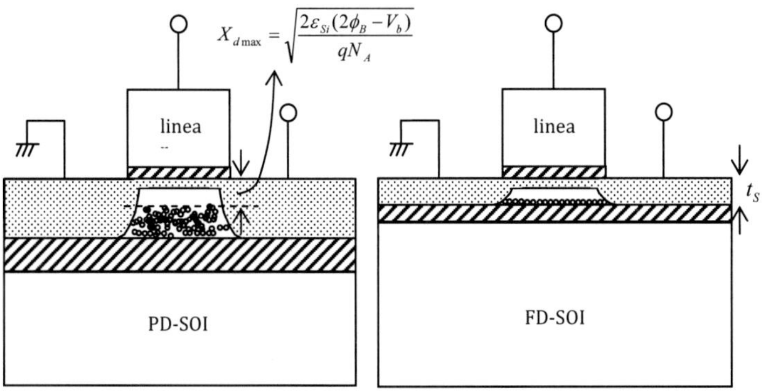

Figure 2.19 Comparison of structures between PD-SOI and FD-SOI MOSFETs.

$$\Delta V_{\mathrm{th}}^{(\mathrm{PD})}=\frac{t_{\mathrm{ox}}}{\varepsilon_{\mathrm{ox}}}\frac{2\varepsilon_{\mathrm{Si}}}{t_{\mathrm{Si}}}[(2\phi_{\mathrm{B}}-V_{\mathrm{b0}})-\sqrt{(2\phi_{\mathrm{B}}-V_{\mathrm{b0}})(2\phi_{\mathrm{B}}-V_{\mathrm{b1}})}]. \tag{2.22}$$

Since

$$\begin{aligned}\sqrt{(2\phi_{\mathrm{B}}-V_{\mathrm{b0}})(2\phi_{\mathrm{B}}-V_{\mathrm{b1}})} &\le \frac{(2\phi_{\mathrm{B}}-V_{\mathrm{b0}})+(2\phi_{\mathrm{B}}-V_{\mathrm{b1}})}{2}\\ &=2\phi_{\mathrm{B}}-\frac{V_{\mathrm{b0}}+V_{\mathrm{b1}}}{2},\end{aligned} \tag{2.23}$$

Eq. 2.22 can be written as

$$\Delta V_{\mathrm{th}}^{(\mathrm{PD})}\ge\frac{t_{\mathrm{ox}}}{\varepsilon_{\mathrm{ox}}}\frac{2\varepsilon_{\mathrm{Si}}}{t_{\mathrm{Si}}}\frac{V_{\mathrm{b1}}-V_{\mathrm{b0}}}{2}=\frac{\varepsilon_{\mathrm{Si}}}{\varepsilon_{\mathrm{ox}}}\frac{t_{\mathrm{ox}}}{t_{\mathrm{Si}}}\Delta V_{\mathrm{b}}=\Delta V_{\mathrm{th}}^{(\mathrm{FD})}, \tag{2.24}$$

where we used Eq. 2.19. Thus, the read signal of the PD FBC is larger than that of the FD-SOI FBC when N_{A} of the PD-SOI is adjusted so that $X_{\mathrm{dmax}}(V_{\mathrm{b0}})$ is equal to t_{Si} of the FD-SOI FBC. In the same way, if we assume

$$X_{\mathrm{dmax}}(V_{\mathrm{b1}})=t_{\mathrm{Si}}, \tag{2.25}$$

for a PD-SOI FBC and an FD-SOI FBC, we have

$$\Delta V_{\mathrm{th}}^{(\mathrm{PD})}\le\Delta V_{\mathrm{th}}^{(\mathrm{FD})}. \tag{2.26}$$

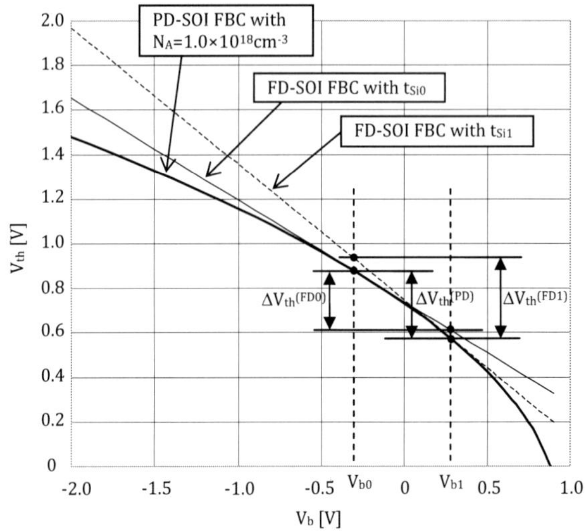

Figure 2.20 The relationship of signals between PD-SOI FBC and FD-SOI FBC.

These two situations can be illustrated in Fig. 2.20 where the body effect curve is described for a PD-SOI FBC with $N_A = 1.0 \times 10^{18}$ cm^{-3} and two body effect straight lines are described for FD-SOI FBCs with two different silicon thickness t_{Si0} and t_{Si1} such that X_{dmax} $(V_{b0}) = t_{Si0}$ and $X_{dmax}(V_{b1}) = t_{Si1}$with signals $\Delta V_{th}^{(FD0)}$ and $\Delta V_{th}^{(FD1)}{}_{t}$, respectively. It is understood that $\Delta V_{th}^{(FD)}$ can be equal to $\Delta V_{th}^{(PD)}$ if we choose t_{Si} somewhere between $X_{dmax}(V_{b0})$ and $X_{dmax}(V_{b1})$.

Figure 2.21 shows TCAD device simulation results for the read signal ΔV_{th} of FBC on SOI as a function of silicon film thickness with its acceptor concentration N_A changed. As are expressed in Eqs. 2.4 and 2.19, the signal ΔV_{th} of PD-SOI FBC is independent of its silicon film thickness and is only depend on its acceptor concentration, while the signal ΔV_{th} of FD-SOI FBC is inversely proportional to its silicon film thickness t_{Si}. Therefore, Fig. 2.21 can be well understood if we

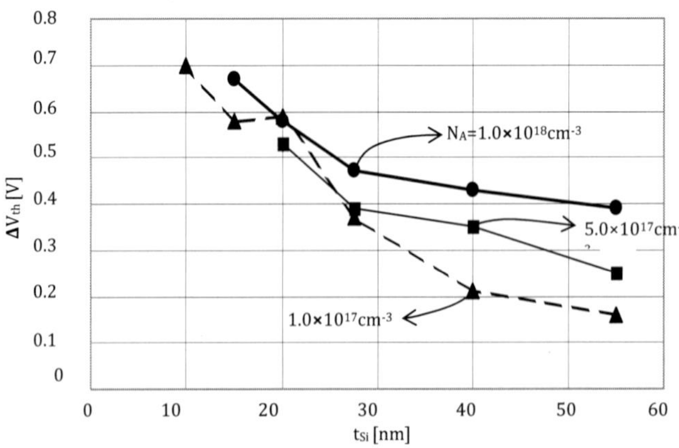

Figure 2.21 Dependence of FBC signal on silicon thickness and acceptor concentration.

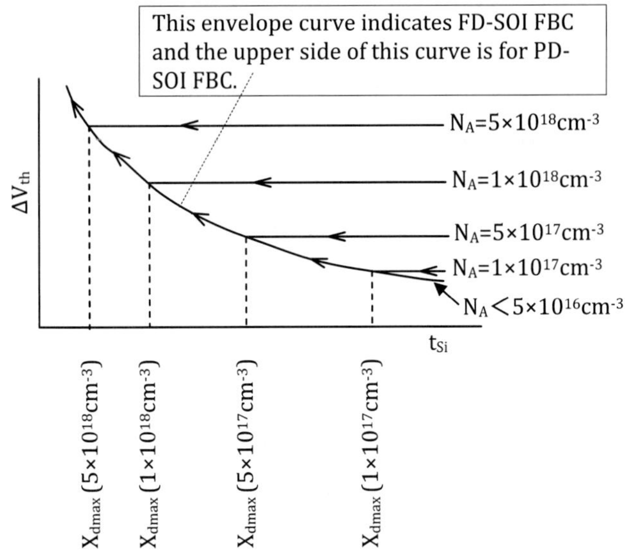

Figure 2.22 Boundary between PD-SOI FBC and FD-SOI FBC

draw an idealized picture of Fig. 2.22. If the silicon film thickness t_{Si} is larger than X_{dmax} at the acceptor concentration N_A, the FBC acts as that on PD-SOI and the signal ΔV_{th} does not depend on the silicon thickness t_{Si} but depends on N_A. However, if t_{Si} is less than X_{dmax} at the acceptor concentration N_A, it acts as that on FD-SOI and the signal ΔV_{th}becomes inversely proportional to t_{Si}. Therefore, under a constant acceptor concentration N_A, the read signal ΔV_{th} is constant in t_{Si} when t_{Si} is large enough. However, it becomes to increase with inversely proportional to t_{Si} when t_{Si}becomes smaller than X_{dmax}:

$$t_{Si} \leq X_{dmax} = \sqrt{\frac{2\varepsilon_{Si}(2\phi_B - V_b)}{qN_A}}. \tag{2.27}$$

Figure 2.23 shows the substrate bias V_{sub}dependence of the threshold voltage V_{th1} of the data "1" cell and of the threshold voltage V_{th0} of the data "0" cell for an FD-SOI FBC calculated in TCAD device simulations. Here, the substrate refers to the p-type silicon under the BOX. Two cases are shown where the BOX is 10 nm thick and 20 nm thick. The gate oxide thickness t_{ox} is 5 nm, the silicon thickness t_{Si} is 10 nm and the boron concentration N_A is 1.0×10^{15} cm^{-3} and the gate length L_g is 70 nm. As is shown in this graph, the read signal ΔV_{th} of the FD-SOI FBC is zero as $V_{th1} = V_{th0}$, when V_{sub} is positive with respect to the source terminal of the FBC. However, ΔV_{th} becomes larger as V_{sub}becomes negative and its absolute value increases. And it saturates to be a certain maximum value for a negative large V_{sub}. The curves can be explained by the substrate bias dependence of the threshold voltage of a PD-SOI FBC and of an FD-SOI FBC. Both a PD-SOI n-type MOSFET and an FD-SOI n-type MOSFET with its back surface accumulated with holes have threshold voltages which are independent of V_{sub}, because the electrical field change at the silicon surface due to the substrate voltage change is shielded by the charges in the bodies. On the other hand, for the FD-SOI n-type MOSFET whose body is completely depleted, the silicon surface electrical potential is changed by the substrate bias change and hence the threshold voltage changes depending on V_{sub}. The rate of body effect γ is calculated as[18] where

$$\gamma = \frac{dV_{th}}{dV_{sub}} = -\frac{\frac{1}{C_{ox}}}{\frac{1}{C_{BOX}} + \frac{1}{C_{Si}}} = -\frac{t_{ox}}{t_{BOX} + \frac{\varepsilon_{ox}}{\varepsilon_{Si}} t_{Si}}, \tag{2.28}$$

C_{Box} stands for the capacitance of the BOX per unit area which is expressed by

$$C_{BOX} = \frac{\varepsilon_{ox}}{t_{BOX}}, \tag{2.29}$$

using the BOX thickness t_{Box}. Equation 2.28 indicates that the threshold voltage for a MOSFET which has no charge accumulated in the body changes along a straight line which has a minus gradient with respect to the substrate bias. These straight lines are observed at the positive V_{sub} region and at the negative V_{sub} region with its absolute value not so large in Fig. 2.23. In fact, γ is calculated from Eq. 2.28 to be −0.37 for t_{ox}= 5 nm, t_{Si} = 10 and t_{Box} = 10 nm and λ = −0.21 for t_{ox} = 5 nm, t_{Si} = 10 nm, and t_{Box} = 20 nm. These calculated slopes are added in Fig. 2.23, showing that the simulated results are well explained by Eq. 2.28. Figure 2.23 can be ideally illustrated by Fig. 2.24 where three regions of different operational modes of FD-SOI FBC are separated as

1. Both "1" and "0" cells operate as FD-SOI MOSFETs (for positive V_{sub} of negative V_{sub} with smaller $|V_{sub}|$),
2. "1" cell operates as PD-SOI MOSFET, while "0" cell operates as FD-SOI MOSFET (for negative V_{sub} with medium $|V_{sub}|$), and
3. Both "1" and "0" cells operate as PD-SOI MOSFETs (for negative V_{sub} with larger $|V_{sub}|$).

And V_{sub} must be positioned so that the operational mode is set in either (2) or (3) for the FD-SOI FBC to exhibit memory function There seems to be several merits for adopting FD-SOI FBCs. As is shown in Eq. 2.19, the read signal of an FD-SOI FBC can be enlarged by thinning the silicon film thickness t_{Si} without increasing the acceptor concentration. Therefore, as will be discussed in detail in Chapter 5, an FD-SOI FBC can be regarded as

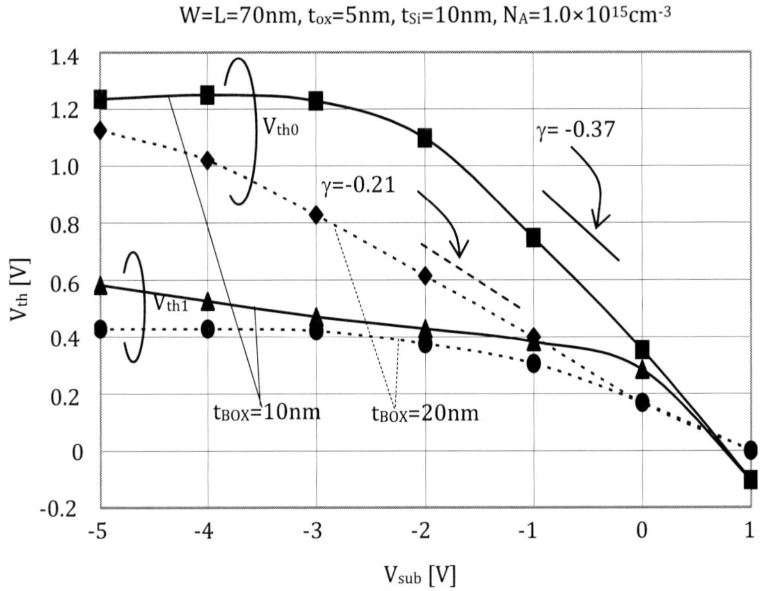

Figure 2.23 Substrate bias dependence of the read signals of two FD-SOI FBCs.

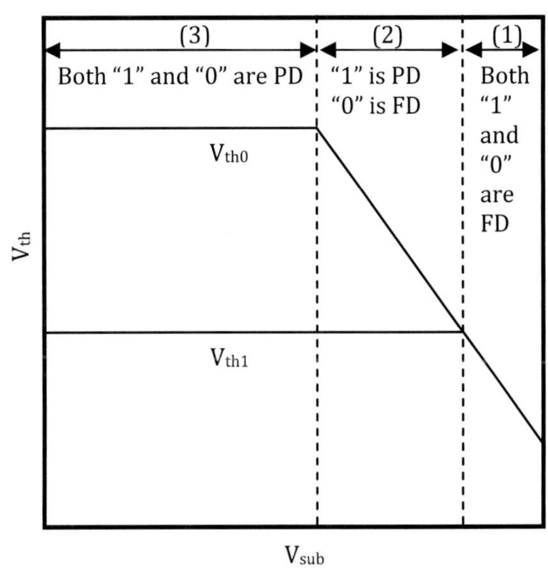

Figure 2.24 Three operational modes of FD-SOI FBC.

more scalable than the PD counterpart, because the threshold voltage variations due to random dopant fluctuations is reasonably suppressed by decreasing the acceptor concentration in the body even for very scaled FBCs. It is very important to decrease the threshold voltage fluctuation for both the "1" and the "0" cells in addition to increasing the separation between the averaged threshold voltage difference between the two data states $<\Delta V_{th}>$ to make a large memory functional, as will be explained quantitatively in Chapter 3.

As is shown in Figs. 2.23 and 2.24, there might be some concerns for malfunction due to the signal variation caused by the substrate voltage fluctuation when $|V_{sub}|$ is small, because the read signal ΔV_{th} is affected by the substrate voltage variation ΔV_{sub}. Especially when the FBC memory is embedded in a logic LSI, the substrate nose which comes from the LSI might affect the memory operation to make it unstable. In those cases, it is recommended to make the FBC memory operate at the mode (3) where $|V_{sub}|$ is so large that the read signal ΔV_{th} is not affected by the substrate voltage change ΔV_{sub}. There is no concern about the signal amount, because the signal ΔV_{th} is maximized there. It is necessary to design the value for V_{sub} with enough margin considered regarding cell-to-cell variations of the characteristics of body effect curves and V_{sub} variation range.

2.6 VARIETY IN WRITE AND READ OPERATIONS OF FBC

In the last section of this chapter, various methods of writing from and reading to FBC which have been proposed are explained. Their merits and demerits are discussed by comparing each other.

2.6.1 "1" Write

Writing "1" is performed by generating holes to be accumulated in the body. Four distinct methods for writing "1" to FBC have been proposed. Table 2.1 summarizes "1" write methods. 2.6.1.1 Impact ionization by MOSFET current[1,2,4,19–24] This is the "1" write method which was explained in sections 2.3 and 2.4. Figure 2.25a shows the polarities of the voltages which are applied to FBC to write "1" by this method. For an n-type FBC, positive voltages are applied to BL and WL to bias the FBC at saturation region, generating electron-hole pairs through impact ionization. The electrons are swept to the BL, while holes are piled up in a potential well built in the drain-body-source (n-p-n) structure. Of course, it will be possible to raise SL higher than BL for biasing FBC at saturation region. The channel current flows between the drain

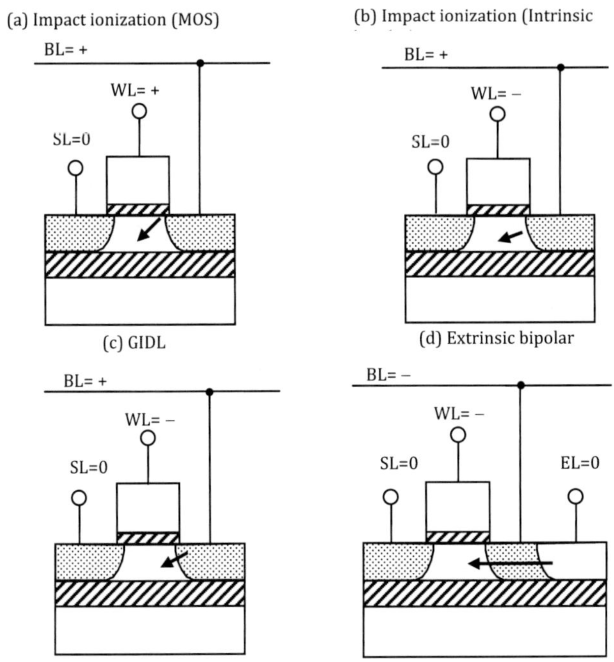

Figure 2.25 Variations of "1" write to FBC.

and the source during the write and hence it has a drawback of consuming large power in writing "1." However, the ionization current is so large that fast writing is possible, making it the most commonly adopted method for writing "1." Since only the cell which exists at the cross point where both WL and BL are high is injected with holes, a selective write can be achieved. As was discussed in section 2.4, the BL "1" disturb is inevitable when this "1" write method is adopted.

2.6.1.2 Impact ionization by intrinsic bipolar current[25]

This method was proposed recently and is depicted in Fig. 2.25b. An n-type MOSFET built on SOI is also regarded as a base floating n-p-n bipolar transistor. In this method, the bipolar transistor is sustained to be turned on with its floating base continuously supplied positive voltage by hole injection caused by impact ionization due to the collector current. As is shown in Fig. 2.25b, the cell is selected by raising WL from a negative voltage to a still-negative voltage which is closer to 0 V and BL is also raised from 0 V to a positive voltage. According to the operations, the potential of the floating body (base) also rises by the capacitive coupling to the gate and the drain (collector), making the p-n junction between the body (base) and the source (emitter) forward biased to flow a base current I_B. A collector current I_C will be induced which is determined by

$I_C = h_{FE}I_B$, where h_{FE} stands for the static common-emitter current gain.[25] The collector current I_C generates electron-hole pairs in the base at the collector edge through impact ionization and holes are being piled up in the base. If the generation rate of holes is larger than the loss rate of holes due to p-n forward current to emitter, a positive feedback loop is built up in which the transistor continues to be turn on and the number of holes accumulated in the base increases until it reaches an equilibrium state where the loss rate gets equal to the generation rate. This is the mechanism of injecting holes into the body to write "1." The parasitic bipolar action was originally intended for the method of reading data from FBC, which will be explained in 2.6.4.2. The read method can be modified to be applied to the "1" write operation by raising WL and/or BL higher than those at read to make both "1" and "0"cells turned on. The collector current flows between the collector and the emitter during the write and hence it has a drawback of consuming large power in writing "1" as is the case 2.6.1.1. The ionization current is so large that fast writing is possible and since only the cell which exists at the cross point where both WL and BL are raised is injected with holes, a selective write can be achieved. This method also brings about the concern of the BL "1" disturb.

2.6.1.3 GIDL (band-to-band tunneling)[27–29]

This is the method to write "1" by the band-to-band tunneling current or so-called gate GIDL.[11,12] Figure 2.25c shows the polarities of applied voltages. A large negative voltage is applied to the gate with respect to the drain by setting the gate voltage negative and the drain voltage positive, leading the silicon surface at the drain edge to a deep depletion state to let electrons in the valence band to tunnel into the conduction band with holes created in the valence band and accumulated in the body. One of important merits of this method is a low power write. Since there is no channel induced at the silicon surface, no drain current flows in this method, providing a very low power "1" write means. By raising BL voltage positive, a selective write is also possible. However, the BL "1" disturb must be taken care of. If the selectivity is sacrificed by remaining BL 0 V when writing "1" with WL voltage negative, the BL "1" disturb can be avoided. In the latter operation, all cells along a selected WL are written "1" at once. Therefore, a read operation using sense amplifier circuits (S/As) is required for all cells along a selected WL prior to the "1" write operation and the "0" data latched in the S/As are need to be written back after the "1" write by using a certain "0" write method which is selective. Thus, BL "1" disturb is avoided at the cost of additional "0" write cycle which is necessary after the "1" write. The selection of write methods, in this manner, depends on the designer's priority on which characteristics or failure modes should be taken most seriously. Though fast write is possible by using GIDL, the reliability issue of the gate dielectrics needs to be taken care of when a large negative voltagc is applied to the gate.

2.6.1.4 Extrinsic bipolar current[30]

As shown in Fig. 2.25d, a p-type emitter region is attached to the drain of an n-type MOSFET FBC and the p-n-p bipolar transistor which consists of the body, the drain, and the emitter is built in outside the n-type MOSFET. Holes are injected into the body of FBC by turning on the p-n-p bipolar transistor to write "1" to the FBC. More precisely, BL is lowered negative while the emitter line EL stay 0 V to turn on the p-n-p bipolar transistor, injecting holes into the body of the FBC. WL can also lowered negative to prevent the silicon surface from being inverted. Though a large p-n diode current flows during the write, the efficiency of charging the body (the hole current to charge the body divided by the diode current) is very high compared with the method using the impact ionization and hence low power write is possible. The time to complete to write "1" by using the extrinsic bipolar current is very short. The BL "1" disturb is avoided because BL is not necessary to be raised high at the "1" write. However, since the emitter region must be added to a MOSFET which cannot be shared with the adjacent cells, the cell size becomes large compared to the FBCs adopting other "1" write methods.

Table 2.1 "1" write variations and their pros and con[c]

"1" write method	WL	BL	SL	EL	Power	Disturb	Selective	Size
Impact ionization (MOS)	+	+	0 V	NA	X	X	O	$6F^2$
Impact ionization (bipolar)	–	+	0 V	NA	X	X	O	$6F^2$
GIDL (1)	–	+	0 V	NA	O	X	O	$6F^2$
GIDL (2)	–	0 V	0 V	NA	O	O	X	$6F^2$
Extrinsic bipolar	–	–	0 V	0 V	X	O	O	$14F^2$

[c] The alphabets X and O in the column Power denote consuming and not consuming DC power, respectively. Those in the column Disturb denote having and not having the concern of BL "1" disturb, respectively. Those in the column Selective denote not having and having the "1" write selectivity, respectively.

2.6.2 "0" Write

Writing "0" is performed by extracting holes from the body. Eight distinct methods for writing "0" to FBC have been proposed. These eight "0" write methods are summarized in Table 2.2.

2.6.2.1 P-N forward with BL minus[1,2,4,19,20,23]

This is what was explained in sections 2.3 and 2.4 as is shown in Fig. 2.26a. After WL is raised high, BL is lowered to a negative voltage with respect to SL, making the p-n junction between the body and the drain of a "1" cell forward biased to eject holes which are accumulated in the body. Though DC power is consumed during write, it is the method most commonly used because it can achieve a selective "1" write operation as was explained in section 2.4.

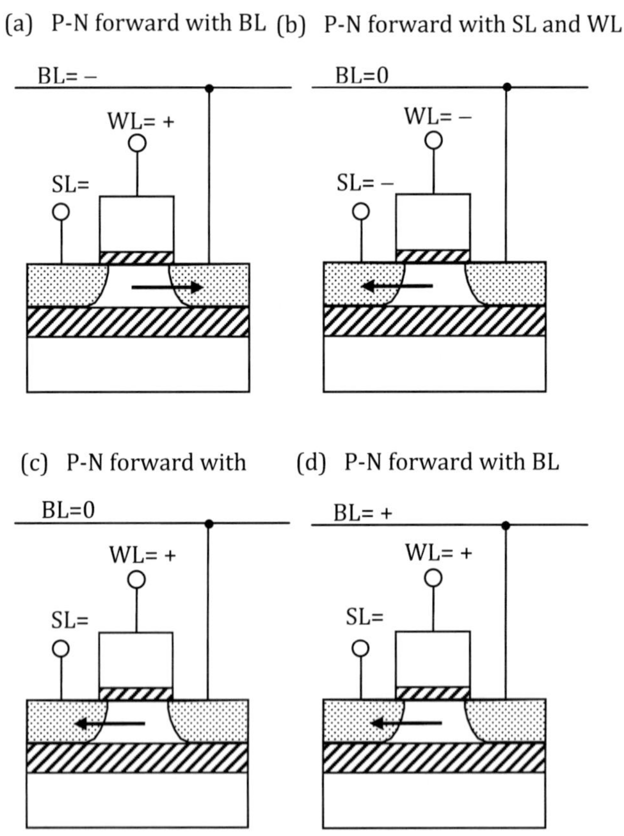

Figure 2.26 Variations of "0" write to FBC (to be continued).

2.6.2.2 P-N forward with SL and WL minus[22]

According to this method which is shown in Fig. 2.26b, holes are extracted to SL by lowering SL to a negative voltage instead of BL which stays 0 V in this method. The p-n junction between the body and the source becomes forward to eject holes from the body. Since BL remains 0 V during the "0" write, there is no concern about BL "0" disturb which cannot be avoided in the method explained in 2.6.2.1. However, all cells along SL are equally written "0."[d] Therefore, as is the case GIDL (2) in Table 2.1, the two-cycle write operation is required, because a read operation using S/As is required for all cells along a selected SL prior to the "0" write operation and the "1" data latched in the S/As are need to be written back after the "0" write by using a certain "1" write method which is selective. Since WL cab be set at the same level as SL or lower, no DC power is consumed. However, cell size needs to be larger because SL cannot be shared with two adjacent cells.

2.6.2.3 P-N forward with SL minus[31]

This is a modification to the method of 2.6.2.2 intended for enhancing the read signal as is shown in Fig. 2.26c. After raising WL high to make the body potential of the data "1" cell high through the gate-body capacitive coupling, SL is lowered to a negative voltage. Since the voltage applied to the p-n junction between the body and the source is larger than that in the case of 2.6.2.2 for the same negative voltage of SL, the efficiency of extracting holes is larger that that of the case of 2.6.2.2, leading to a larger read signal than the previous case of 2.6.2.2. This method consumes DC power during write and has no selectivity in writing along the SL. However, since the selected WL differentiates the body potential of the selected cell from those of other unselected cells along BLs, SLs can be designed to be shared with two adjacent cell, achieving a small cell size.

2.6.2.4 P-N forward with BL plus[32]

As is shown in Fig. 2.26d, this method is different from those previously explained from the view point of the direction of extracting holes relative to the driving terminal. In the previous three cases, holes are extracted into the terminal to which a negative voltage is applied. In this method, on the other hand, holes are extracted to the terminal opposite to one to which a positive voltage is applied. Holes are "pulled in" for the cases 2.6.2.1 to 2.6.2.3, while they are "pushed out" for this case 2.6.2.4 as it were. WL is raised high to increase the efficiency of write. The WL level must be higher than the BL level plus the threshold voltage of the cell for the cell to be biased at linear

---

[d] This type of collective write is sometimes called "purge" probably after flash memory operation of simultaneous extraction of electrons from the floating gates of all cells in a block.

region. DC power is consumed during write and the selectivity of write is guaranteed.

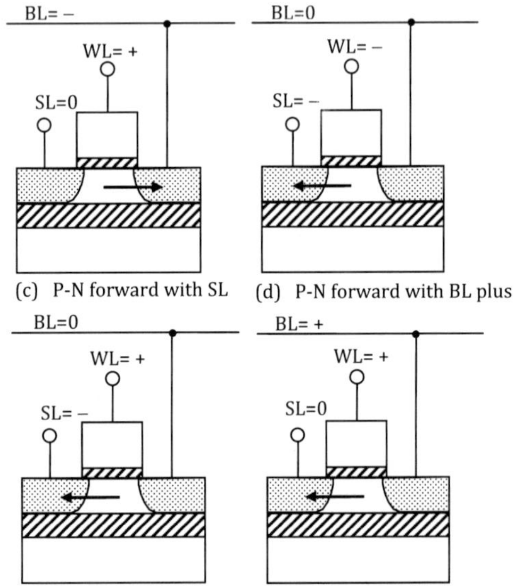

Figure 2.26 Variations of "0" write to FBC (continued from the previous figure).

2.6.2.5 P-N forward with BL plus and WL 0 V

This is a modification to the method of 2.6.2.4 intended for reducing power which is shown in Fig. 2.26e. Since WL is 0 V to prevent the silicon surface from being inverted, the channel current between the source and the drain does not flow. However, signal will be smaller than that of 2.6.2.4, because the body potential does not rise as high as that in the case of 2.6.2.4 through the capacitive coupling between the gate and the body.

2.6.2.6 P-N forward with BL and back word line plus[33–35]

This method which is illustrated in Fig. 2.26f is applied to an FBC having a back-gate whose voltage is controllable in addition to a front-gate like the structure of a double-gate SOI MOSFET. Holes accumulated in a "1" cell are ejected by raising both WL formed by the front-gate and back word line (BWL) formed by the back-gate higher than their respective hold levels. This operation is a purge in a sense that holes in all "1" cells along the WL and the BWL are evacuated simultaneously. Therefore, the two-cycle write is required where a selective "1" write using S/As which are latching data stored in all cells along the WL follows this operation of "0" write. Since the channel current doesn't flow, a low power write is achieved. In addition, there is no concern about the BL "0" disturb.

2.6.2.7 Charge pumping[36]

This method is what was explained in Fig. 1.13 in section 1.3. The voltage application is shown in Fig. 2.26g. The gate voltage is essentially a positive pulse or positive pulses. When a single or several positive voltage pulses are applied to an n-type MOSFET, the data "0" is written with holes extracted from the body in making use of the charge pumping phenomenon. A state-of-the-art MOSFET has its channel so short that the rate of electron-hole pair recombination is very low for residual channel electrons.[37] Therefore, the charge pumping which recombines the accumulated holes with residual channel electrons when V_{gs} goes from high to low cannot be used for the method of writing "0." And, unfortunately, another charge pumping mechanism which recombines the accumulated holes with electrons which are trapped in the interface states at the gate oxide when V_{gs} goes from high to low cannot be used either for the method of writing "0." The reason will be explained as follows. The interface state density of a state-of-the-art FBC is in the order of 10^{10} cm^{-2}[38,39]; therefore, only a few electrons can be trapped at the interface states when the channel is formed whose area is 100 × nm × 100 nm = 10^{-10} cm^2. The number of holes which are accumulated in the body of the "1" cell of this size is more than several hundreds as will be shown in Chapter 5. Therefore, it takes more than a few dozen cycles for the "1" cell to be changed to a "0" cell. Furthermore, the number of the interface states fluctuates largely according to Poisson distribution as will be shown in sections 6.6 and 7.5. It seems difficult for the charge pumping mechanism to be used as a method of writing "0" to state-of-the-art FBCs.

2.6.2.8 Gate direct tunneling[40]

It is possible to write "0" by the direct tunneling current of the gate dielectrics of a MOSFET which is shown in Fig. 2.26h. The gate dielectrics thickness needs to be around 3 nm or thinner. No channel current flows in this method, providing a means for low power write. The BL "0" disturb does not take place, since BL can remain at 0 V. If this method is combined with the "1" write by GIDL (1) in Table 2.1, both data "1" and "0" can be written simultaneously depending on the level of BLs which are 0 V for "0" write and high voltage for "1" write., achieving a low power and high speed write.[40]

2.6.3 Criteria for Choosing Write Methods

As for write operations of FBC, a suitable combination between a "1" write method from Table 2.1 and a "0" write method form Table 2.2 can be chosen.

Either the "1" write method or the "0" write method chosen must be equipped with the selectivity of write. One cycle write can be possible only when a combination of a "1" write method and a "0" write method both are selective is chosen. If you chose a "1" write method which is not selective, the "1" write operation must be preceded by a read operation for the data of all cells selected by a WL to be latched in S/As and must be followed by a selective "0" write according to the latched data and vice versa, requiring the two-cycle write.

The DC power which is consumed in a cell during write affects the total write power of a memory and it is important to reduce it for a low power operation. However, it is worth noting that the total write power of a memory is also affected by the BL and WL charging power which consumes when the large capacitances of the BLs and WLs are to be charged when they are biased from their holding voltages. Especially, since the number of BLs which are to be charged is very large, the AC power due to the BL charging is much larger than the AC power of WL charging and can not be neglected even compared with the DC power of cells.

Table 2.2 "0" write variations and their pros and cons[e]

"0" write method	WL	BL	SL	BWL	Power	Disturb	Selective	Size
P-N forward (BL minus)	+	–	0 V	NA	X	X	O	$6F^2$
P-N forward (SL&WL minus)	–	0 V	–	NA	O	O	X	$10F^2$
P-N forward (SL minus)	+	0 V	–	NA	X	O	X	$6F^2$
P-N forward (BL plus)	+	+	0 V	NA	O	X	O	$6F^2$
P-N forward (BL plus/WL = 0 V)	0 V	+	0 V	NA	O	X	O	$6F^2$
P-N forward (WL&BWL plus)	+	0 V	0 V	+	O	O	X	$6F^2$
Charge pumping	+	0 V	0 V	NA	O	O	X	$6F^2$
Gate tunneling	–	0 V	0 V	NA	O	O	X or O[f]	$6F^2$

[e] The meanings of the alphabets X and O in the columns *Power*, *Disturb*, and *Selective* are the same as those in Table 2.1.

[f] Though this write doesn't have the selectivity of write in general, it can have it only when it is combined with the *GIDL* (1) in Table 2.1.

However, the most important criterion for choosing write methods is to maximize the read signal which the chosen write operations can achieve. The read signal of an FBC ΔV_{th} changes depending on the write methods to be adopted. And it is known that different structures of FBC have their own optimum methods of write which maximize the signal. Referring to "0" write methods, for example, the "pull in" write operations such as Fig. 2.26a–c are suitable for an FBC having a thin silicon film and a thin BOX, while the "push out" write operations such as Fig. 2.26d,e are suitable for an FBC having a thick silicon film and a thick BOX from a view point of maximizing the read signal. This is because the body potential change caused by the potential change of the drain (source) depends on the ratio of the body-drain (source) capacitance to the body-substrate capacitance. If this ratio is small, the "pull in" "0" write methods can eject holes more effectively from the body. If this ratio is large, the "push out" "0" write methods can eject holes more effectively from the body.

Another important issue which should be considered is the BL disturbs which certainly shorten a static retention time of an FBC. Referring to Tables 2.1 and 2.2, it is noticed that there are only two methods which are disturb-free and, at the same time, selective. One is the "1" write method by extrinsic bipolar, the other is the "0" write method by gate direct tunneling which is combined with the GIDL "1" write with BL high. Since the latter must be combined with the "1" write method which is not BL disturb-free, there remains only one method which can provide an overall disturb-free writing system, i.e., external bipolar "1" write. This "1" write method can be combined with any "0" write methods which are disturb-free without write selectivity. A demerit of this strategy is having to adopt a large cell. It is a big challenge to search for a disturb-free selective write method without sacrificing the small cell size.

2.6.4 Read

So far there are proposed two distinct methods for reading data from FBC as are shown in Fig. 2.27.

2.6.4.1 MOS current[36]

This read method is to distinguish the "1" cell MOS current from the "0" cell MOS current due to their threshold voltage difference caused by the difference of hole numbers between the "1" cell and the "0" cell through the body effect as was explained in section 2.3. This is a classical method[36] and is exclusively used until a novel read method which will be explained in 2.6.4.2 was recently proposed.

2.6.4.2 Bipolar current[25]

This new read method makes use of a sustained collector current of a floating base parasitic bipolar transistor formed in a MOSFET structure of an FBC. The difference of the number of holes accumulated in the body of an FBC affects the sensitivity for the base floating bipolar transistor to make the collector current sustained in a latched state. Figure 2.28 explains the principle of the bipolar current read. The body potential of a "1" cell which

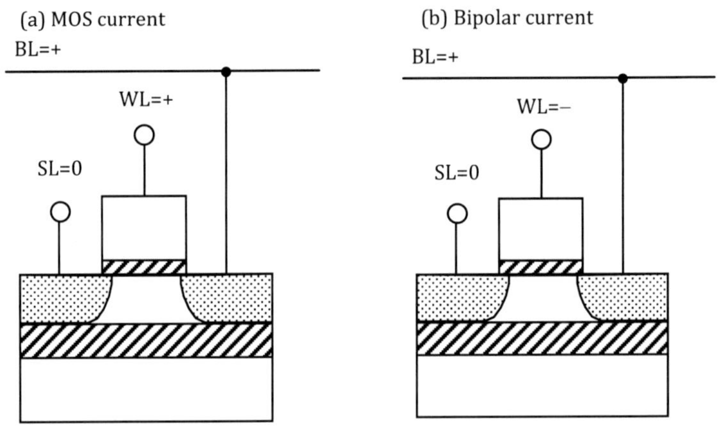

Figure 2.27 Variations of read from FBC.

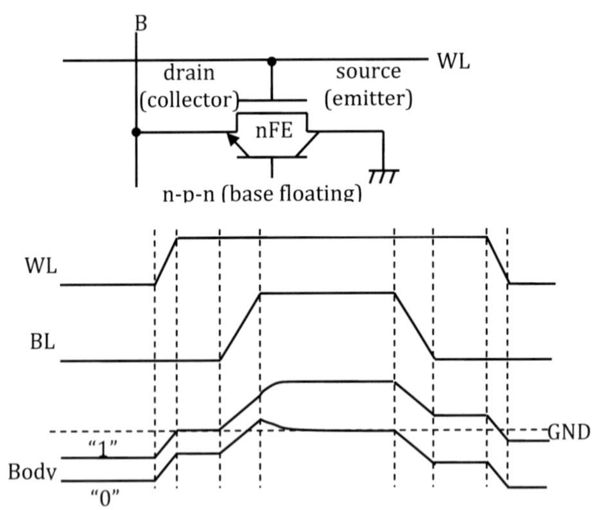

Figure 2.28 Principle of bipolar current read.

is selected by a WL is raised higher than that of a "0" cell which is also selected, because of their initial potential difference. After that, BL is raised to a high level, making the body potentials of both the "1" cell and the "0" cell forward to the emitter. The forward bias voltage between the base and the emitter of the "1" cell is larger than that of the "0" cell, making the base current of the "1" cell I_{B1} larger than that of the "0" cell I_{B0}. If the static common-emitter current gain is denoted by h_{FE}, the "1" cell and the "0" cell flow the collector currents $I_{C1} = h_{FE}I_{B1}$ and $I_{C0} = h_{FE}I_{B0}$, respectively. These collector currents generate electron-hole pairs in the base by impact ionization, letting electrons swept into the collector and holes accumulated in the base. Since $I_{C1} > I_{C0}$, the generation rate of holes of the "1" cell is larger than that of the "0" cell. On the other hand, since the p-n junction between the base and the emitter is forward, holes are flowing out of the base. If the rate of generating holes is larger than the rate of flowing out of the base, a positive feedback is latched and the collector current continues to flow with the body potential sustained strongly forward biased. If the rate of generating holes is smaller than the rate of flowing out of the base, the collector current decreases to zero with the body potential decays to GND. It is possible to tune the base potentials of the "1" cell and the "0" cells by changing the WL and the BL high voltages so that the border from which the latched state is separated by the unlatched one is positioned between the "1" cell and the "0" cell. Thus, by choosing the BL and the WL high voltages properly, the "1" cell turns on the collector current with the "0" cell remained turned off. This is the principle of the bipolar current read. As was mentioned in 2.6.1.2, this principle can also be made use of for writing "1" where the high voltages of BL and/or WL is set higher than those at read to make both the floating base bipolar transistors of the "1" cell and the "0" cell turned on. According to this method, the WL high level can be minus with respect to the source (emitter) terminal potential, because the silicon channel needs not to be formed as is the case of the MOS current read method. Therefore, an FD-SOI FBC having a thin silicon film can hold holes under the front-gate forming WL at read, eliminating the necessity of applying a negative voltage at the back-gate to make the FD-SOI FBC function. This is a big advantage over the MOS current read method, because FBC can be constructed by a simple FinFET and a surrounded gate transistor (SGT) in addition to a thin silicon film FD-SOI transistor.[26]

When data is read by the MOS current, these transistors must have a second gate which makes a potential well to hold holes at read with the first gate biased at a positive voltage, i.e., thin BOX or double-gate structure is mandatory for a planar FD-SOI MOSFET and the independent-gate structure is also required for a FinFET. It might be difficult for a simple SGT to be used as FBC unless the bipolar current read method is adopted.

The read current ratio between the "1" cell and the "0" cell I_1/I_0 is very large and the large I_1 may have a merit in designing S/As to detect the data in a short period of time. However, the challenge the bipolar current read faces is a relatively high BL voltage to be applied at read for turning on the parasitic bipolar transistor. This high voltage inevitably deteriorates the data "0" state of unselected cells during hold, another disturb which is named BL read disturb. The feasibility of this read method hinges on whether the turn on BL voltage can be lowered to a reasonable level.

References

1. T. Ohsawa, "Semiconductor memory device with floating storage bulk region and method of manufacturing the same," *U.S. Patent 6,621,725*, Sept. 16, 2003.
2. T. Ohsawa, K. Fujita, T. Higashi, Y. Iwata, T. Kajiyama, Y. Asao and K. Sunouchi, "Memory design using one-transistor gain cell on SOI," *ISSCC Dig. Tech. Papers*, pp. 152–153, Feb. 2002.
3. S. M. Sze, *Physics of Semiconductor Devices*, 2nd Edition, p. 442, John Wiley & Sons, New York, 1981.
5. T. Ohsawa, K. Fujita, K. Hatsuda, T. Higashi, M. Morikado, Y. Minami, T. Shino, H. Nakajima, K. Inoh, T. Hamamoto and S. Watanabe, "An 18.5ns 128Mb SOI DRAM with a floating body cell," *ISSCC Dig. Tech. Papers*, pp. 458–459, Feb. 2005.
6. T. Ohsawa, K. Fujita, K. Hatsuda, T. Higashi, T. Shino, Y. Minami, H. Nakajima, M. Morikado, K. Inoh, T. Hamamoto, S. Watanabe, S. Fujii and T. Furuyama, "Design of a 128-Mb SOI DRAM using the floating body cell (FBC)," *IEEE J. Solid-State Circuits*, vol. 41, no. 1, pp. 135–145, Jan. 2006.
4. T. Ohsawa, K. Fujita, T. Higashi, Y. Iwata, T. Kajiyama, Y. Asao and K. Sunouchi, "Memory design using a one-transistor gain cell on SOI," *IEEE J. Solid-State Circuits*, vol. 37, no. 11, pp. 1510–1522, Nov. 2002.
7. S. M. Sze, *Physics of Semiconductor Devices*, 2nd Edition, p. 452, John Wiley & Sons, New York, 1981.
8. S. Tam, P.-K. Ko and C. Hu, "Luck-electron model of channel hot-electron injection in MOSFET's," *IEEE Trans. Electron Devices*, vol. ED-31, no. 9, pp. 1116–1125, Sept. 1984.
9. S. M. Sze, *Physics of Semiconductor Devices*, 2nd Edition, p. 89, John Wiley & Sons, New York, 1981.
10. M. R. Tack, M. Gao, C. L. Claeys and G. J. Declerck, "The multistable charge-contro-lled memory effect in SOI MOS transistors at low temperatures," *IEEE Trans. Electron Devices*, vol. ED-37, no. 5, pp. 1373–1382, May 1990.
11. C. Chang and J. Lien, "Corner-field induced drain leakage in thin oxide MOSFETs," *IEDM Tech. Dig.*, pp. 714–717, Dec. 1987.
12. T. Y. Chang, J. Chen, P. K. Ko and C. Hu, "The impact of gate-induced drain leakage current on MOSFET scaling," *IEDM Tech. Dig.*, pp. 718–721, Dec. 1987.
13. T. Ohsawa, T. Higashi, K. Fujita, K. Hatsuda, N. Ikumi, T. Shino, H. Nakajima, Y. Minami, N. Kusunoki, A. Sakamoto, J. Nishimura, T. Hamamoto and S. Fujii, "A 128Mb floating body RAM (FBRAM) on SOI with multi-averaging scheme of dummy cell," *2006 Symposium on VLSI Circuits Dig. Tech. Papers*, pp. 224–225, June 2006.
14. T. Ohsawa, "Semiconductor memory device," *U.S. Patent 6,617,651*, Sept. 9, 2003.

15. J.-P. Colinge, *Silicon-on-Insulator Technology: Materials to VLSI*, 3rd Edition, p. 203, Kluwer Academic Publishers, Boston, 2004.
16. J.-P. Colinge, *Silicon-on-Insulator Technology: Materials to VLSI*, 3rd Edition, p. 161, Kluwer Academic Publishers, Boston, 2004.
17. C. Kuo, T.-J. King and C. Hu, "A capacitorless double-gate DRAM cell," *IEEE Electron Device Lett.*, vol. 23, no. 6, pp. 345–347, June 2002.
18. J.-P. Colinge, *Silicon-on-Insulator Technology: Materials to VLSI*, 3rd Edition, p. 165, Kluwer Academic Publishers, Boston, 2004.
19. S. Okhonin, M. Nagoga, J. M. Sallese and P. Fazan, "SOI capacitor-less 1T-DRAM concept," *IEEE Int. SOI Conf.*, pp. 153–154, Oct. 2001.
20. S. Okhonin, M. Nagoga, J. M. Sallese and P. Fazan, "A capacitor-less 1T-DRAM cell," *IEEE Electron Device Lett.*, vol. 23, no. 2, pp. 85–87, Feb. 2002.
21. A. Villaret, R. Ranica, P. Mazoyer, P. Candelier, F. Jacquet, S. Cristoloveanu and T. Skotnicki, "Triple-well nMOSFET evaluated as a capacitor-less DRAM cell for nanoscale low-cost & high-density applications," *2003 Silicon Nanoelectronics Workshop*, pp. 40–41, June 2003.
22. R. Ranica, A. Villaret, P. Malinge, P. Mazoyer, D. Lenoble, P. Candelier, F. Jacquet, P. Masson, R. Bouchakour, R. Fournel, J. P. Schoellkopf and T. Skotnicki, "A one transistor cell on bulk substrate (1T-Bulk) for low-cost and high-density eDRAM," *2004 Symposium on VLSI Technology Dig. Tech. Papers*, pp. 128–129, June 2004.
23. I. Ban, U. E. Avci, U. Shah, C. E. Barns, D. L. Kencke and P. Chang, "Floating body cell with independently-controlled double gates for high-density memory," *IEDM Tech. Dig.*, pp. 573–576, Dec. 2006.
24. C. Kuo, T.-J. King and C. Hu, "A capacitorless double-gate DRAM cell," *IEEE Electron Device Lett.*, vol. 23, no. 6, pp. 345–347, June 2002.
25. S. Okhonin, M. Nagoga, E. Carman, R. Beffa and E. Faraoni, "New generation of Z-RAM," *IEDM Tech. Dig.*, pp. 925–928, Dec. 2007.
26. S. M. Sze, *Physics of Semiconductor Devices*, 2nd Edition, p.141, John Wiley & Sons, New York, 1981.
27. F. Horiguchi, T. Ohsawa, Y. Iwata and T. Yamada, "Semiconductor memory device," *U.S. Patent 6,548,848*, Apr. 15, 2003.
28. E. Yoshida and T. Tanaka, "A design of a capacitorless 1T-DRAM cell using gate-induced drain leakage (GIDL) current for low-power and high-speed embedded memory," *IEDM Tech. Dig.*, pp. 913–916, Dec. 2003.
29. E. Yoshida and T. Tanaka, "A capacitorless 1T-DRAM technologiy using Gate-Induced Drain Leakage (GIDL) current for low-power and high-speed embedded memory," *IEEE Trans. Electron Devices,* vol. 53, no. 4, pp. 692–697, Apr. 2006.
30. T. Ikehashi and T. Ohsawa, "Semiconductor integrated circuit device," *U.S. Patent 6,825,524*, 2004.
31. T. Ohsawa, "Semiconductor memory device," *U.S. Patent Application Publication US 2008/0130379 A1*, June 5, 2008.

32. P. Fazan, S. Okhonin, M. Nagoga and J.-M. Sallese, "A simple 1-transistor capacitor-less memory cell for high performance embedded DRAMs," *Proc. IEEE 2002 Custom Integrated Circuits Conf.*, pp. 99–102, 2002.

33. C. Kuo, T.-J. King and C. Hu, "A capacitorless double-gate DRAM cell," *IEEE Electron Device Lett.*, vol. 23, no. 6, pp. 345–347, June 2002.

34. C. Kuo, T.-J. King and C. Hu, "A capacitorless double-gate DRAM cell design for high-density applications," *IEDM Tech. Dig.*, pp. 843–846, Dec. 2002.

35. C. Kuo, T.-J. King and C. Hu, "A capacitorless double-gate DRAM technology for sub-100-nm gate embedded and stand-alone memory applications." *IEEE Trans. Electron Devices*, vol. 50, no. 12, pp. 2408–2416, Dec. 2003.

36. N. Sasaki, "Charge pumping in SOS-MOS transistors," *IEEE Trans. Electron Devices*, vol. ED-28, no. 1, pp. 48–52, Jan. 1981.

37. J. S. Brugler and P. G. A. Jespers, "Charge pumping in MOS devices," *IEEE Trans. Electron Devices*, vol. ED-16, no. 3, pp. 297–302, March 1969.

38. G. Groeseneken, H. E. Maes, N. Beltran and R. F. De Keersmaecker, "A reliable approach to charge-pumping measurements in MOS transistors," *IEEE Trans. Electron Devices*, vol. ED-31, no. 1, pp. 42–53, Jan. 1984.

39. S. Okhonin, M. Nagoga and P. Fazan, "Principles of transient charge pumping on partially depleted SOI MOSFETs," *IEEE Electron Device Lett.*, vol. 23, no. 5, pp. 279–281, May 2002.

40. S. Okhonin and M. Nagoga, "Memory cell having an electrically floating body transistor and programming technique therefor," *U.S. Patent 7, 476, 939*, 2009.

Chapter 3

Signal of Floating Body Cell

Optimum structures of floating body cell (FBC) for maximizing the signal are discussed. Two kinds of signal-to-noise ratios (SNRs) of FBC, one with respect to threshold voltage difference and the other with respect to cell current difference, are defined. It is shown that the latter generally provides more accurate index for predicting the yield of a memory using FBC. The minimum SNR which is necessary for making a multiple-gigabit memory functional is calculated. Hardware measurement results of SNR for a state-of-the-art FBC memory chip which demonstrate the validity of the theory and also satisfy the minimum SNR criteria will be given in Chapter 4.

3.1 IMPORTANT FACTORS DETERMINING MAGNITUDE OF FLOATING BODY CELL SIGNAL

Binary data stored in a floating body cell (FBC) is distinguished by the cell read current difference between the "1" cell and the "0" cell. The signal of FBC, therefore, can be defined by the threshold voltage difference $\Delta V_{th} = V_{th0} - V_{th1}$ between the threshold voltage of the "0" cell V_{th0} and that of the "1" cell V_{th1} or by the read current difference $\Delta I_{cell} = I_{cell1} - I_{cell0}$ between the cell read current of the "1" cell I_{cell} and that of the "0" cell I_{cell0}. The former definition is more fundamental, because it is essentially independent of bias voltages and device dimensions such as channel width. The latter, on the other hand, depends on bias conditions and measured device sizes, though this must be the ultimate physical quantity which will be detected by sense amplifier (S/A) circuits. The validity of the signal definitions for FBC will be discussed in section 3.2 in detail. In the meantime, let us presume that the magnitude of FBC signal is proportional to the body voltage difference between the "1" cell and the "0" cell at read operation.

The factors which determine the magnitude of FBC signal will be studied. Figure 3.1 shows the process in which the read current difference is determined staring from writing data to reading it through data holding. In reality, data may degrade during being held without disturb (static retention) and with disturb (dynamic retention). However, in this section, the data

Floating Body Cell: A Novel Capacitor-less DRAM Cell
By Takashi Ohsawa and Takeshi Hamamoto

www.panstanford.com

degradation due to retention is not considered. The first principle signal of FBC which is read just after being written will be discussed. As is shown in this figure, if V_{WLHW} is applied to word line (WL) and V_{BLH} or V_{BLL} is applied to bit line (BL) depending on writing the data "1" or "0" during t_w for an FBC with certain characteristics of drain current drivability, impact ionization current, and p-n junction current, the body voltages of the data "1" V_{bw1} and the data "0" V_{bw0} and their difference $\Delta V_{bw} = V_{bw1} - V_{bw0}$ are determined. The data is held in the state where WL is V_{WLL} which is set negative with respect to SL and BL is set 0 V. This hold state defines the body voltages of the both data V_{bh1} and V_{bh0} at hold and their difference $\Delta V_{bh} = V_{bh1} - V_{bh0}$ at hold. To read the data, WL is raised to V_{WLHR} and BL is biased at V_{BLR}, where V_{WLHR} is equal to or lower than V_{WLHW} and V_{BLR} is as small as 0.2 V, for example. This read state defines the body voltages of the both data V_{br1} and V_{br0} at read and their difference $\Delta V_{br} = V_{br1} - V_{br0}$ at read. The body voltage difference at read brings about the threshold voltage difference ΔV_{th} at read by body effect and the cell drain current characteristics determine the cell read current difference ΔI_{cell}.

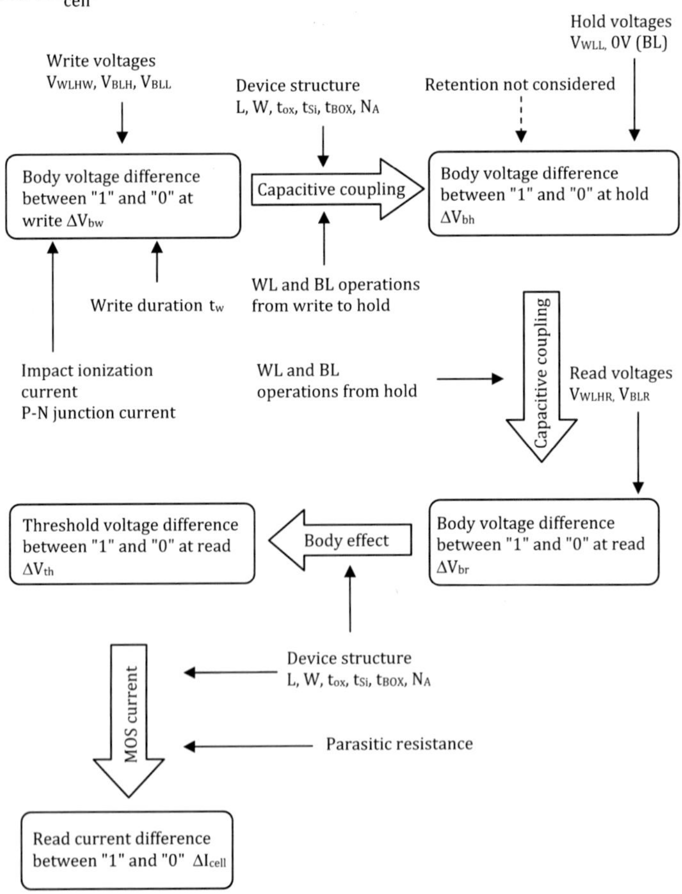

Figure 3.1 Factors which determine the FBC signal ΔI_{cell}.

To increase the body voltage of the data "1" at write, it is effective to increase the impact ionization current of the cell. The body voltage of the data "1" V_{bw1} is determined by the balance between the impact ionization current I_{sub} and the p-n junction forward current between the body and the source I_{fwd}. When the saturation current of the p-n junction between the body and the source is written as I_s, the balance condition to determine V_{bw1} is expressed by

$$I_{sub} = I_s\left[\exp\left(\frac{q \cdot V_{bw1}}{kT}\right) - 1\right] \tag{3.1}$$

Since I_{sub} is a complicated function of V_{bw1}, it is difficult to solve Eq. 3.1 for V_{bw1} analytically. However, when Eq. 3.1 is expressed as

$$V_{bw1} = \frac{kT}{q}\ln\left(\frac{I_{sub}}{I_s} + 1\right), \tag{3.2}$$

it can be understood that V_{bw1} is balanced at a higher voltage as the ratio I_{sub}/I_s is larger. Therefore, to increase $\Delta V_{bw} = V_{bw1} - V_{bw0}$, it is important to optimize the WL and the BL voltages at "1" write for the impact ionization current I_{sub} to increase. The impact ionization current I_{sub} increases, when the BL voltage during writing "1" V_{BLH} becomes large. However, there is an upper limit because the channel current increase with the larger V_{BLH} leads to the increase the power and the BL "1" disturb becomes the more severe as was explained in section 2.4.

To decrease the body voltage of the data "0" at write V_{bw0} for enhancing ΔV_{bw}, it is effective to lower the BL "0" write level V_{BLL} as was expressed in Eq. 2.9. However, there is also a lower limit, because the BL "0" disturb becomes the more severe as was explained in section 2.4. And since the data "0" is defined not by its equilibrium point as is the case of the data "1" write but by the write time t_w, V_{bw0} becomes lower as t_w increase. However, there is in an upper limit for t_w which comes from the product spec for write time.

The subject to be discussed in this section is not related to the method to increase ΔV_{bw} but to the method to keep the original signal ΔV_{bw} in the process from write to read through hold. As you can see in Fig. 2.4a in Chapter 2, the body voltage difference between the data "1" cell and the data "0" cell at hold ΔV_{bh} is smaller than the original difference just after write ΔV_{bw} and the body voltage difference at read ΔV_{br} which is directly affects the signal is smaller than that at hold ΔV_{bh}. Here, we discuss the reason why the body voltage difference decreases as it undergoes the transition from write to read through hold and the method how to minimize the signal degradation.

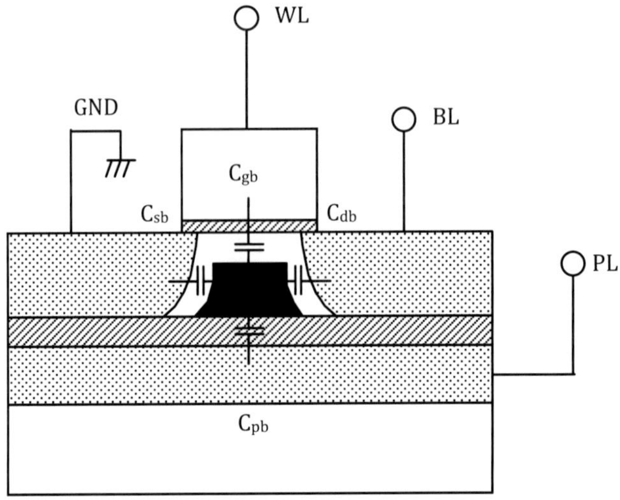

Figure 3.2 Capacitive couplings between the body neutral region and other FBC's terminals at cut-off region.

The body voltage at write changes from write to read through hold according to the capacitive coupling between the body and the gate and between the body and the drain. First, let us define several capacitive coupling coefficients which govern the body voltage shift by the gate and drain voltage changes. Figures 3.2–3.4 illustrate capacitive couplings between the body neutral region and four other terminals (gate, drain, source, and substrate) and a channel for an SOI-MOSFET at cut-off region, linear region, and saturation region, respectively. We define the capacitive coupling coefficient between the gate and the body at cut-off region as

$$\lambda_g = \frac{C_{gb}}{C_{gb} + C_{db} + C_{sb} + C_{pb}}, \tag{3.3}$$

where C_{gb}, C_{db}, C_{sb}, and C_{pb} are the capacitances between the gate and the body, the drain and the body, the source and the body, and substrate (plate) and the body, respectively. In the same way, we define the capacitive coupling coefficient between the drain and the body at cut-off region as

$$\lambda_d = \frac{C_{db}}{C_{gb} + C_{db} + C_{sb} + C_{pb}}. \tag{3.4}$$

For linear region, we define the capacitive coupling coefficient between the drain and the body as

$$\lambda_{dl} = \frac{C_{db} + C_{dep}/2}{C_{dep} + C_{db} + C_{sb} + C_{pb}}, \tag{3.5}$$

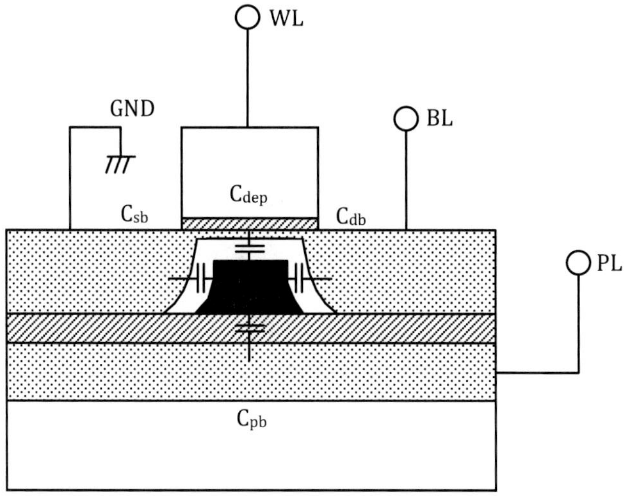

Figure 3.3 Capacitive couplings between the body neutral region and other FBC's terminals at linear region.

where C_{dep} represents the capacitance of depletion region formed under the channel. For saturation region, we define the capacitive coupling coefficient between the drain and the body as

$$\lambda_{\text{ds}} = \frac{C_{\text{db}}}{C_{\text{dep}} + C_{\text{db}} + C_{\text{sb}} + C_{\text{pb}}}. \tag{3.6}$$

These coefficients are the capacitances between the gate and the body or the drain and the body divided by the body total capacitance in three different operating states. The capacitances and therefore the coefficients are all functions of body voltage and other terminal voltages. However, we treat them as constants which are regarded as averaged values over the voltage changes, for we seek for intuitive understanding which sheds light on the mechanism of the signal degradation from write to read through hold rather than rigorous treatments with complexity. The definitions and calculations which follow can be applied to both partially depleted silicon-on-insulator (PD-SOI) FBC and fully depleted silicon-on-insulator (FD-SOI) FBC equally.

We discuss the variations of the body voltages of the data "1" cell and the data "0" cell from write to read through hold for the case where WL is set to its hold state (V_{WLL}) prior to BL (0 V) in the transition from write to hold separately from the case where BL (0 V) is set to its hold state prior to WL (V_{WLL}) in the transition.

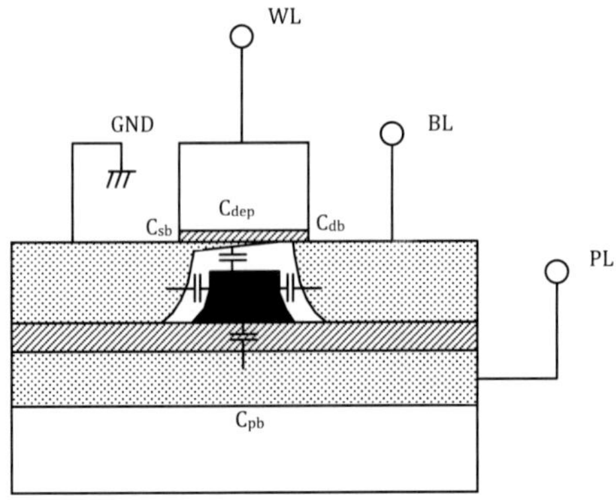

Figure 3.4 Capacitive couplings between the body neutral region and other FBC's terminals at saturation region.

3.1.1 WL Is Set to Its Hold Level Prior to BL

Figure 3.5 illustrates operational waveforms of FBC from write to read through hold for the case where WL is set to its hold state prior to BL. The externally applied voltages of WL and BL and the response of body voltage are shown for the data "1" write, hold and read in Fig. 3.5a and for the data "0" write, hold, and read in Fig. 3.5b.

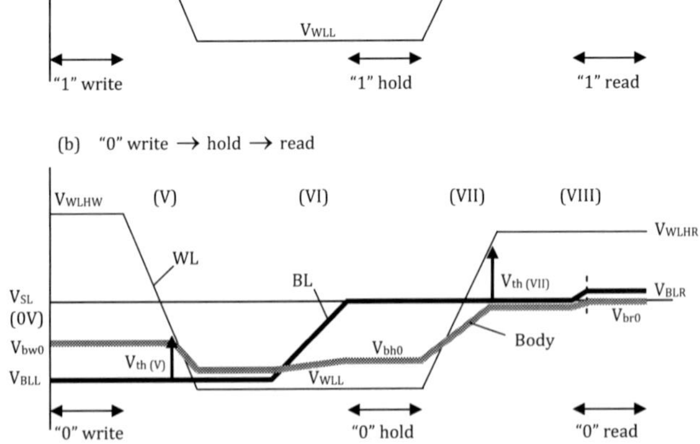

Figure 3.5 WL and BL waveforms in write, hold, and read and the body voltage response illustrated for the case where WL is set to its hold state prior to BL in the transition from write to hold. (a) "1" write, (b) "0" write.

Let us analyze the case of the data "1" at first. In the region (I) when WL is lowered from the write level V_{WLHW} to the hold level V_{WLL}, the body voltage V_{bw1} which is the "1" write level decreases due to the gate to the body capacitance C_{gb} in Fig. 3.2 or according to the capacitive coupling coefficient λ_g in Eq. 3.3 after WL level becomes smaller than the threshold voltage of the cell at the region (I) $V_{th(I)}$ to cut off the channel. In the region (II) when the voltage of BL is decreasing, the body voltage decreases again due to the drain to the body capacitance C_{db} in Fig. 3.2 or the capacitive coupling coefficient λ_d. in Eq. 3.4. As a result, the body voltage of the "1" cell at hold V_{bh1} is expressed by

$$V_{bh1} = V_{bw1} - \left(V_{th(I)} + |V_{WLL}|\right) \cdot \lambda_{g(I)} - V_{BLH} \cdot \lambda_{d(II)}, \tag{3.7}$$

where $\lambda_{g(I)}$ is the coefficient in Eq. 3.3 defined at cut-off region and averaged over the region (I) and $\lambda_{d(II)}$ is the coefficient in Eq. 3.4 defined at cut-off region and averaged over the region (II). For reading the data, WL is raised to V_{WLHR} followed by the BL rising to V_{BLR} as is shown in Fig. 3.5a regions (III) and (IV). In the region (III), the body voltage rises due to the capacitance C_{gb} in Fig. 3.2 or according to the capacitive coupling coefficient λ_g in Eq. 3.3. In the region (IV), the body voltage rises due to the capacitance C_{db} in Fig. 3.3 or according to the capacitive coupling coefficient λ_{dl} in Eq. 3.5. Therefore, the body voltage at read "1" V_{br1} is written as

$$V_{br1} = V_{bh1} + \left(V_{th(III)} + |V_{WLL}|\right) \cdot \lambda_{g(III)} + V_{BLR} \cdot \lambda_{dl(IV)}, \tag{3.8}$$

where $V_{th(III)}$ is the threshold voltage of the cell at the region (III), $\lambda_{g(III)}$ is the coefficient in Eq. 3.3 defined at cut-off region and averaged over the region (III), and $\lambda_{dl(IV)}$ is the coefficient in Eq. 3.5 defined at linear region and averaged over the region (IV).

Next, we consider the case of the data "0." In the region (V) in Fig. 3.5b, if WL becomes smaller than V_{BLL} + $V_{th(V)}$, the body voltage decreases due to the capacitance C_{gb} in Fig. 3.2 or the capacitive coupling coefficient λ_g in Eq. 3.3 averaged over the region (V), because the source voltage of the cell is not 0 V but V_{BLL} which is negative in this region. $V_{th(V)}$ is the threshold voltage of the cell for the "0" write body voltage V_{bw0}. However, $V_{th(V)}$ may be very close to $V_{th(I)}$, because the source voltage of the cell is lower than V_{bw0}. And when BL is raised to 0 V in the region (VI), the body voltage increases due to the capacitance C_{db} in Fig. 3.2 or the capacitive coupling coefficient λ_d in Eq. 3.4. As a result, the BL voltage of the data "0" is held at V_{bh0} which is expressed by

$$V_{bh0} = V_{bw0} - \left(V_{th(V)} - |V_{BLL}| + |V_{WLL}|\right) \cdot \lambda_{g(V)} + |V_{BLL}| \cdot \lambda_{d(VI)} \tag{3.9}$$

where $\lambda_{g(V)}$ is the coefficient in Eq. 3.3 defined at cut-off region and averaged over the region (V) and $\lambda_{d(II)}$ is the coefficient in Eq. 3.4 defined at cut-off region and averaged over the region (VI). In the same way as Eq. 3.8, the body voltage at read for the data "0" V_{br0} is written as

$$V_{br0} = V_{bh0} + \left(V_{th(VII)} + \left|V_{WLL}\right|\right) \cdot \lambda_{g(VII)} + V_{BLR} \cdot \lambda_{dl(VIII)}, \tag{3.10}$$

where in the region (VII) the body is coupled to the gate through the capacitance C_{gb} in Fig. 3.2 or the capacitive coupling coefficient λ_g in Eq. 3.3 averaged over the region (VII) and in the region (VIII) the body is coupled to the drain through the capacitance C_{db} in Fig. 3.3 or the capacitive coupling coefficient λ_{dl} in Eq. 3.5 averaged over the region (VIII).

From Eqs. 3.7 and 3.9, the body voltage difference at hold ΔV_{bh} is written as

$$\begin{aligned}\Delta V_{bh} &= V_{bh1} - V_{bh0} \\ &= \Delta V_{bw} - \left(V_{th(I)} + \left|V_{WLL}\right|\right) \cdot \lambda_{g(I)} - V_{BLH} \cdot \lambda_{d(II)} \\ &\quad + \left(V_{th(V)} - \left|V_{BLL}\right| + \left|V_{WLL}\right|\right) \cdot \lambda_{g(V)} - \left|V_{BLL}\right| \cdot \lambda_{d(VI)}\end{aligned} \tag{3.11}$$

where ΔV_{bw} stands for the body voltage difference at write which is defined by

$$\Delta V_{bw} = V_{bw1} - V_{bw0}. \tag{3.12}$$

Equation 3.11 indicates that the body voltage difference at hold is smaller than that at write by the amount $\Delta\Delta V_{bwh}$ which is

$$\begin{aligned}\Delta\Delta V_{bwh} &= \Delta V_{bw} - \Delta V_{bh} \\ &= \left(V_{th(I)} + \left|V_{WLL}\right|\right) \cdot \lambda_{g(I)} + V_{BLH} \cdot \lambda_{d(II)} \\ &\quad - \left(V_{th(V)} - \left|V_{BLL}\right| + \left|V_{WLL}\right|\right) \cdot \lambda_{g(V)} + \left|V_{BLL}\right| \cdot \lambda_{d(VI)}.\end{aligned} \tag{3.13}$$

The effects of drain voltage variations V_{BLH} and V_{BLL} are trivial and can be neglected especially for a thin film FD-SOI FBC, because λ_d is much smaller than λ_g. And, by assuming $\lambda_{g(I)} = \lambda_{g(V)} = \lambda_g$ and $V_{th(I)} = V_{th(I)}$, we have from Eq. 3.13

$$\Delta\Delta V_{bwh} = \left|V_{BLL}\right| \cdot \lambda_g. \tag{3.14}$$

This result shows that the loss of the signal from write to hold is related to the method of writing "0" with BL voltage negative (V_{BLL}) and holding the data by cutting the cell transistor off with BL voltage negative. In other words, the WL cut-off point of the "1" cell is higher than that of the "0" cell by $|V_{BLL}|$, making C_{gb} for the "1" cell appears earlier than for the "0" cell when WL is falling to hold the data. This makes the amplitude of WL shift during which body voltage is coupled to the gate is larger for the "1" cell than for the "0" cell. Thus, the voltage of the "1" cell's body migrates a longer distance than the voltage of the "0" cell's body does at the transition from write to hold, bringing about the major signal loss from write to hold.

From Eqs. 3.8 and 3.10, the body voltage difference at read ΔV_{br} is written as

$$\begin{aligned}\Delta V_{\text{br}} &= V_{\text{br1}} - V_{\text{br0}} \\ &= \Delta V_{\text{bh}} + \left(V_{\text{th(III)}} + \left|V_{\text{WLL}}\right|\right)\cdot \lambda_{\text{g(III)}} + V_{\text{BLR}} \cdot \lambda_{\text{dl(IV)}} \\ &\quad - \left(V_{\text{th(VII)}} + \left|V_{\text{WLL}}\right|\right)\cdot \lambda_{\text{g(VII)}} - V_{\text{BLR}} \cdot \lambda_{\text{dl(VIII)}}.\end{aligned} \tag{3.15}$$

Equation 3.15 indicates that the body voltage difference at read is smaller than that at hold by the amount $\Delta\Delta V_{\text{bhr}}$ which is

$$\begin{aligned}\Delta\Delta V_{\text{bhr}} &= \Delta V_{\text{bh}} - \Delta V_{\text{br}} \\ &= -\left(V_{\text{th(III)}} + \left|V_{\text{WLL}}\right|\right)\cdot \lambda_{\text{g(III)}} - V_{\text{BLR}} \cdot \lambda_{\text{dl(IV)}} \\ &\quad + \left(V_{\text{th(VII)}} + \left|V_{\text{WLL}}\right|\right)\cdot \lambda_{\text{g(VII)}} + V_{\text{BLR}} \cdot \lambda_{\text{dl(VIII)}}.\end{aligned} \tag{3.16}$$

If we assume $\lambda_{\text{dl(IV)}} = \lambda_{\text{dl(VIII)}}$ and $\lambda_{\text{g(III)}} = \lambda_{\text{g(VII)}} = \lambda_{\text{g}}$, we have from Eq. 3.16

$$\Delta\Delta V_{\text{bhr}} = \left(V_{\text{th(VII)}} - V_{\text{th(III)}}\right)\cdot \lambda_{\text{g}} = \Delta V_{\text{th}} \cdot \lambda_{\text{g}}, \tag{3.17}$$

where ΔV_{th} is the threshold voltage difference observed at read between the "1" cell and the "0" cell. The last equation shows that the loss of the signal from hold to read is due to the threshold voltage difference itself between the data "1" cell and the data "0" cell. The silicon surface of the data "1" cell which has a lower threshold voltage will be inverted earlier than that of the data "0" cell when WL is rising to read the cell, making the "0" cell's body voltage migrate a longer distance than the "1" cell's body does. Though this is essential in the event of reading the cell as a MOSFET current difference, the loss $\Delta\Delta V_{\text{bhr}}$ is in general smaller than another loss $\Delta\Delta V_{\text{bwh}}$ forl $|V_{\text{BLL}}| > \Delta V_{\text{th}}$.

Now, let us consider how to minimize these signal losses. As is indicated in Eqs. 3.14 and 3.17, both $\Delta\Delta V_{\text{bwh}}$ and $\Delta\Delta V_{\text{bhr}}$ are proportional to λ_{g} which is defined in Eq. 3.3. To make λ_{g} small, C_{db}, C_{sb}, or C_{pb} need to become large compared to C_{gb}. However, increasing C_{db} must be avoided, because this makes λ_{d} larger and a harmful effect of the BL swings becomes large, making the approximation which was used to derive Eq. 3.14 invalid. This means that C_{sb} cannot be increased either, because a MOSFET is symmetrical for the drain and the source. The only way to make λ_{g} small, therefore, is to increase C_{pb} with other components unchanged.

3.1.2 BL Is Set to Its Hold Level Prior to WL

In 3.1.1, we found that the signal loss of FBC from write to read through hold is approximately expressed by Eq. 3.14 and is caused by the asymmetry of C_{gb} seen in the "1" cell and in the "0" cell when WL is falling in the transition from write to hold for the case where WL is set to its hold level prior to

BL. Since this asymmetry is due to the difference of the source levels by $|V_{BLL}|$between the "1" cell and the "0" cell when WL is falling, the signal loss may be eliminated if we can make C_{gb} seen symmetry for the both data polarities when the cell transits from write to hold. If we change the order of WL and BL in the transition from write to hold, the source voltages of the "1" cell and the "0" cell coincide and the asymmetry seems to be eliminated to keep the signal at hold as large as the original signal at write. Therefore, we now analyze the case where BL is set to its hold level prior to WL.

Figure 3.6 illustrates operational waveforms of FBC from write to read through hold for the case where BL is set to its hold state prior to WL. The externally applied voltages of WL and BL and the response of body voltage are shown for the data "1" write, hold, and read in Fig. 3.6a and for the data "0" write, hold, and read in Fig. 3.6b.

Let us consider the case of the data "1" cell at first based on Fig. 3.6a. In the region (I) when BL falls from the "1" write voltage V_{BLH} to the hold voltage 0 V, the body voltage of the data "1" cell decreases from its "1" write level V_{bw1} due to the capacitive coupling between the drain and the body. When the BL voltage is larger than $V_{WLHW} - V_{th(I)}$, FBC is in saturation region and the coupling is governed by C_{db} in Fig. 3.4 or the capacitive coupling coefficient λ_{dl} in Eq. 3.6. In contrast, when the BL voltage decreases to less than $V_{WLHW} - V_{th(I)}$, FBC is in linear region and the coupling is governed by C_{db} in Fig. 3.3 or the capacitive coupling coefficient λ_{dl} in Eq. 3.5. After that, when WL is falling from the write voltage V_{WLHW} to the hold voltage V_{WLL} (region (II)), the body voltage is decreasing due to the capacitance C_{gb} defined in Fig. 3.2 or according to the capacitive coupling coefficient λ_g in Eq. 3.3. All the coefficients are those averaged over the corresponding regions. Therefore, the hold voltage of the "1" cell's body V_{bh1} is written as

$$V_{bh1} = V_{bw1} - \Big[\left(V_{th(II)} + |V_{WLL}|\right) \cdot \lambda_{g(II)} + \left(V_{BLH} - V_{WLHW} + V_{th(I)}\right) \cdot \lambda_{ds(I)} + \left(V_{WLHW} - V_{th(I)}\right) \cdot \lambda_{dl(I)}\Big] \quad (3.18)$$

In the same way, the "1" cell's body voltage at read V_{br1} is written as

$$V_{br1} = V_{bh1} + \left(V_{th(II)} + |V_{WLL}|\right) \cdot \lambda_{g(II)} + V_{BLR} \cdot \lambda_{dl(III)}. \quad (3.19)$$

Next, we consider the case of the data "0" cell based on Fig. 3.6b. In the region (IV) when BL rises from V_{BLL} to 0 V, the body voltage of the "0" cell also rises due to the capacitance at linear region C_{db} in Fig. 3.3 or according to the capacitive coupling coefficient λ_{dl} in Eq. 3.5 averaged over the region (IV). After that, in the region (V) when WL falls from V_{WLHW} to V_{WLL}, the body voltage also falls due to the capacitance at cut-off region C_{gb} in Fig. 3.2 or according to the capacitive coupling coefficient λ_g in Eq. 3.3 averaged over the region (V). As a result, the body voltage of the "0" cell at hold V_{bh0} is expressed by

$$V_{\mathrm{bh0}} = V_{\mathrm{bw0}} - \left(V_{\mathrm{th(V)}} + \left|V_{\mathrm{WLL}}\right|\right)\cdot \lambda_{\mathrm{g(V)}} + \left|V_{\mathrm{BLL}}\right|\cdot \lambda_{\mathrm{dl(IV)}}. \tag{3.20}$$

In the same way, the body voltage of the "0" cell at read V_{br0} is expressed by

$$V_{\mathrm{br0}} = V_{\mathrm{bh0}} + \left(V_{\mathrm{th(V)}} + \left|V_{\mathrm{WLL}}\right|\right)\cdot \lambda_{\mathrm{g(V)}} + V_{\mathrm{BLR}}\cdot \lambda_{\mathrm{dl(VI)}}. \tag{3.21}$$

Here, it should be noted that the p-n junction between the body and the drain in a latter part of the region (I) and in almost all part of the region (IV) is forward biased. These situations have never been experienced in the analyses and the capacitance C_{db} averaged for the BL voltage from $V_{\mathrm{WLHW}} - V_{\mathrm{th(I)}}$ to 0 V in the region (I) and that averaged for the BL voltage from V_{BLL} to 0 V in the region (IV) are very large, making both $\lambda_{\mathrm{dl(I)}}$ in Eq. 3.18 and $\lambda_{\mathrm{dl(IV)}}$ in Eq. 3.20 very large. Therefore, V_{bw1} migrates a long distance downward when the data "1" is to be held and at the same time V_{bw0} migrates a long distance upward when the data "0" is to be held, leading to a significant signal loss from write to hold.

From Eqs. 3.18 and 3.20, the body voltage difference between the "1" cell and the "0" cell at hold ΔV_{bh} is expressed as

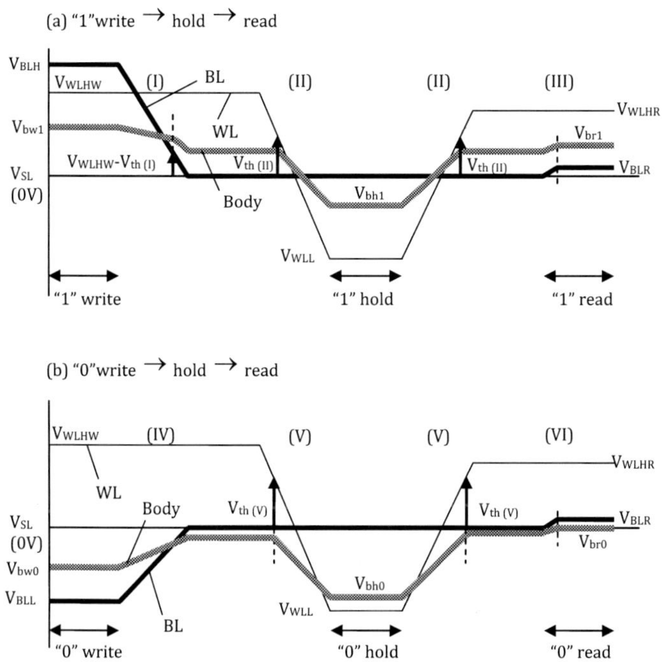

Figure 3.6 WL and BL waveforms in write, hold, and read and the body voltage response illustrated for the case where BL is set to its hold state prior to WL in the transition from write to hold. (a) "1" write, (b) "0" write.

$$\begin{aligned}\Delta V_{\text{bh}} &= \Delta V_{\text{bw}} + \left(V_{\text{th(V)}} \cdot \lambda_{\text{g(V)}} - V_{\text{th(II)}} \cdot \lambda_{\text{g(II)}}\right) \\ &+ \left(\lambda_{\text{g(V)}} - \lambda_{\text{g(II)}}\right) \cdot \left|V_{\text{WLL}}\right| - \left(V_{\text{BLH}} - V_{\text{WLHW}} + V_{\text{th(I)}}\right) \cdot \lambda_{\text{ds(I)}} \\ &- \left(V_{\text{WLHW}} - V_{\text{th(I)}}\right) \cdot \lambda_{\text{dl(I)}} - \left|V_{\text{BLL}}\right| \cdot \lambda_{\text{dl(IV)}}.\end{aligned} \tag{3.22}$$

Therefore, the signal loss from write to hold $\Delta\Delta V_{\text{bwh}}$ in this operation is written as

$$\begin{aligned}\Delta\Delta V_{\text{bwh}} &= \Delta V_{\text{bw}} - \Delta V_{\text{bh}} \\ &= \left(V_{\text{BLH}} - V_{\text{WLHW}} - V_{\text{th(I)}}\right) \cdot \lambda_{\text{ds(I)}} + \left(V_{\text{WLHW}} - V_{\text{th(I)}}\right) \cdot \lambda_{\text{dl(I)}} \\ &+ \left|V_{\text{BLL}}\right| \cdot \lambda_{\text{dl(IV)}} - \left(V_{\text{th(V)}} \cdot \lambda_{\text{g(V)}} - V_{\text{th(II)}} \cdot \lambda_{\text{g(II)}}\right) \\ &- \left(\lambda_{\text{g(V)}} - \lambda_{\text{g(II)}}\right)\left|V_{\text{WLL}}\right|.\end{aligned} \tag{3.23}$$

Here, if we assume that $\lambda_{\text{ds(I)}}$ is negligible compared with $\lambda_{\text{dl(I)}}$ and $\lambda_{\text{dl(IV)}}$ and that $\lambda_{\text{g(II)}} = \lambda_{\text{g(V)}} = \lambda_{\text{g}}$, Eq. 3.23 can be simpified as,

$$\Delta\Delta V_{\text{bwh}} = \left(V_{\text{WLHW}} - V_{\text{th(I)}}\right) \cdot \lambda_{\text{dl(I)}} + \left|V_{\text{BLL}}\right| \cdot \lambda_{\text{dl(IV)}} - \Delta V_{\text{th}} \cdot \lambda_{\text{g}}, \tag{3.24}$$

where ΔV_{th} is the threshold voltage difference observed to read between the cell "1" and the cell "0" which is defined by

$$\Delta V_{\text{th}} = V_{\text{th(V)}} - V_{\text{th(II)}}. \tag{3.25}$$

In the same way, from Eqs. 3.19 and 3.21, the body voltage difference between the "1" cell and the "0" cell at read ΔV_{br} is expressed as

$$\begin{aligned}\Delta V_{\text{br}} &= V_{\text{br1}} - V_{\text{br0}} \\ &= \Delta V_{\text{bh}} + \left(V_{\text{th(II)}} + \left|V_{\text{WLL}}\right|\right) \cdot \lambda_{\text{g(II)}} \\ &- \left(V_{\text{th(V)}} + \left|V_{\text{WLL}}\right|\right) \cdot \lambda_{\text{g(V)}} + V_{\text{BLR}} \cdot \left(\lambda_{\text{dl(III)}} - \lambda_{\text{dl(VI)}}\right).\end{aligned} \tag{3.26}$$

Therefore, the signal loss from hold to read $\Delta\Delta V_{\text{bhr}}$ in this operation is written as

$$\begin{aligned}\Delta\Delta V_{\text{bhr}} &= \Delta V_{\text{bh}} - \Delta V_{\text{br}} \\ &= \left(V_{\text{th(V)}} + \left|V_{\text{WLL}}\right|\right) \cdot \lambda_{\text{g(V)}} - \left(V_{\text{th(II)}} + \left|V_{\text{WLL}}\right|\right) \cdot \lambda_{\text{g(II)}} \\ &- V_{\text{BLR}} \cdot \left(\lambda_{\text{dl(III)}} - \lambda_{\text{dl(VI)}}\right).\end{aligned} \tag{3.27}$$

Here, if we assume that $\lambda_{\text{g(II)}} = \lambda_{\text{g(V)}} = \lambda_{\text{g}}$ and $\lambda_{\text{dl(VI)}} = \lambda_{\text{d(III)}}$, Eq. 3.27 is simplified as

$$\Delta\Delta V_{\text{bhr}} = \Delta V_{\text{th}} \cdot \lambda_{\text{g}}. \tag{3.28}$$

Now, we newly define the body voltage difference decrease from write to read $\Delta\Delta V_{\mathrm{bwr}}$ which is expressed by adding Eq. 3.24 and Eq. 3.28 as

$$\begin{aligned}\Delta\Delta V_{\mathrm{bwr}} &= \Delta\Delta V_{\mathrm{bwh}} + \Delta\Delta V_{\mathrm{bhr}} \\ &= \left(V_{\mathrm{WLHW}} - V_{\mathrm{th(I)}}\right)\cdot\lambda_{\mathrm{dl(I)}} + \left|V_{\mathrm{BLL}}\right|\cdot\lambda_{\mathrm{dl(IV)}}.\end{aligned} \quad (3.29)$$

The last equation indicates that the signal loss from write to read in the case when BL is set to its hold level prior to WL is due to the body voltage bounces during the BL swings caused by large capacitance coupling in the p-n forward junction between the body and the drain.

To minimize the signal loss, it is important to make the capacitive coupling coefficient λ_{dl} defined in Eq. 3.5 as small as possible. The only way to minimize the signal loss in this case is, again, to increase the capacitance between the body and the substrate (plate) C_{pb} with other components unchanged.

3.1.3 TCAD Simulation Results

As was predicted analytically in 3.1.1 and 3.1.2, it is important to increase the capacitance between the body and substrate C_{pb} for minimizing the signal loss from write to read regardless the order of BL and WL to be set to their hold levels.[a]

Figure 3.7 shows TCAD simulation results in which the body voltage response to the external voltages of BL and WL is calculated for FBC with gate length L_{g} = 150 nm, gate oxide thickness t_{ox} = 6 nm, silicon thickness t_{Si} = 55 nm, buried oxide (BOX) thickness t_{Box} = 25 nm, and acceptor concentration of the body $N_{\mathrm{A}} = 1.0 \times 10^{18}\ \mathrm{cm^{-3}}$ which is constant. Figure 3.7a shows the operational waveforms and Fig. 3.7b shows the cell read current for the "1" cell I_{cell} observed from 26 ns to 28 ns and the cell read current for the "0" cell I_{cell0} observed from 46 ns to 48 ns. This simulation corresponds to the case where WL is set to its hold level prior to BL and is similar to Fig. 2.4 in Chapter 2 except for differences of device parameters and driving voltages. Although C_{pb} is increased by thinning BOX to 25 nm from the conventionally used SOI, the effect of the asymmetry of C_{gb} is observed for the both data in the transition from write to hold, i.e., from 20 ns to 22 ns for the data "1" and from 40 ns to 42 ns to for the data "0." If we use an SOI wafer with BOX which is thicker than 25 mn, this asymmetry becomes larger and the signal of FBC will become smaller than the result obtained in Fig. 3.7b.

Figure 3.8 shows TCAD similar simulation results to be performed for the same FBC as Fig. 3.7. This simulation corresponds to the case where BL is set to its hold level prior to WL.

---

[a] Generally, the capacitances between the body and other nodes whose voltages remain unchanged from write to read need to be enlarged for minimizing the signal loss.

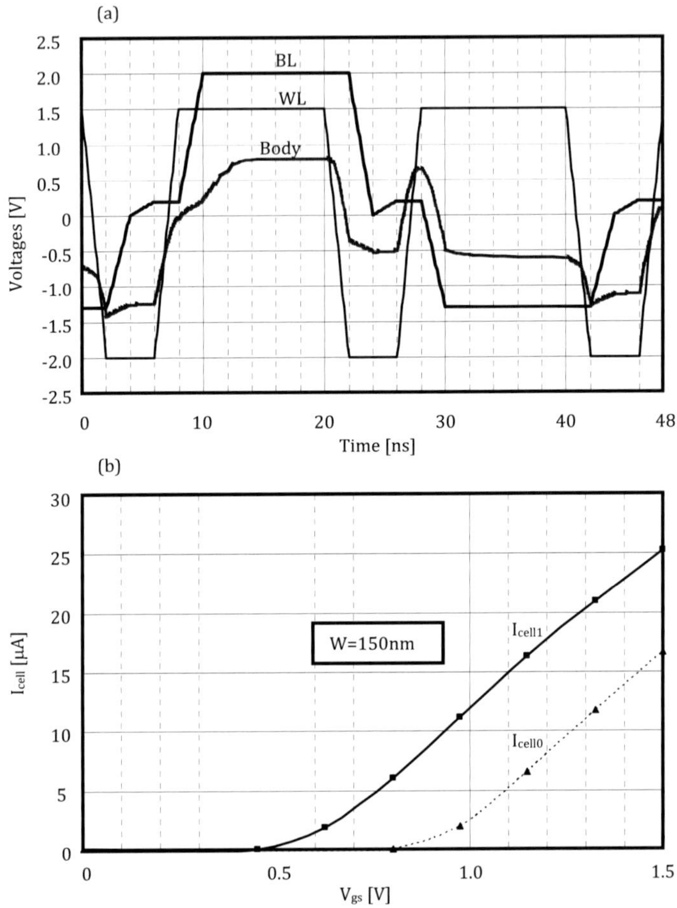

Figure 3.7 TCAD simulation results of write, hold, and read for the case where WL is set to its hold level prior to BL. (a) Operational wave forms and (b) cell read currents observed from 26 ns to 28 ns for the "1" data and from 46 ns and 48 ns for the "0" data. The cell's device parameters are L = 150 nm, t_{ox} = 6 nm, t_{Si} = 55 nm, t_{BOX} = 25 nm, N_A = 1.0 × 10^{18} cm^{-3}.

The asymmetry effect observed in Fig. 3.7 during the WL falling is eliminated, as is shown in Fig. 3.8, form 22 ns to 24 ns for the "1" data and form 42 ns to 44 ns for the "1" data. However, large downward migration of the body voltage for the data "1" when BL is set to its hold level is observed between 20 ns and 22 ns. Similarly, large upward migration of the body voltage for the data "0" when BL is set to its hold level is observed between 40 ns and 42 ns. As a result, the difference in the body voltage at read between the data "1" and the data "0" in Fig. 3.8 is no larger than that in Fig. 3.7 and the difference in the threshold voltage for the cell read currents in Fig. 3.8 is almost the same as that in Fig. 3.7, namely about 0.3 V.

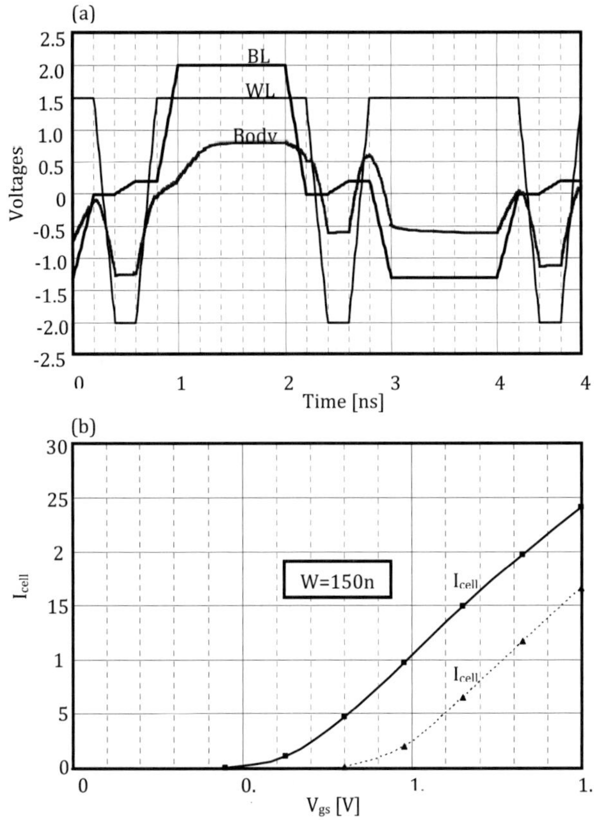

Figure 3.8 TCAD simulation results of write, hold, and read for the case where BL is set to its hold level prior to WL. (a) Operational wave forms and (b) cell read currents observed from 26 ns to 28 ns for the "1" data and from 46 ns and 48 ns for the "0" data. The cell's device parameters are L = 150 nm, t_{ox} = 6 nm, t_{Si} = 55 nm, t_{BOX} = 25 nm, and $N_A = 1.0 \times 10^{18}$ cm^{-3}.

Actually, if the charge in the body conserves from write to read, the body voltages at read for both the "1" data V_{br1} and the "0" data V_{br0} and hence the body voltage difference at read ΔV_{br} should be identical for the both cases 3.1.1 and 3.1.2, being independent of the order of resetting WL and BL, though Eqs. 3.8, 3.10, and 3.15 appear very different from Eqs. 3.19, 3.21, and 3.26, respectively, with the capacitive coupling coefficients averaged at different regions to different values.

Lastly, let us examine the effect of C_{pb} on the FBC signal. Figure 3.9 shows the device structure for which we performed TCAD simulation to check the effect. The value of C_{pb} is changed by shifting the position of the back gate up and down, i.e., by changing the distance t_{ox2} from the back surface of the silicon body to the top surface of the back gate. The top gate oxide thickness simulations. Figure 3.10a shows the TCAD simulation ead signal ΔV_{th} as a function of the ratio of t_{ox2} to t_{ox1}. $_{BLL}$ to write the data "0" is also changed depending on the as is shown in Fig. 3.10b. This is because the body voltage of at hold becomes larger as C_{pb} becomes larger by decreasing the s was explained in section 2.4, the BL "0" disturb becomes severer n the "1" cell's body voltage at hold is larger, since the difference in ne p-n junction forward voltage between the body and the drain for the selected and the unselected cells is smaller. Therefore, the data "0" is written more weakly for the cell with larger C_{pb} to avoid the BL "0" disturb. In this simulation, WL is set to its hold level prior to BL. However, the results will be the same as the case where BL is set to its hold level prior to WL, as was commented above. According to Fig. 3.10a, the signal of FBC increases as the ratio t_{ox2}/t_{ox1} decreases, in other words, as C_{pb} increases with other capacitors kept constant, verifying the analytical predictions we made in 3.1.1.

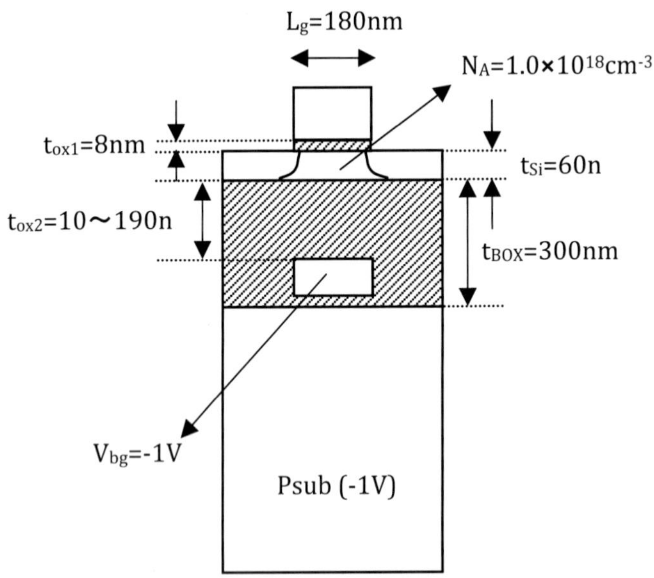

Figure 3.9 Device structure we used for performing TCAD simulations which check the effect of C_{pb} on the FBC read signal. The value of C_{pb} is changed by shifting the back gate position up and down.

The capacitor C_{pb} was named a stabilizing capacitor in this sense and was originally formed by poly-silicon pillars which are built in shallow trench isolations and biased with a constant voltage as is shown in Fig. 3.11.[1,2] Since the capacitance between the body and the pillars becomes small as the device is scaled with the silicon film thinner, the stabilizing capacitor is now formed by thinning the BOX of SOI in which FBC is built as a more natural and simpler way instead of building the pillars.[3] A large stabilizing capacitance can also be achieved by applying a constant voltage either to the back gate of the double-gate SOI transistor FBC[4] or to the separated gate of the independently controlled double-gate FBC.[5] These methods are also effective for making the absolute value of the negative bias voltage small which is to be applied in the back gate to form a potential well for holes to be accumulated in the body of an FD-SOI FBC.[6,7] Apart from FBC, thin BOX technologies are recently gathering attention as effective means to enhance the static-noise margin of SRAM[8,9] and also to change the threshold voltages of MOSFETs between the active mode and the standby mode to achieve

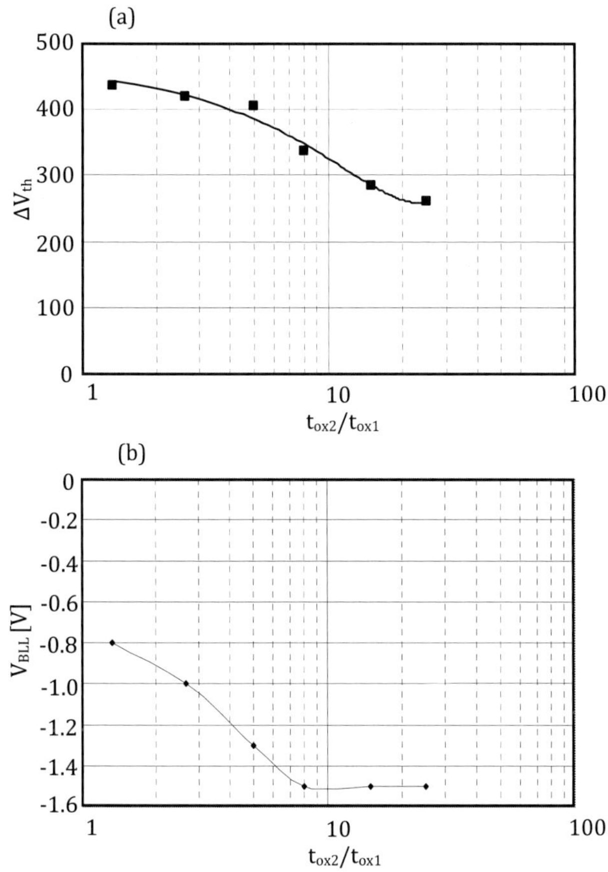

Figure 3.10 (a) The TCAD simulationr results of the threshold voltage difference for read currents of the data "1" cell and the date "0" cell. (b) The BL Voltage V_{BLL} in writing the data "0" is changed depending on the ration t_{ox2}/t_{ox1} to void the BL "0" disturb in the simulation. The simulated device is shown in Fig. 3.10.

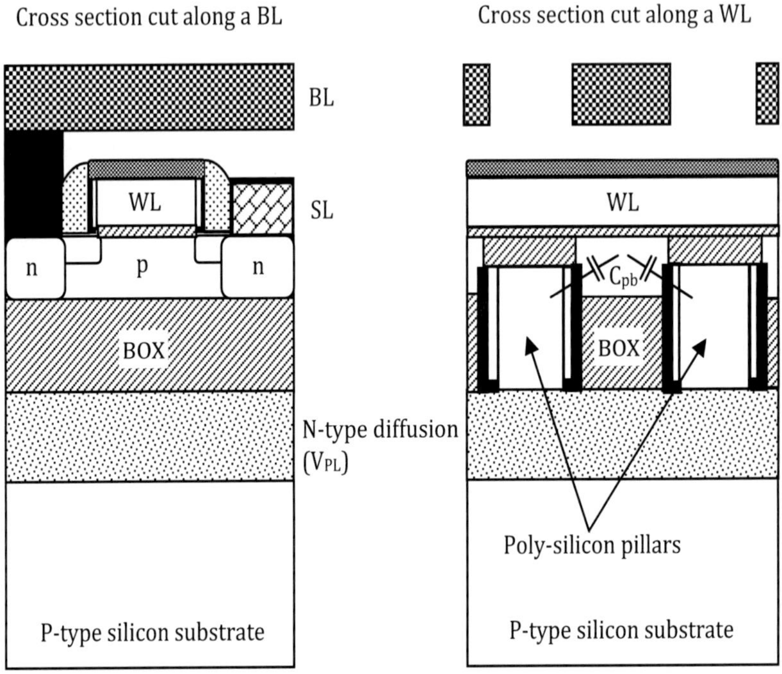

Figure 3.11 The original method to provide stabilizing capacitors by building poly-silicon pillars within shallow trench isolations (STIs).

high-speed and low-power LSIs[10] both by controlling the back gate of an FD-SOI MOSFET through a thin BOX. Responding these situations, major SOI makers are actively developing the thin BOX technologies.

3.2 SIGNAL-TO-NOISE RATIO OF FBC AND THE EFFECT OF PARASITIC RESISTANCE OF FBC ON THE SIGNAL-TO-NOISE RATIO

Figure 3.12a schematically shows the distributions of the threshold voltages for the data "1" cells V_{th1}, the data "0" cells V_{th0}, and dummy cells V_{thref}. Dummy cell is a cell which flows a certain current between the "1" cell current and the "0" cell current to distinguish the data. In general, V_{th1} in an array of an FBC memory fluctuates according to a normal distribution with average $<V_{th1}>$ and standard deviation σV_{th1}. In the same way, V_{th0} in the array fluctuates according to a normal distribution with average $<V_{th0}>$ and standard deviation σV_{th0}. If we assume that the dummy cell consists of a pair of "1" cell and "0" cell and that its current is the average of the "1" cell's

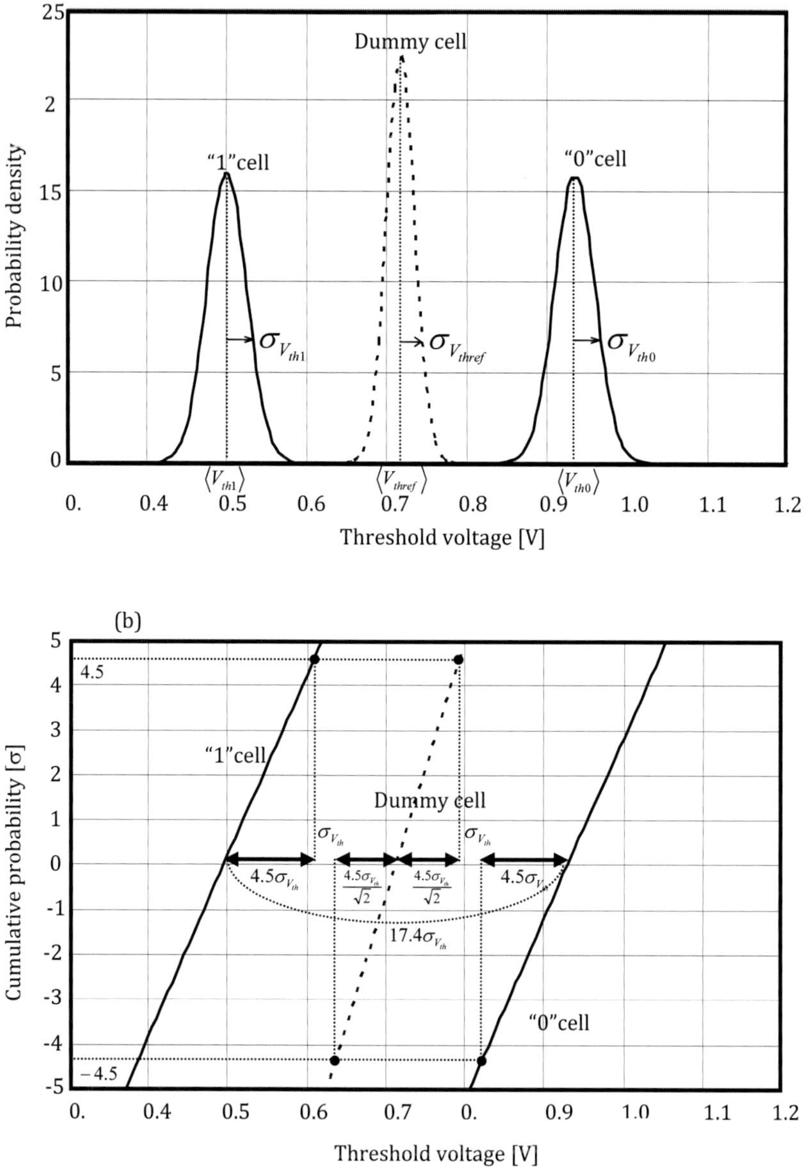

Figure 3.12 (a) Probability density distribution of treshold voltage for "1" cells, "0" cells, and dummy cells within a memory array and (b) their cumulative probability represented by standard deviation.

current I_{cell} and the "0" cell's one I_{cell0}, i.e. $(I_{cell1} + I_{cell0})/2$, V_{thref} in the array will fluctuate according to the normal distribution with the average $<V_{thref}>$

$= (<V_{\text{th1}}> + <V_{\text{th0}}>)/2$ and the standard deviation $\sigma_{V_{\text{thref}}} = \frac{\sqrt{\sigma_{V_{\text{th1}}^2} + \sigma_{V_{\text{th0}}^2}}}{2}$ due to the probability theory.[b]

Now we estimate the minimum signal which is necessary for a certain size of FBC memory functions by using a simplified mathematical model represented in Fig. 3.12b. Figure 3.12b shows the cumulative distributions which correspond to the distributions in Fig. 3.12a in the linear scale of standard deviation. A memory cell array of FBC contains 2 Mbit cells and redundancy which can fix up to 8 bit fails. Although the number of dummy cells will be much smaller than that of cells in actual designs, we assume the numbers are equal for simplicity and for the worst case estimation. Since up to 8 bits fails are permitted in 2 Mbit, the maximum failure rate permitted is 8/2 M = 3.8 ppm.. This corresponds to $\pm 4.5\sigma_{V_{\text{thi}}}$ a normal distribution where $\sigma_{V_{\text{th}}}$ is a standard deviation of the threshold voltage distribution in the FBC memory array. Here we assume that the standard deviation of the threshold voltage distribution of the data "1" cells and that of the data "0" cell is the same $\sigma_{V_{\text{th1}}} = \sigma_{V_{\text{th9}}} = \sigma V_{\text{th}}$. The standard deviation of the dummy cell's threshold voltage distribution is expressed as $\sigma_{V_{\text{thref}}} = \sigma_{V_{\text{th}}}/\sqrt{2}$. And the necessary margin for S/As to sense data correctly is assumed to be equal to σ_{Vth}. Under these assumptions, the threshold voltage difference $\Delta V_{\text{th}} = <V_{\text{th0}}> - V_{\text{th1}}>$ must satisfy the inequality

$$\Delta V_{\text{th}} \geq \left(4.5\sigma_{V_{\text{th}}} + \frac{4.5\sigma_{V_{\text{th}}}}{\sqrt{2}} + \sigma_{V_{\text{th}}}\right) \times 2 = 17.4\sigma_{V_{\text{th}}}, \tag{3.30}$$

for the array to function. The necessary condition for the signal-to-noise ratio (SNR) will be derived from Eq. 3.30 as

$$(\text{SNR})_{V_{\text{th}}} \equiv \Delta V_{\text{th}} / \sigma_{V_{\text{th}}} \geq 17.4, \tag{3.31}$$

where $(\text{SNR})_{V_{\text{th}}}$ stands for the SNR which is measured as the threshold voltage of cells. If an actual large memory consists of many of the cell arrays, the minimum $(\text{SNR})_{V_{\text{th}}}$ required for the memory to function will be larger than 17.4. As is shown in Fig. 3.13, the SNR which is measured by the cell read current is subject to the same constraint for the functionality of the 2 Mbit array:

[b] Detailed designs of dummy cell will be explained in Chapter 6.

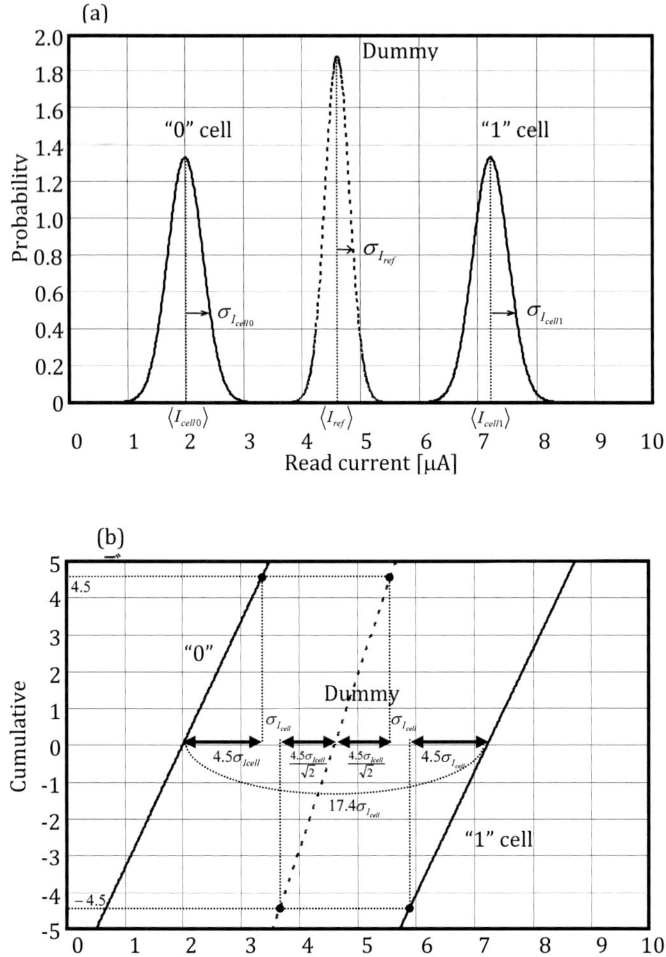

Figure 3.13 (a) Probability density distribution of read current for "1" cells, "0" cells, and dummy cells within a memory array and (b) their cumulative probability represented by standard deviation.

$$(\mathrm{SNR})_{I_{\mathrm{cell}}} \equiv \Delta I_{\mathrm{cell}} / \sigma_{I_{\mathrm{cell}}} \geq 17.4, \tag{3.32}$$

S/A circuit detects the current difference between the data "1" or the data "0" cell and the reference cell. In this sense, the memory chip yield may have a stronger correlation with $(\mathrm{SNR})_{I_{\mathrm{cell}}}$ than with $(\mathrm{SNR})_{V_{\mathrm{th}}}$. However, $(\mathrm{SNR})_{V_{\mathrm{th}}}$ is more convenient for us to handle than $(\mathrm{SNR})_{I_{\mathrm{cell}}}$, because the definition of the cell read current I_{cell} requires more parameters than that of the cell threshold voltage V_{th}. Actually, the former depends on many parameters regarding the operational condition and the device structure such as the gate voltage, the drain voltage, the channel length, and the channel width in contrast with the latter which is independent of those parameters in the zero-order approximation. We will discuss the relation between the two SNRs $(\mathrm{SNR})_{V_{\mathrm{th}}}$ and $(\mathrm{SNR})I_{\mathrm{cell}}$ in taking the parasitic resistance of the cell's channel current into consideration. FBC usually has lightly doped n-type diffusion regions under the gate spacers of the cell transistor as is shown in Fig. 3.14 for the purpose of relaxing the electrical field strength within the body to reduce the p-n junction leakage current for improving the data retention time as well as improving the immunity against the threshold voltage instability due to the hot carrier injection. This is the so-called lightly doped drain-source structure.[11] In this cell structure, parasitic resistances R are introduced at the source and the drain edges in the channel. Let us analyze the effect of R on the drain-source current I_{ds} and its fluctuation.

As is shown in Fig. 3.15, the voltage applied between the gate and the source $\hat{V}_{\mathrm{gs}}$ and that applied between the drain and the source $\hat{V}_{\mathrm{ds}}$ will be decreased by ΔV_{gs} and ΔV_{ds} from their applied voltages V_{gs} and V_{ds}, respectively:

$$\hat{V}_{\mathrm{gs}} = V_{\mathrm{gs}} - \Delta V_{\mathrm{gs}}, \tag{3.33}$$

$$\hat{V}_{\mathrm{ds}} = V_{\mathrm{ds}} - \Delta V_{\mathrm{ds}}, \tag{3.34}$$

where

$$\Delta V_{\mathrm{gs}} = I_{\mathrm{ds}} \cdot R, \tag{3.35}$$

$$\Delta V_{\mathrm{ds}} = 2 \cdot I_{\mathrm{ds}} \cdot R. \tag{3.36}$$

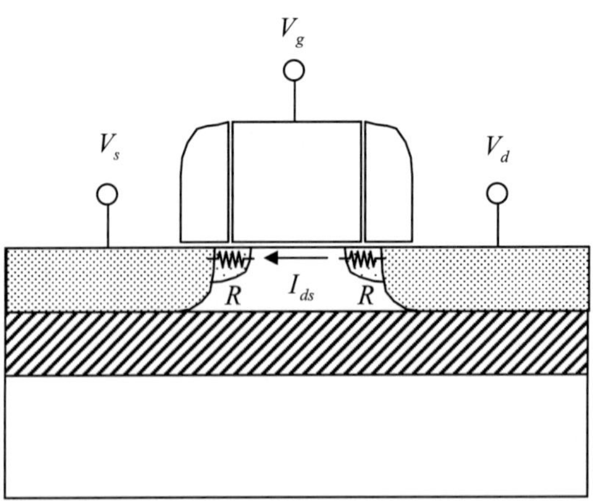

Figure 3.14 Parasitic resistances appearing under the FBC's gate spacers for the purpose of relaxing the electrical field strength within the body to reduce the pn junction leak current for improving the data retention time.

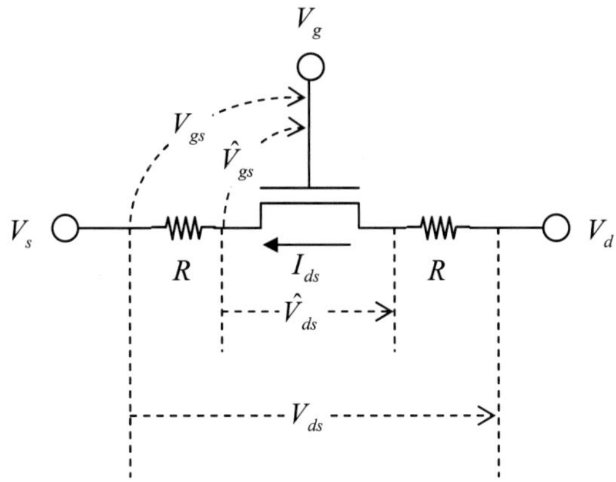

Figure 3.15 Equivalent circuit of FBC in Fig. 3.14.

If the cell is operated at linear region, i.e.,

$$\hat{V}_{gs} > \hat{V}_{ds} + V_{th}, \tag{3.37}$$

the current which flows between the drain and the source I_{ds} is expressed as

$$I_{ds} = \beta.\left(\hat{V}_{gs} - V_{th} - \frac{\hat{V}_{ds}}{2}\right).\hat{V}_{ds} \tag{3.38}$$

where V_{th} is the threshold voltage of the cell and β is expressed as

$$\beta = \mu_e \cdot \frac{\varepsilon_{ox}}{t_{ox}} \cdot \frac{W}{L_{eff}}, \tag{3.39}$$

where μ_e is the mobility of electron, ε_{ox} is the permittivity of SiO_2, t_{ox} is the thickness of the gate oxide, W is the channel width, and L_{eff} is the effective channel length. Substituting Eqs. 3.33 and 3.34 into Eq. 3.37, the condition for the cell to operate at linear region is expressed as

$$V_{gs} > V_{ds} - I_{ds} \cdot R + V_{th}, \tag{3.40}$$

where Eqs. 3.35 and 3.36 are used. Substituting Eqs. 3.33–3.36 into Eq. 3.38 and assuming that ΔV_{ds} and ΔV_{gs} are small enough compared with V_{ds} and V_{gs}, respectively, for the quadratic expressions with respect to ΔV_{ds} and ΔV_{gs} to be neglected in I_{ds}, we have

$$I_{ds} = \frac{V_{ds}}{2 \cdot R + \dfrac{1}{\beta \cdot (V_{gs} - V_{th} - 0.5 \cdot V_{ds})}}. \tag{3.41}$$

At saturation region, on the other hand, i.e., for

$$\hat{V}_{gs} \leq \hat{V}_{ds} + V_{th}, \tag{3.42}$$

or

$$V_{gs} \leq V_{ds} - I_{ds} \cdot R + V_{th}, \tag{3.43}$$

the current between the drain and the source I_{ds} will be expressed as

$$I_{ds} = \frac{1}{2}\beta\,(\hat{V}_{gs} - V_{th})^2. \tag{3.44}$$

Substituting Eqs. 3.33 and 3.35 into Eq. 3.34 and neglecting the term which is proportional to ΔV_{gs}^2, we have

$$I_{ds} = \frac{V_{gs} - V_{th}}{2 \cdot R + \dfrac{1}{0.5 \cdot \beta\,(V_{gs} - V_{th})}}. \tag{3.45}$$

Next, we consider the fluctuation in I_{ds} which is caused by the variation in the device parameters. At first, the linear region expressed by Eq. 3. 41 will be analyzed. We consider I_{ds} as a function of β, R, and V_{th} with applied voltages V_{ds} and V_{gs} constant:

$$I_{ds} = I_{ds}(\beta\,,\ R, V_{th}). \tag{3.46}$$

The total differential of this equation will be

$$dI_{ds} = \frac{\partial I_{ds}}{\partial \beta} d\beta + \frac{\partial I_{ds}}{\partial R} dR + \frac{\partial I_{ds}}{\partial V_{th}} dV_{th}, \tag{3.47}$$

where each partial differential coefficient can be expressed as follows:

$$\frac{\partial I_{ds}}{\partial \beta} = \frac{V_{ds} \cdot (V_{gs} - V_{th} - 0.5 \cdot V_{ds})}{[1 + 2 \cdot R \cdot \beta \cdot (V_{gs} - V_{th} - 0.5 \cdot V_{ds})]^2}, \tag{3.48}$$

$$\frac{\partial I_{ds}}{\partial R} = -\frac{2 \cdot \beta^2 \cdot V_{ds} \cdot (V_{gs} - V_{th} - 0.5 \cdot V_{ds})^2}{[1 + 2 \cdot R \cdot \beta \cdot (V_{gs} - V_{th} - 0.5 \cdot V_{ds})]^2}, \tag{3.49}$$

$$\frac{\partial I_{ds}}{\partial V_{th}} = -\frac{\beta \cdot V_{ds}}{[1 + 2 \cdot R \cdot \beta \cdot (V_{gs} - V_{th} - 0.5 \cdot V_{ds})]^2}. \tag{3.50}$$

From Eq. 3.47, the standard deviation $\sigma_{I_{ds}}$ of the fluctuation in I_{ds} will be expressed by σ_β, σ_R, and $\sigma_{V_{th}}$ which are, respectively, the standard deviations of the variation in β, R, and V_{th} as follows:

$$\sigma_{I_{ds}}^2 = \left(\frac{\partial I_{ds}}{\partial \beta}\right)^2 \cdot \sigma_\beta^2 + \left(\frac{\partial I_{ds}}{\partial R}\right)^2 \cdot \sigma_R^2 + \left(\frac{\partial I_{ds}}{\partial V_{th}}\right)^2 \cdot \sigma_{V_{th}}^2. \tag{3.51}$$

Substituting Eqs. 3.48–3.50 into Eq. 3.51, we have

$$\sigma_{I_{ds}}^{2}=\frac{V_{ds}^{2}\cdot\left[V_{gc}^{2}\cdot\sigma_{\beta}^{2}+4\cdot\beta^{4}\cdot V_{gc}^{4}\cdot\sigma_{R}^{2}+\beta^{2}\cdot\sigma_{V_{th}}^{2}\right]}{(1+2\cdot R\cdot\beta\cdot V_{gc})^{4}},\tag{3.52}$$

where V_{gc} is the averaged voltage between the gate and the annel which induces the channel electrons defined by

$$V_{gc}\equiv V_{gs}-V_{th}-0.5\cdot V_{ds}.\tag{3.53}$$

Equation 3.52 is for linear region. For saturation region, in the same way, we have

$$\sigma_{I_{ds}}^{2}=\frac{V_{g}^{2}\cdot\left[0.25\cdot V_{g}^{2}\cdot\sigma_{\beta}^{2}+0.25\cdot\beta^{4}\cdot V_{g}^{4}\cdot\sigma_{R}^{2}+\beta^{2}\cdot(1+0.5\cdot R\cdot\beta\cdot V_{g})2\cdot\sigma_{V_{th}}^{2}\right]}{(1+R\cdot\beta\cdot V_{g})^{4}},\tag{3.54}$$

where we defined the gate overdrive voltage as

$$V_{g}\equiv V_{gs}-V_{th}.\tag{3.55}$$

Here, we calculate the SNR with respect to the cell current of FBC. When FBC is read at linear region, the read current of the data "1" cell I_{cell} and that of the data "0" cell I_{cell0} are, respectively, written from Eq. 3.41 as

$$I_{cell1}=\frac{V_{BLR}}{2\cdot R+\dfrac{1}{\beta\cdot(V_{WLHR}-V_{th1}-0.5\cdot V_{BLR})}},\tag{3.56}$$

$$I_{cell0}=\frac{V_{BLR}}{2\cdot R+\dfrac{1}{\beta\cdot(V_{WLHR}-V_{th0}-0.5\cdot V_{BLR})}},\tag{3.57}$$

where V_{th1} and V_{th0} are the respective threshold voltages and the voltage between the gate and the source and that between the drain and the source both at read are denoted by V_{WLHR} and V_{BLR}, respectively. Therefore, the read current difference between the "1" cell and the "0" cell ΔI_{cell} is expressed by

$$\Delta I_{cell}=\frac{\beta\cdot\Delta V_{th}\cdot V_{BLR}}{(1+2\cdot R\cdot\beta\cdot V_{gc1})\cdot(1+2\cdot R\cdot\beta\cdot V_{gc0})},\tag{3.58}$$

where

$$\Delta V_{th}=V_{th0}-V_{th1},\tag{3.59}$$

$$V_{gc1}=V_{WLHR}-V_{th1}-0.5\cdot V_{BLR},\tag{3.60}$$

$$V_{gc0}=V_{WLHR}-V_{th0}-0.5\cdot V_{BLR}.\tag{3.61}$$

From Eq. 3.52, the standard deviation of the fluctuation in the "1" cell read current $\sigma_{I_{\text{cell}}}$ and that of the fluctuation in the "0" cell read current $\sigma_{I_{\text{cell0}}}$ are, respectively, expressed as

$$\sigma_{I_{\text{cell1}}} = \frac{\beta \cdot V_{\text{BLR}} \cdot \sqrt{\beta^{-2} \cdot V_{\text{gc1}}^2 \cdot \sigma_\beta^2 + 4 \cdot \beta^2 \cdot V_{\text{gc1}}^4 \cdot \sigma_R^2 + \sigma_{V_{\text{th1}}}^2}}{(1 + 2 \cdot R \cdot \beta \cdot V_{\text{gc1}})^2}, \tag{3.62}$$

$$\sigma_{I_{\text{cell0}}} = \frac{\beta \cdot V_{\text{BLR}} \cdot \sqrt{\beta^{-2} \cdot V_{\text{gc0}}^2 \cdot \sigma_\beta^2 + 4 \cdot \beta^2 \cdot V_{\text{gc0}}^4 \cdot \sigma_R^2 + \sigma_{V_{\text{th0}}}^2}}{(1 + 2 \cdot R \cdot \beta \cdot V_{\text{gc0}})^2}. \tag{3.63}$$

Now we define the SNR of FBC read current as

$$(\text{SNR})_{I_{\text{cell}}} = \frac{\Delta I_{\text{cell}}}{\sqrt{\sigma_{I_{\text{cell1}}} \cdot \sigma_{I_{\text{cell0}}}}}. \tag{3.64}$$

Substituting Eqs. 3.58, 3.62, and 3.63 into Eq. 3.64, we have
$(\text{SNR})_{I_{\text{cell}}} =$

$$\frac{(\text{SNR})_{V_{\text{th}}}}{\sqrt[4]{1 + \left(\frac{V_{\text{gc1}}}{\beta} \cdot \frac{\sigma_\beta}{\sigma_{V_{\text{th1}}}}\right)^2 + \left(2 \cdot \beta \cdot V_{\text{gc1}}^2 \cdot \frac{\sigma_R}{\sigma_{V_{\text{th1}}}}\right)^2} \sqrt[4]{1 + \left(\frac{V_{\text{gc0}}}{\beta} \cdot \frac{\sigma_\beta}{\sigma_{V_{\text{th0}}}}\right)^2 + \left(2 \cdot \beta \cdot V_{\text{gc0}}^2 \cdot \frac{\sigma_R}{\sigma_{V_{\text{th0}}}}\right)^2}}, \tag{3.65}$$

where $(\text{SNR})_{V_{\text{th}}}$ is the SNR with respect to the cell's threshold voltage which is defined by

$$(\text{SNR})_{V_{\text{th}}} = \frac{\Delta V_{\text{th}}}{\sqrt{\sigma_{V_{\text{th1}}} \cdot \sigma_{V_{\text{th0}}}}}. \tag{3.66}$$

Equation 3.65 indicates that if $\sigma_\beta = \sigma_R = 0$, then

$$(\text{SNR})_{I_{\text{cell}}} = (\text{SNR})_{V_{\text{th}}}. \tag{3.67}$$

Thus, if the fluctuation in the FBC read current is determined by the variation in the threshold voltage and if the variations in β and R are neglected, the SNR with respect to FBC's threshold voltage is identical to that with respect to FBC's read current. However, in general cases where β and R are supposed to vary cell-to-cell in a memory array, the SNR with respect to FBC's read current is smaller than that with respect to FBC's threshold voltage:

$$(\text{SNR})_{I_{\text{cell}}} < (\text{SNR})_{V_{\text{th}}}. \tag{3.68}$$

When FBC is read at saturation region, the SNR is calculated in the same way by using Eqs. 3.45 and 3.54.

As was calculated in this section, when both β which is the cell's transconductance divided by the drain voltage at read V_{BLR} and the cell's parasitic resistance R are not fluctuating within cells in a memory, the SNR defined by the threshold voltage coincides with that defined by the read current. However, if the fluctuation in parasitic resistance of the cell and/or that in the eletron mobility of the cell (under the assumption that t_{ox}, L_{eff}, and W are aligned for all cells) are not negligible, the SNR defined by the cell read current is smaller than that defined by the cell threshold voltage. Therefore, to predict the memory functionality, it is necessary to calculate $(\text{SNR})_{I_{\text{cell}}}$ which will directly correlate with the memory chip yield rather than $(\text{SNR})_{V_{\text{th}}}$, though the latter which is more fundamental parameter with a clearer definition can be used to check the ideally maximum value of the SNR without fluctuation in parasitic resistances.

3.3 NECESSARY SNR OF FBC FOR HIGH-DENSITY MEMORIES

In the previous section, we roughly estimated the necessary SNR of FBC for functioning high-density memories. However, the treatment is not rigorous but rather intuitive. In this section, we calculate SNR of FBC starting from the first principle.

Figure 3.16 illustrates schematically the probability density distributions of cell current for the "1" cell $f_1(I_{\text{cell}})$, the "0" cell $f_0(I_{\text{cell}})$, and the dummy cell $f_{\text{R}}(I_{\text{cell}})$. These distributions represent the fluctuations of corresponding cell currents in a memory cell array. Here, we assume that all three distribution functions are symmetrical with respect to the averages $<I_{\text{cell1}}>$, $<I_{\text{cell0}}>$, and $<I_{\text{ref}}>$, respectively. Furthermore, $f_1(I_{\text{cell}})$ and $f_0(I_{\text{cell}})$ are placed symmetrically with respect to $f_{\text{R}}(I_{\text{cell}})$. In the assumption,

$$< I_{\text{ref}} > = \frac{< I_{\text{cell1}} > + < I_{\text{cell0}} >}{2} \tag{3.69}$$

and the bit yield which is a probability for an arbitrary cell to be sensed correctly by a S/A circuit is independent of the data pattern in the cell array. Therefore, we assume that all cells have "0" data for simplicity. As is shown in Fig. 3.16, the probability for a dummy cell to have its reference current between I_{cell} and $I_{\text{cell}} + dI_{\text{cell}}$ is expressed by

$$f_{\text{R}}(I_{\text{cell}})dI_{\text{cell}}. \tag{3.70}$$

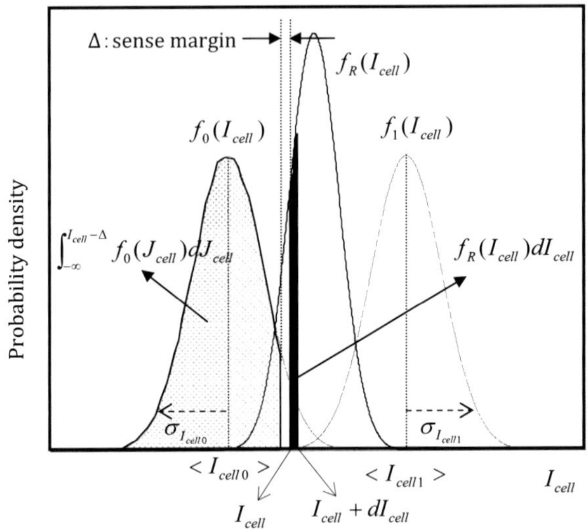

Figure 3.16 The probability density distributions of cell current for the "1" cell , the "0" cell, and the dummy cell to be used for calculating the bit yield in a memory array.

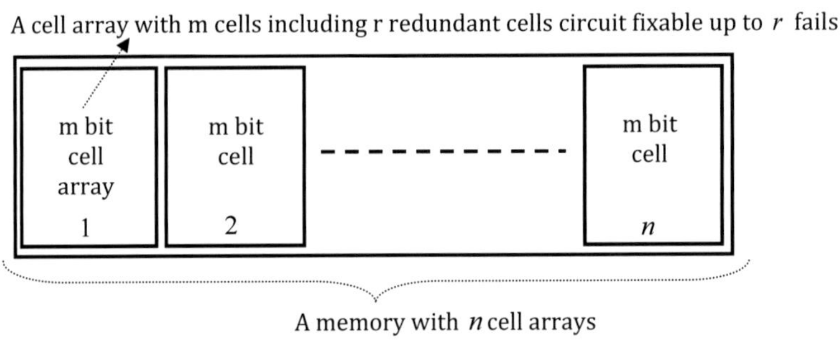

Figure 3.17 General definition of a hierarchical memory structure which consists of n independent cell arrays each includes m cells with r redundant cells to fix r random fails.

When a S/A senses a "0" data cell by using the dummy cell, the cell's current must be $I_{\text{cell}} - \Delta$ or smaller for being sensed correctly, where Δ stands for a sense margin which is the minimum difference between the data cell current and the reference cell one for the cell to be sensed correctly by the S/A. Therefore, the probability for a "0" cell to be sensed correctly by using the dummy cell is expressed by

$$\int_{-\infty}^{I_{\text{cell}}-\Delta} f_0(J_{\text{cell}})dJ_{\text{cell}}. \tag{3.71}$$

Therefore, the probability for a dummy cell to flow the reference current between I_{cell} and $I_{\text{cell}} + dI_{\text{cell}}$ with successful sensing is expressed by

$$f_R(I_{cell})dI_{cell}\int_{-\infty}^{I_{cell}-\Delta} f_0(J_{cell})dJ_{cell}. \tag{3.72}$$

Since dummy cell current distributes from $-\infty$ to ∞, the bit yield or the probability p for a cell to be sensed correctly is written as

$$p=\int_{-\infty}^{\infty} f_R(I_{cell})dI_{cell}\int_{-\infty}^{I_{cell}-\Delta} f_0(J_{cell})dJ_{cell}. \tag{3.73}$$

The memory chip which we will calculate the yield for is assumed to have n identical cell arrays each of which has m cells plus r redundant cells which can fix up to r random fails as shown in Fig. 3.17. The memory chip yield Y_{SM} regarding cell functionality due to signal margin[c] is written by

$$Y_{SM}=\left[\sum_{i=0}^{r} {}_m C_i \cdot p^{m-i}\cdot(1-p)^i\right]^n. \tag{3.74}$$

We assume that $f_1(I_{cell})$, $f_0(I_{cell})$, and $f_R(I_{cell})$ are the normal distributions with the averages $<I_{cell}>$, $<I_{cell0}>$, and $<I_{ref}>$, respectively. The cell distributions $f_1(I_{cell})$ and $f_0(I_{cell})$have the same standard deviation $\sigma_{I_{cell}}$, while the dummy cell distribution $f_R(I_{cell})$ has the standard deviation $\sigma_{I_{cell}}/\sqrt{2}$ corresponding to the method to generate the reference current by averaging the currents of a pair of the "1" cell and the "0" cell. Therefore, we have

$$f_0(I_{cell})=\frac{1}{\sqrt{2\pi}\cdot\sigma_{I_{cell}}}\exp\left[-\frac{(I_{cell}-<I_{cell0}>)^2}{2\cdot\sigma_{I_{cell}}^2}\right], \tag{3.75}$$

$$f_1(I_{cell})=\frac{1}{\sqrt{2\pi}\cdot\sigma_{I_{cell}}}\exp\left[-\frac{(I_{cell}-<I_{cell1}>)^2}{2\cdot\sigma_{I_{cell}}^2}\right], \tag{3.76}$$

$$f_R(I_{cell})=\frac{1}{\sqrt{\pi}\cdot\sigma_{I_{cell}}}\exp\left[-\frac{\left(I_{cell}-\frac{<I_{cell0}>+<I_{cell1}>}{2}\right)^2}{\sigma_{I_{cell}}^2}\right]. \tag{3.77}$$

By substituting Eqs. 3.75–3.77 into Eq. 3.73, we can calculate the bit yield p. And by substituting the calculated p into Eq. 3.74, we can calculate the memory chip yield Y_{SM} regarding to cell functionality due to the signal margin as a function of the SNR $(SNR)_{I_{cell}} = \Delta I_{cell}/\sigma_{I_{cell}} = (<I_{cell1}> - <I_{cell0}>)/\sigma I_{cell}$. Figure 3.18 shows the calculated memory chip yield Y_{SM} regarding to cell functionality due to the signal margin as a function of the SNR $(SNR)I_{cell}$ for $m = 2 \times 1024 \times 1024$ (2 Mbitarray),

[c] The eventual memory chip yield Y will be expressed by $Y = Y_{SM} \times Y_D \times Y_{peri} \times ...$, where Y_D and Y_{peri} represent the memory chip yield loss caused by defects due to, for example, particles introduced when being processed and the peripheral circuit yield, respectively.

$r = 8$, and $\Delta = \sigma_{I\text{cell}}$. We consider three cases: $n = 512$(1 Gbit), $n = 2048$ (4Gbit), and $n = 8192$ (16 Gbit). These results indicate that the necessary minimum value of the SNR will be about 18 or larger, a little bit larger than what we have estimated in section 3.2.

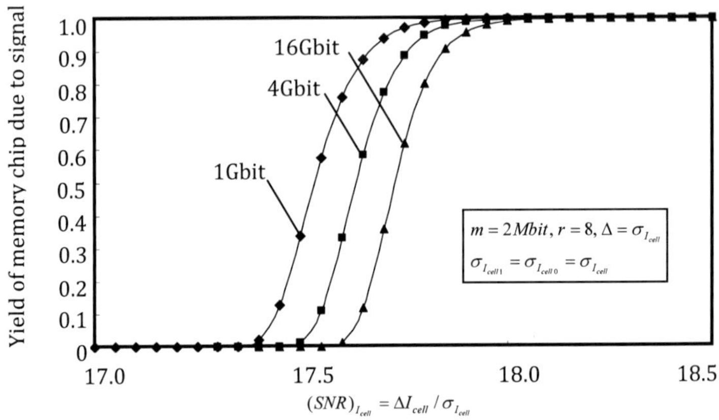

Figure 3.18 The relationship between the memory chip yields regarding cell functionality due to signal margin and the signal-to-noise ratio SNR with respect to cell read current.

In the above method for generating the reference current, the cell read currents of a pair of dummy FBC cells with opposite data are averaged. In general, it is possible to design the dummy cell system by averaging *N* pairs of dummy FBC cells to generate an accurate reference current with a smaller standard deviation which contributes to increasing the signal margin yield, where *N* is a natural number such as 128, 256, or larger.[12,13] In fact, according to the probability theory, the standard deviation of the reference current generated by averaging *N* pairs of FBC dummy cells with opposite data will be expressed by

$$\sigma_{I_{\text{ref}}} = \sigma_{I_{\text{cell}}} / \sqrt{N}, \tag{3.78}$$

where $\sigma_{I_{\text{cell}}}$ stands for the standard deviation of the cell's read current under the assumption that the extent to which the "1" cell's current fluctuates is the same as that to which the "0" cell's current does. The simulation and experimental results on the reference current generation by multi-pair dummy cell averaging will be described in Chapter 6 in detail.

References

1. T. Ohsawa, K. Fujita, T. Higashi, Y. Iwata, T. Kajiyama, Y. Asao and K. Sunouchi, "Memory design using one-transistor gain cell on SOI," *ISSCC Dig. Tech. Papers*, pp. 152–153, Feb. 2002.
2. T. Ohsawa, K. Fujita, T. Higashi, Y. Iwata, T. Kajiyama, Y. Asao and K. Sunouchi, "Memory design using a one-transistor gain cell on SOI," *IEEE J. Solid-State Circuits*, vol. 37, no. 11, pp. 1510–1522, Nov. 2002.

3. T. Shino, T. Higashi, K. Fujita, T. Ohsawa, Y. Minami, T. Yamada, M. Morikado, H. Nakajima, K. Inoh, T. Hamamoto and A. Nitayama, "Highly scalable FBC (Floating Body Cell) with 25nm BOX structure for embedded DRAM applications," *2004 Symposium on VLSI Technology Digest of Technical Papers*, pp. 132–133, June 2004.
4. C. Kuo, T.-J. King and C. Hu, "A capacitorless double-gate DRAM technology for sub-100-nm gate embedded and stand-alone memory applications." *IEEE Trans. Electron Devices*, vol. 50, no. 12, pp. 2408–2416, Dec. 2003.
5. I. Ban, U. E. Avci, U. Shah, C. E. Barns, D. L. Kencke and P. Chang, "Floating body cell with independently-controlled double gates for high density memory," *IEDM Tech. Dig.*, pp. 573–576, Dec. 2006.
6. T. Shino, T. Higashi, N. Kusunoki, K. Fujita, T. Ohsawa, N. Aoki, H. Tanimoto, Y. Minami, T. Yamada, M. Morikado, H. Nakajima, K. Inoh, T. Hamamoto and A. Nitayama, "Fully-depleted FBC (Floating body cell) with enlarged signal window and excellent logic process compatibility," *IEDM Tech. Dig.*, pp. 281–284, Dec. 2004.
7. I. Ban, U. E. Avci, D. L. Kencke and P. L. D. Chang, "A scaled floating body cell (FBC) memory with high-k+metal gate on thin-silicon and thin-box for 16-nm technology node and beyond," *2008 Symposium on VLSI Technology Digest of Technical Papers*, pp. 92–93, June 2008.
8. M. Yamaoka, K. Osada, R. Tsuchiya, M. Horiuchi, S. Kimura and T. Kawahara, "Low power SRAM menu for SoC application using yin-yang-feedback memory cell technology," 2004 Symposium on VLSI Circuits Digest of Technical Papers, pp. 288–291, June 2004.
9. M. Yamaoka, R. Tsuchiya and T. Kawahara, "SRAM circuit with expanded operating margin and reduced stand-by leakage current using thin-box FD SOI transistors," *IEEE J. Solid-State Circuits*, vol. 41, no. 11, pp. 2366–2372, Nov. 2006.
10. R. Tsuchiya, M. Horiuchi, S. Kimura, M. Yamaoka, T. Kawahara, S. Maegawa, T. Ipposhi, Y. Ohji and H. Matsuoka, "Silicon on thin BOX: a new paradigm of the CMOSFET for low-power and high-performance application featuring wide-range back-bias control," *IEDM Tech. Dig.*, pp. 631–634, Dec. 2004.
11. S. Ogura, P. J. Tsang, W. W. Walker, D. L. Critchlow and J. F. Shepard, "Design and characteristics of the lightly doped drain-source (LDD) insulated gate field-effect transistor," *IEEE Trans. Electron Devices*, vol. ED-27, no. 8, pp. 1359–1367, Aug. 1980.
12. T. Ohsawa, K. Fujita, K. Hatsuda, T. Higashi, T. Shino, Y. Minami, H. Nakajima, M. Morikado, K. Inoh, T. Hamamoto, S. Watanabe, S. Fujii and T. Furuyama, "Design of a 128-Mb SOI DRAM using the floating body cell (FBC)," *IEEE J. Solid-State Circuits*, vol. 41, no. 1, pp. 135–145, Jan. 2006.
13. T. Ohsawa, T. Higashi, K. Fujita, K. Hatsuda, N. Ikumi, T. Shino, H. Nakajima, Y. Minami, N. Kusunoki, A. Sakamoto, J. Nishimura, T. Hamamoto and S. Fujii, "A 128Mb floating body RAM (FBRAM) on SOI with multi-averaging scheme of dummy cell," *2006 Symposium on VLSI Circuits Digest of Technical Papers*, pp. 224–225, June 2006.

Chapter 4

128 Mbit Floating Body RAM on Silicon-on-Insulator

To verify the extendibility of floating body cell (FBC) for embedded macros, a 128 Mbit floating body RAM (FBRAM) has been designed and fabricated. The memory cell design is based on 90 nm technology node. The operation of FBC has been designed as fully depleted MOSFET on silicon-on-insulator. Based on the measurement results of the 128 Mbit FBRAM, many kinds of important characteristics of FBC have been clarified. FBC has an adequate signal for the single cell operation. The data retention can be controlled by the electrical filed of the depletion layer, which is almost same as the conventional dynamic random access memory. As for the disturb issue, the effect of two kinds of disturb has been clarified. The one is bit line disturb, the other is bipolar disturb.

4.1 MEMORY CELL DESIGN

Figure 4.1 shows a schematic view of the fabricated floating body cell (FBC) structure. The memory cell consists of a one-transistor memory cell on silicon-on-insulator (SOI) substrate with thin buried oxide (BOX). P^- diffused layer on the substrate is used as a plate electrode, which can stabilize the body potential. Salicide process is introduced to reduce the parasitic resistance inside the array device, which is indispensable for high-speed write operation.[1] The memory cell layout and the design parameters are shown in Fig. 4.2 and Table 4.1, respectively. The unit cell area is $6.2F^2$ (0.17 μm^2). The thicknesses of SOI and BOX are 55 nm and 25 nm, respectively. To reduce the parasitic resistance, Co salicide is formed on the source, the drain, and the gate. M1 (Cu wiring) was used both for the source line (SL) and the pad electrode for the bit line (BL) contact. M2 (Cu wiring) was used for the BL. Figure 4.3 shows a cross-sectional picture of the fabricated memory cell array in the BL direction.[2]

Floating Body Cell: A Novel Capacitor-less DRAM Cell
By Takashi Ohsawa and Takeshi Hamamoto

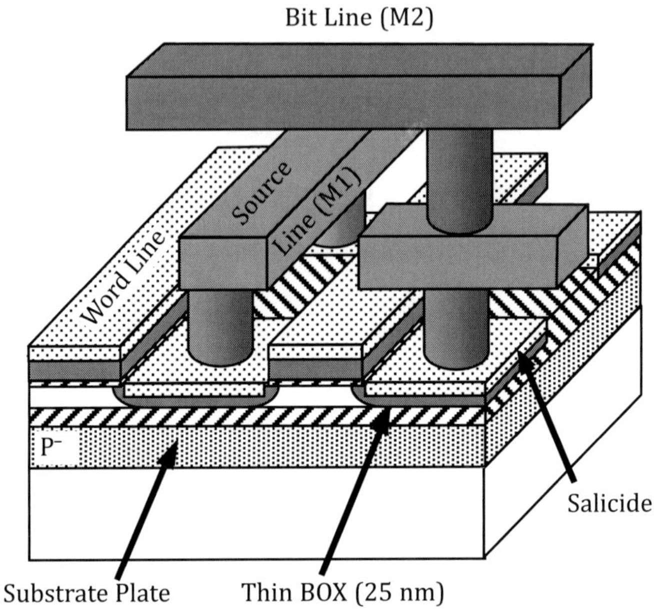

Figure 4.1 Schematic view of the fabricated FBC structure.

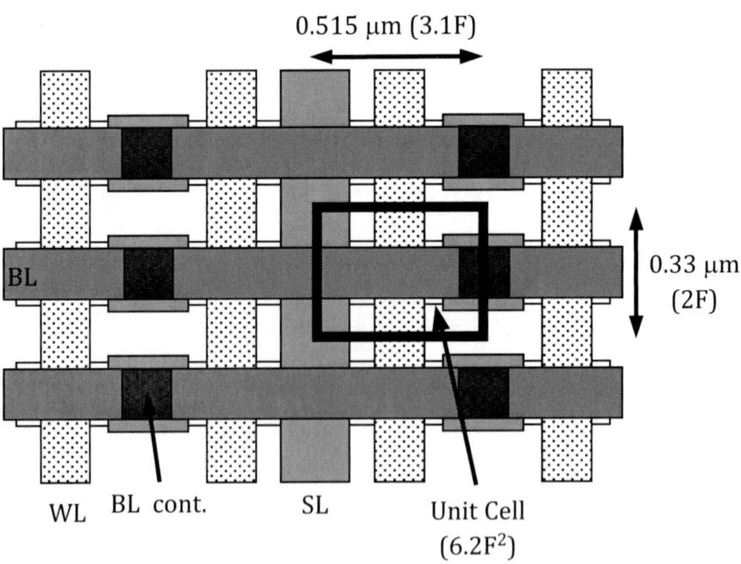

Figure 4.2 The memory cell layout.

Table 4.1 The design parameters

Structure	Cell size (μm^2)	0.17
	Cell size (F^2)	6.2
	BL pitch (μm)	0.33
	WL pitch (μm)	0.515
	L_{gate} (μm)	0.145
	Silicon thickness (nm)	55
	Tox (nm)	6
	BOX thickness (nm)	25
Bias	V_{WLH} (V)	1.5
	V_{WLL} (V)	–2
	V_{BLH} (V)	2.5
	V_{BLL} (V)	–1.6
	V_{PL} (V)	–2.5
Process	Technology node	90 mm (CMOS IV)
	Additional process	Deep Nwell
	Salicide	$CoSi_2$
	BEOL	Cu wiring

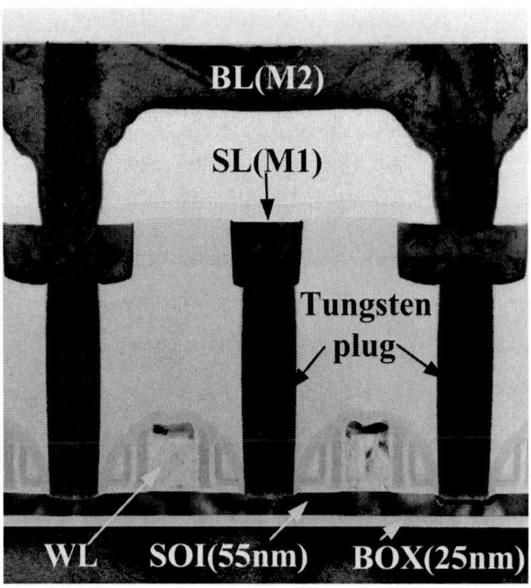

Figure 4.3 Cross-sectional picture of the fabricated memory cell array in the BL direction.

4.2 MEASUREMENT RESULTS OF FBC SIGNAL

In this section, we will describe fully depleted (FD) operation of the fabricated FBC. Then, we also describe the measured results of the signal-to-noise ratio (SNR) along with the memory chip functionality of a 128 Mbit floating body RAM (FBRAM).[3,4] After that we will present an optimized FBC device structure which balances the functionality and the retention time characteristics by comparing two kinds of cell structures with different gate spacer width. The theory presented in the previous section will also be verified in the comparison.

4.2.1 FD Operation

The most important issue for realizing FBRAM is to achieve an adequate signal. In the case of FD operation, the negative plate bias makes much improvement of the threshold voltage difference of "1" state and "0" state.[5] The holes, which are generated by the impact ionization at "1" state write, are accumulated at the interface between body and BOX by the negative plate bias. Figure 4.4a shows the threshold voltages as a function of the plate bias. The threshold voltage of "1" state is nearly constant, which indicates generated holes are stored and the back surface is in the accumulation state. On the other hand, the threshold voltage of "0" state significantly increases by applying the plate bias in the range from 0 to −2 V. In this region, the back surface is depleted. Due to the so-called back gate effect, the increase of the negative plate bias makes the increase of the threshold voltage of "0" state. At a plate bias lower than −2 V, the threshold voltage is saturated. Similar to "1" state, the back surface is accumulated. The electrical field from the plate electrode is shielded by the accumulated layer.

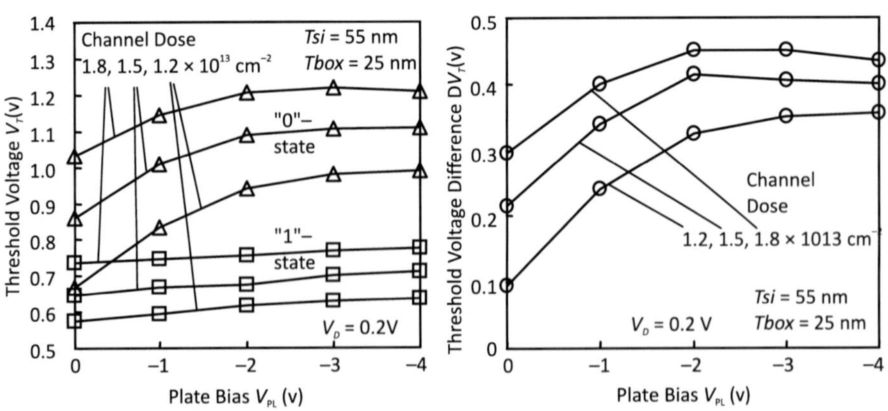

Figure 4.4 Plate bias dependence of (a) the threshold voltage and (b) the threshold voltage differences, as a parameter of the channel dose.

The back gate effect, therefore, no longer occurs. The dependence of the threshold voltage on the plate bias indicates that the operation of FBC is the mixture of FD and partially depleted (PD). In case of "1" state and "0" state of the plate bias lower than −2 V, the channel region can be regarded as PD because the channel potential is not modulated by the plate bias. However, in case of "0" state of the plate bias between 0 V and −2 V, the channel region can be regarded as FD because the channel potential is modulated by the plate bias.

The threshold voltage differences, which is the signals (ΔV_T), as a function of the plate bias are plotted in Fig. 4.4b. The signal can be enlarged by applying the negative plate bias. In both Fig. 4.4a,b, the threshold voltages for the different channel implant doses are also shown. The threshold voltage difference of 0.4 V is sufficient to get a fully functional die assuming the redundancy system of the 128 Mbit FBRAM.[3,4]

4.2.2 Measured SNR

In this subsection, we will disclose the measurement results of the SNR of cells for a state-of-the-art cell array. However, before showing the measurement results, we will explain the method for measuring the signal of FBC. All memory cells in the 128 Mbit FBRAM are designed to be accessed for measuring their cell currents after entering into a test mode with an external PAD directly connected to the corresponding BL. Therefore, a memory tester can measure the cell current from the PAD by selecting the cell with the corresponding row and column addresses entered. However, an ammeter built in the standard memory testers cannot afford to measure the current in a period as short as a few nanoseconds, since it outputs the current averaged over several hundreds microseconds. In the case of FBC memory current measurements, measuring in a short period is very important, because the signal degrades in a range of several hundreds microseconds, as will be explained later. Therefore, we developed a special technique for memory cell read current to be measured in a short period.[6]

Figure 4.5 explains the method to measure the read current I_{cell} of a cell in a memory cell array. The cells are written to "1" or "0" in the normal write cycle prior to entering into a test mode which changes the circuit configuration to what is to be adapted to the cell read current measurement as is shown in Fig. 4.5. A load current is to be supplied to a cell selected by the associated word line (WL) which is raised to V_{WLHR} through the left p-type MOSFET of the current mirror circuit whose drivability is determined by its gate voltage. This current mirror circuit exists per a BL pair originally and consists of a part of the sense amplifier (S/A) circuit implemented in

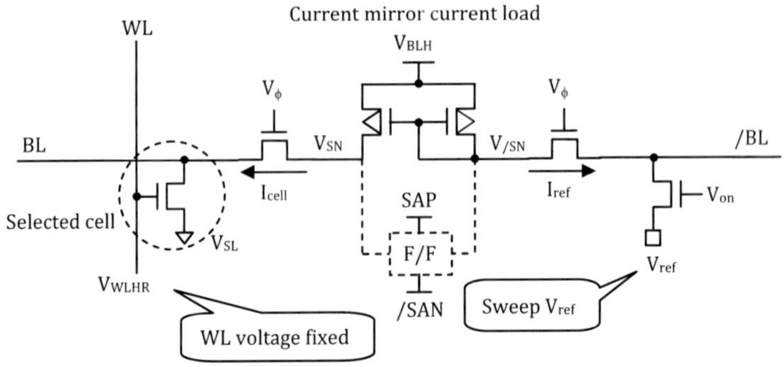

Figure 4.5 The method to measure the read current of cells in an array in a short period.

the memory chip. The gate voltage of the left p-type MOSFET is determined by the reference current I_{ref} which is drawn through the right p-type MOSFET to the external PAD which a reference voltage V_{ref} is being applied to. And the sense node voltage V_{SN} is determined in the balance between the cell current and the load current. The current mirror circuit automatically controls the current mirror node voltage $V_{/SN}$ so as to become smaller than V_{SN} if I_{ref} is larger than I_{cell} and vice versa. Therefore, if the dynamic flip-flop (F/F) circuit latches the voltage difference between V_{SN} and $V_{/SN}$ for various I_{ref} or V_{ref} to notify outside the chip of pass or fail state of the S/A read, we can predict I_{cell} as I_{ref} which is monitored at the transition point from either pass to fail or fail to pass point. This measurement can be performed in a period of a range of several hundreds nanoseconds. This time scale is limited by the signal development time between a BL pair with very small current difference $|I_{cell} - I_{ref}|$ and inevitably longer than a typical sensing time scale of a few nanoseconds with relatively larger current difference.

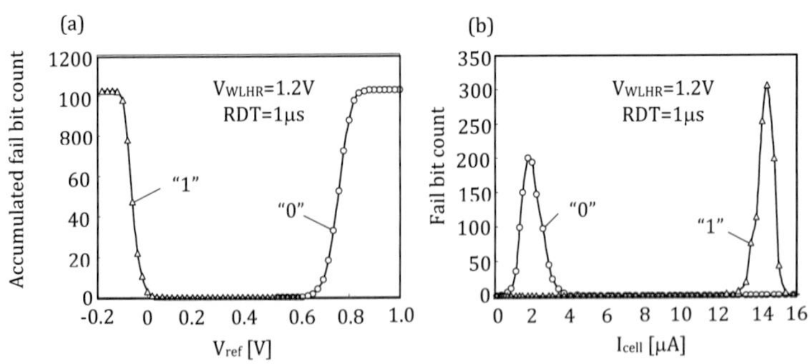

Figure 4.6 (a) Accumulated fail bit counts as a function of V_{ref}. (b) Fail bit count as a function of I_{cell} which is equal to I_{ref} corresponding to V_{ref}.

Figure 4.6a shows the accumulated fail bit count as a function of V_{ref} measured for both the data "1" and the data "0" for WL voltage V_{WLHR} = 1.2 V and the read duration time RDT = 1 μs. The RDT is defined by the delay time from WL activation to the F/F latch. The external voltage V_{ref} can be converted into the reference current I_{ref} by using a table which was made beforehand to correspond between V_{ref} and I_{ref}. This I_{ref} can be regarded as the cell read current I_{cell}. And by differentiating the accumulated fail bit count, we can convert the *y*-axis to obtain the distribution of the cell read current for cells in an array as Fig. 4.6b where V_{WLHR} = 1.2 V and RDT = 1 μs. This graph shows that the averaged cell read current of the data "0" is around 2 μA and that of the data "1" is around 14.5 μA for the read conditions. And it is also shown that the fluctuations of the both currents are well suppressed and that a large SNR is obtained.

Figure 4.7 shows the variation of the cell read current distribution for different RDTs. As RDT becomes larger from 10 μs, 100 μs, to 1 ms, the median of the "1" data read current decreases to approach the "0" data read current and at the same time its fluctuation becomes enhanced. This phenomenon can also be observed in TCAD simulation. Figure 4.8a,b show TCAD device simulation results on the body voltage which is represented by the hole quasi-Fermi potential and on the cell read current, respectively, both as a function of the RDT. The FBC structure for which we

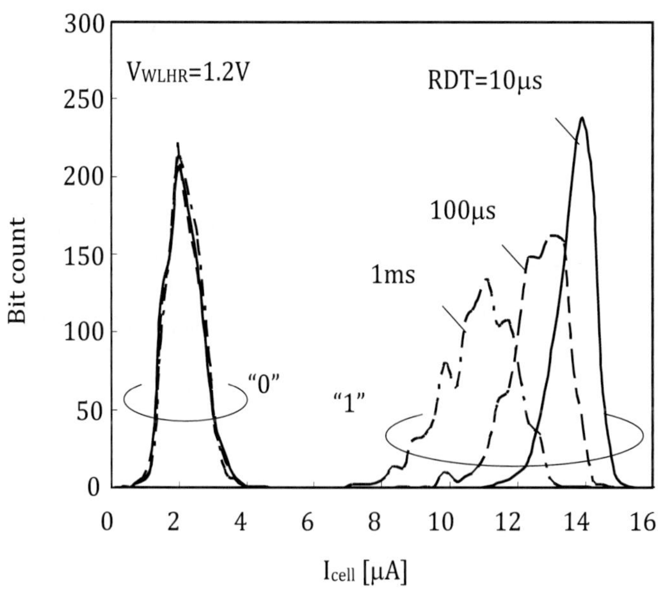

Figure 4.7 Changes in the cell read current distributions for various read duration time RDT.

performed the TCAD simulations is the same as what we used for the measurements. The cells which were written either the data "1" or "0" are being read during RDT with WL 1.5 V and BL 0.2 V, 0.4 V, and 0.6 V. As is shown in Fig. 4.8, the body voltage of the data "1" cell decreases as RDT increases, making the cell read current I_{cell1} decreasing. On the other hand, Fig. 4.8 shows that there is no change for the data "0" cell's body voltage and the cell read current I_{cell0}. The reason will be explained by the forward bias of the p-n junction between the body voltage of the data "1" cell and the grounded source at read in which holes are flowing out of the body gradually to the source terminal. Although the body voltage of the "0" cell might also be positive, the forward bias is too weak to exhibit a noticeable change. The increase of the extent to which the "1" cell current distribution fluctuates as RDT increases can be attributed to the p-n junction forward current variations among various cells which will make the body voltage variation enhanced. Since the actual RDT from the activation of WL to the activation of S/A is about a few nanoseconds or less, this kind of degradation of the "1" cell read current has no effect on the memory performance.

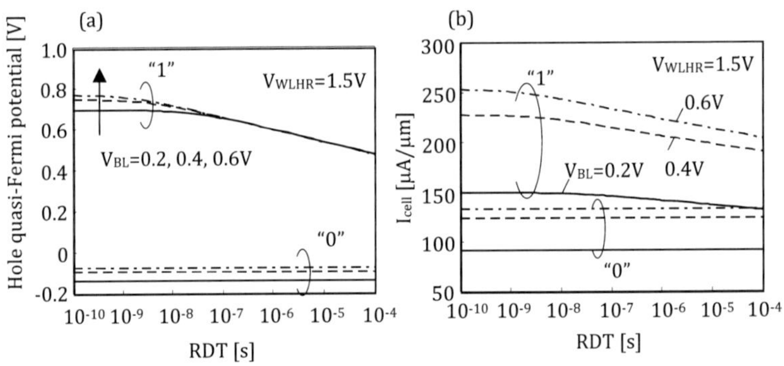

Figure 4.8 TCAD simulation results for (a) the body voltage variation and (b) the cell read current on the read duration time RDT.

However, it is necessary to measure the signal at an RDT as short as a few nanoseconds to know the "real" signal, making the measuring technique very difficult. In fact, though this method can measure the signal for RDT as short as a several hundreds nanoseconds, it is impossible to measure the signal for RDT as short as a few nanoseconds, because the signal development speed will be slowed down when the current mirror load circuit is operated at the critical point where I_{ref} is close to I_{cell}. Therefore, we will extrapolate the measurement results up to 1 μs into a few nanoseconds area to predict the "real" signal as will be explained next.

Figure 4.9 shows (a) the medians of the cell read currents and (b) their standard deviations for the "1" cells and the "0" cells as a function of RDT. As for the standard deviations, two kinds of groups of cells are considered: (I) cells on a BL and (II) cells on a WL. As was explained, this method cannot measure the cell current for RDT = 500 ns or shorter. In Fig. 4.9a, $\langle I_{cell1}\rangle$ and $\langle I_{cell0}\rangle$ stand for the median of the "1" cell read current I_{cell1} and the "0" cell read current I_{cell0}, respectively, and $\langle \Delta I_{cell}\rangle$ means the median of the difference $I_{cell1} - I_{cell0}$. In Fig. 4.9b, $\sigma_{I_{cell0}}$ and $\sigma_{I_{cell1}}$ stand for the standard deviation of the "1" cell read currents and the "0" cell read currents, respectively. As was mentioned, it can be seen that $\langle I_{cell1}\rangle$ decreases and $\sigma_{I_{cell1}}$ increases as RDT increases. Both $\langle I_{cell0}\rangle$ and $\sigma_{I_{cell0}}$ are unchanged. As a consequence, $\langle \Delta I_{cell}\rangle$ decreases as RDT increases. As is shown in Fig. 4.9b, the standard deviation of I_{cell1} for (I) 1024 cells on a BL is smaller than that for (II) 512 cells on a WL. The difference can be explained by the performance variations in S/A, because the

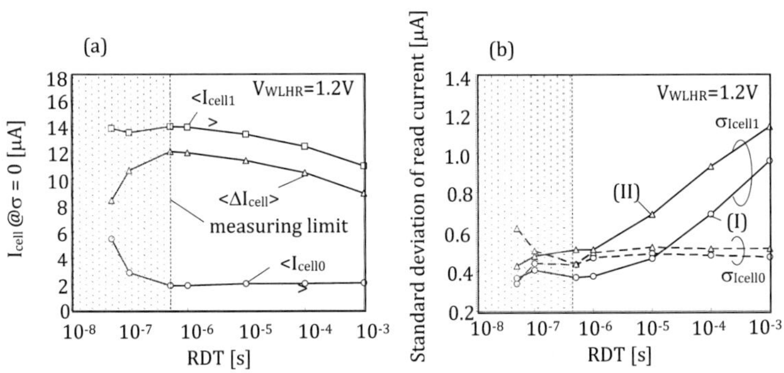

Figure 4.9 Measured data for (a) the medians of the cell current distributions and (b) the standard deviations for the data "1" cells and the data "0" cells. We measured two kinds of standard deviations, i.e., (I) that for 1024 cells on a BL and (II) that for 512 cells on a WL.

measurement results of (I) are obtained by using only one S/A, while those of (II) are obtained by using 512 different S/As. In fact, if we denote the S/A fluctuation in terms of the cell read current by $\sigma_{S/A}$, the standard deviation of (I) $\sigma^{(I)}_{I_{cell1}}$ can be expressed as $\sigma^{(I)}_{I_{cell1}} = \sqrt{\sigma^{(II)^2}_{I_{cell1}} - \sigma^2_{S/A}}$, where $\sigma^{(II)}_{I_{cell1}}$ denotes the standard deviation of (II). Therefore, the intrinsic standard deviation of cell current for the data "1" is expressed by $\sigma^{(I)}_{I_{cell1}}$.

Figure 4.10 shows the SNR with respect to the cell current $(SNR)_{I_{cell}}$ calculated from the measured data in Fig. 4.9. Both ratios for (I) cells on a BL and (II) cells on WL are shown. The intrinsic $(SNR)_{I_{cell}}$ will be indicated in the case (I). If we extrapolate the calculated value to a few nanoseconds area from the data which are reliable, we can estimate $(SNR)_{I_{cell}}$ might be as large as 30 or larger. Since the minimum necessary $(SNR)_{I_{cell}}$ for functioning a multi-gigabit memory by using FBC is 18 as was estimated in section 3.2, the state-of-the-art memory cell array of the L_g = 150 nm generation has an enough potential to function a multi-gigabit memories.

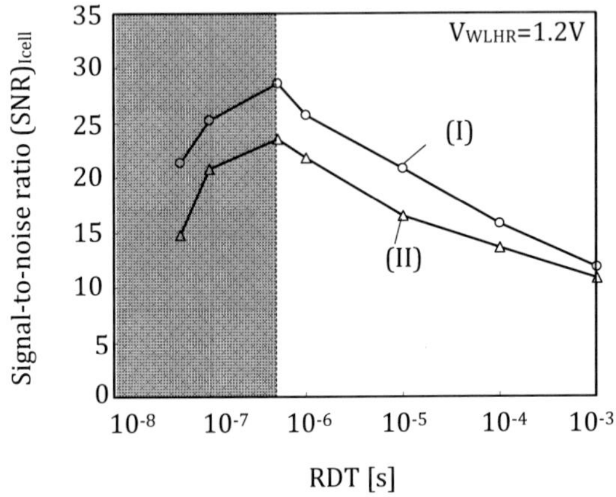

Figure 4.10 Measured signal-to-noise ratio with respect to the cell current for (I) 1024 cells on a BL and (II) 512 cells on a WL.

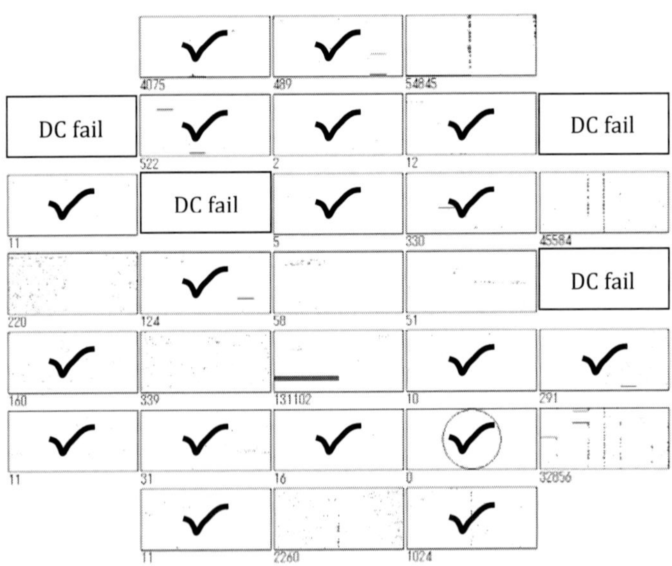

Figure 4.11 Wafer map of function results a 16 Mbit cell array in the 128 Mbit floating body RAM (FBRAM). The number written at the lower left of each chip indicates that of fails in the chip.

Figure 4.11 is a wafer map of functional results for a 16 Mbit cell array in the 128 Mbit FBRAM.[3,4] The four chips which are written "DC fail" indicate the chips whose standby current is very large and do not function at all due to the short-circuited power lines. These chips are considered to be affected by some particles to cause defects leading to the power line shorts. The 18 chips which are checked are fixable chips by redundancy (64 fails are fixable in the 16 Mbit array), though they include multiple fails except for one circled chip. The yield of the 16 Mbit array in this wafer is $Y = 18/27 = 66.7\%$. However, the signal margin yield Y_{SM} must be larger than this value, because we can recognize some systematic pattern in failed bit cells, which means these fails are not due to signal margin loss but due to fails occurred at S/A, row, and column decoders caused by particles or some other defects. Anyway, this high array yield supports the measured results for the large SNR of the FBC.

4.2.3 Optimizing Cell Structure

As was predicted in section 3.3, the lightly doped drain-source (LDD) structure[7] which introduces parasitic resistance at the channel edges may degrade the SNR with respect to the cell current $(\mathrm{SNR})_{I_{\mathrm{cell}}}$ which will directly correlate with the memory chip yield. On the other hand, the LDD structure relaxes the electric field in the body, which will enhance the data retention time.[8–10] Thus, the introduction of LDD or the drain/source engineering in general will be an important issue to optimize FBC structure which can balance between the functionality and the retention time. We will be answering the question to some extent by comparing the functionality and the retention time distribution of two FBCs with different drain/source impurity profiles.

Figure 4.12 shows the cross-sectional transmission electron microscopy (TEM) images of two FBCs, type A and type B. While type A has 120 nm spacers which exist besides the gate, type B has 85 nm spacers. Other parameters are the same for the both types: gate length L_g = 150 nm, channel width W = 180 nm, silicon thickness t_{Si} = 48 nm, gate dielectric thickness t_{ox} = 6 nm, BOX thickness t_{Box} = 25 nm. The FBC structure of type B is exactly the same as that used as the logic n-type MOSFET made by a standard CMOS process. The FBC structure of type A, on the other hand, is made by an extra ion implantation of n-type impurities over the FBC-specific gate spacers whose dose is relatively low prior to the usual n-type ion implantation over the standard CMOS n-type MOSFET gate spacers. The type A, therefore, has a double LDD structure where the source/drain impurity profile is more graded than

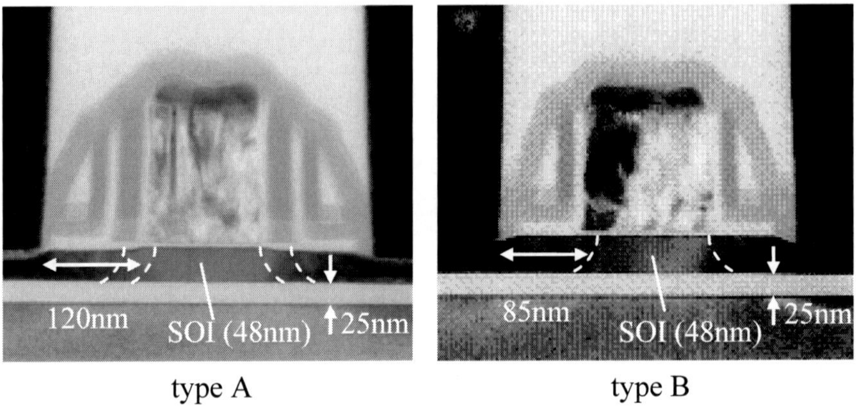

Figure 4.12 TEM images for the two FBC structures (type A and type B) to be compared for memory functionality and retention time distribution. Type A has wider gate spacers for relaxing the electrical field in the body, expected to contribute to improved retention time. Type B has lower parasitic resistance, expected for better functionality.

that of type B. Thus, type A FBC is expected to have more relaxed electrical filed in the body, while type B is expected to have smaller parasitic resistance inserted under the spacers. In other words, type A FBC is designed on a priority basis of retention time characteristics, while type B FBC is designed on a priority basis of cell SNR characteristics.

Figure 4.13 shows the threshold voltage distributions for the cell arrays made by the both types of FBC measured at RDT = 10 μs, and 1 ms. The method of measuring the threshold voltage of FBC at short RDT is the same as described in the previous subsection 3.4.1 for measuring the cell read current except for sweeping V_{WLHR} with I_{ref} fixed at the threshold current instead of sweeping I_{ref} with V_{WLHR} fixed at the WL voltage at read. Type A (Fig. 4.13a) has signal sense margin of 218 mV at ± 4.5σ at RDT = 10 μs, while type B (Fig. 4.13b) has signal sense margin of 242 mV at ± 4.5σ at RDT = 10 μs. Although type B seems to have a little bit larger signal sense margin, no significant difference is recognized.

Figure 4.14 shows the cell read current distributions for the cell arrays made by the both types of FBC measured at RDT 1 μs. Type B has clearly better performance with larger signal sense margin of 5.8 μA at ± 4.5s compared with that of 3.2 μA at ± 4.5s for type A. The differences between the averages $<I_{cell1}> - <I_{cell0}>$ are, respectively, 8.0 μA and 10.3 μ A for type

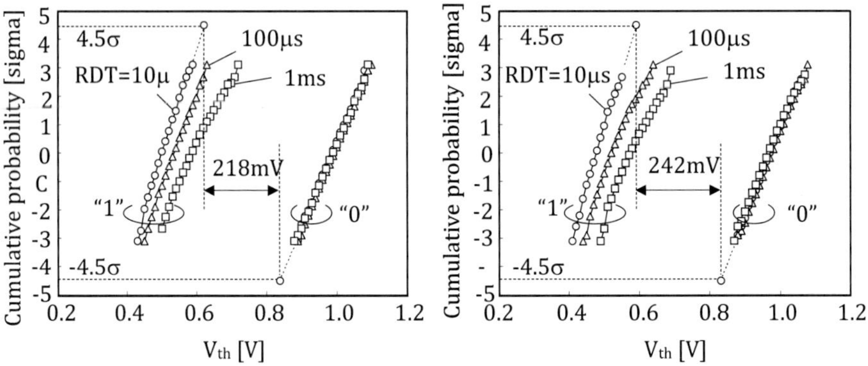

Figure 4.13 Threshold voltage distribution in the array of (a) the type A cells and (b) the type B cells for various RTDs.

A and type B. Furthermore, the distribution of type B cell current is sharper with standard deviation for the data "1" $\sigma_{I_{\text{cell}}}$ = 0.83 μA. As for the data "0" current distribution, the standard deviation of type B is a little bit larger than that of type A. However, since the values themselves are smaller than those for the data "1," they will not affect their SNRs. Therefore, $(\text{SNR})_{I_{\text{cell}}}$ of type B cell is clearly larger than that of type A. If we assume that FBC of type type A has larger parasitic resistance with larger standard deviation of its fluctuation among cells in the array than FBC of type B, this observation was predicted by Eq. 3.65.

Figure 4.15 shows the comparison of functional results for a 16 Mbit cell array using FBC type A and that using type B. The array of type A cell includes 301 fail bits. However, that of type B includes only 10 fail bits. This difference reflects the difference in $(\text{SNR})_{I_{\text{cell}}}$ between the two types of cell. These experiments show that the memory chip functionality cannot be predicted by $(\text{SNR})_{V_{\text{th}}}$ but by $(\text{SNR})_{I_{\text{cell}}}$.

A concern of reducing the parasitic resistance by thinning the gate spacers is retention time degradation due to increased electrical field strength within the floating body. To see the extent to which the type B cell structure degrades its retention time from the value of type A, we measured the retention time distributions for the cell arrays made by both types of cells. Figure 4.16 shows the comparison results. Although the retention times of several tail bits in the array of type B are about the half of those in the array of type A, the two distributions exhibit almost the same performance.

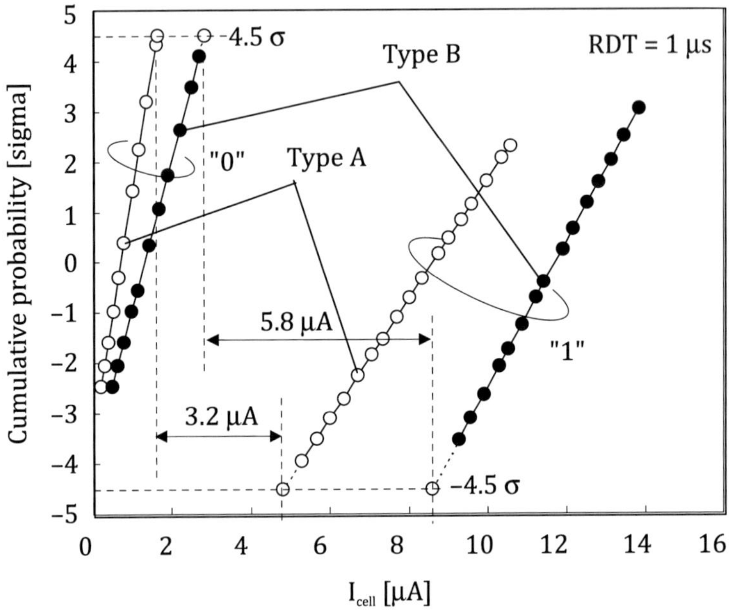

Figure 4.14 Cell read current distribution in the array of the type A cells and the type B cells for RTD = 1 μs.

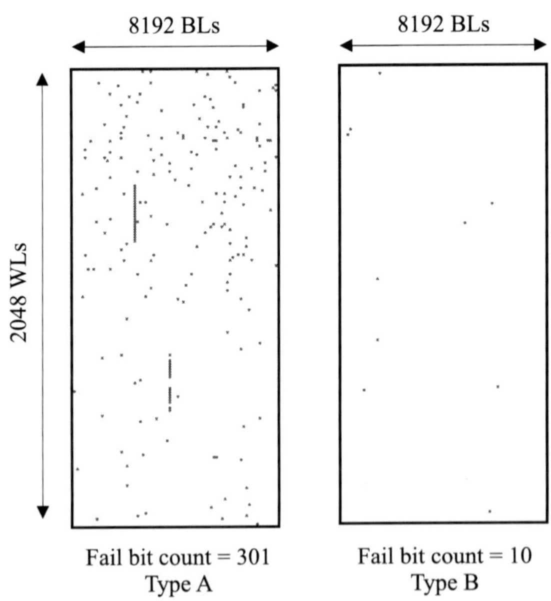

Figure 4.15 Comparison between the functional results of 16 Mbit cell arrays using type A cells and type B cells.

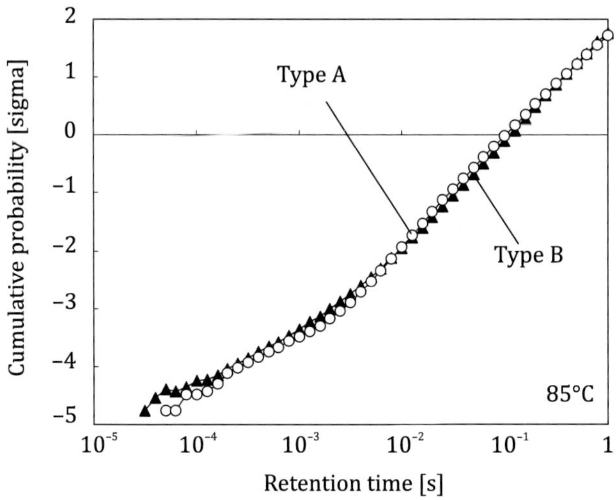

Figure 4.16 Comparison between the retention time distribution measurements of 16 Mbit cell arrays using type A cells and type B cells.

As a consequence of the comparison experiments, FBC which is designed and fabricated according to a standard CMOS process exhibits superior characteristics of chip functionality with large SNR and the degradation of retention time from values of FBC which is designed and fabricated according to a standard CMOS process plus an extra special process steps for widening the gate spaces by about 35 nm for L_g = 150 nm is within an allowable range. This implication brings about a great advantage in implementing the FBC memory into logic LSIs, because the process to make FBCs can be the same as one to make the logic transistors without additional process cost and logic LSI's performance degradation.

4.3 DATA RETENTION

Data retention which is usually represented by the lapse time until which the written data keeps its integrity to be read correctly by S/A is one of the most important features which characterizes usability as well as standby current of FBRAM just like dynamic random access memory (DRAM). Though the retention time distributions are explained in Fig. 4.16, we will discuss data retention in some detail in this section.

As is indicated in Figs. 2.4, 3.7, and 3.8, the p-n junction between the p-type floating body and the n-type source/drain is reverse biased when retaining data regardless its data polarity. The situations of Fig. 2.4 are illustrated in Fig. 4.17 where the body voltage of the data "0" cell and the data "1" cell at hold are −0.65 V and −0.1 V, respectively. Therefore, both data are held by reverse bias between p-n junctions, suppressing the current flowing through the junction to the level of the saturation current of the diode which is quite low. In this hold condition, only the data "0" is degraded with the data "1" state unchanged, because any leak current including the p-n junction current is in the direction to increase the number of holes in the body. Figure 4.18 is one of typical measurement results for data retention characteristics for an FBC at room temperature. The read current of the "0" data cell increases to that of the "1" data cell from 1 s to 10 s approximately.[a]

The retention time of actual FBC is dominated both by the stored signal and the leakage current from the body. To improve the retention time characteristics, there are two countermeasures: (1) increase the stored signal and (2) reduce the leakage current. These countermeasures are basically same as that of the conventional DRAM cell.[8] In addition to the leakage current from the metallurgical junction, the leakage current from the body/BOX interface is important in the case of FD FBC. Figure 4.19 shows retention time cumulative distributions for cells within an array built on PD-SOI with silicon thickness t_{Si} = 150 nm for various doses of channel Boron impurity atoms.[10] The data were measured at 85°. Other device parameters than the silicon thickness are gate length L_g = 150 nm, channel width W = 165 nm, BOX thickness t_{BOX} = 25 nm, and gate dielectric thickness t_{ox} = 6 nm. Four different doses of Boron atoms were compared:

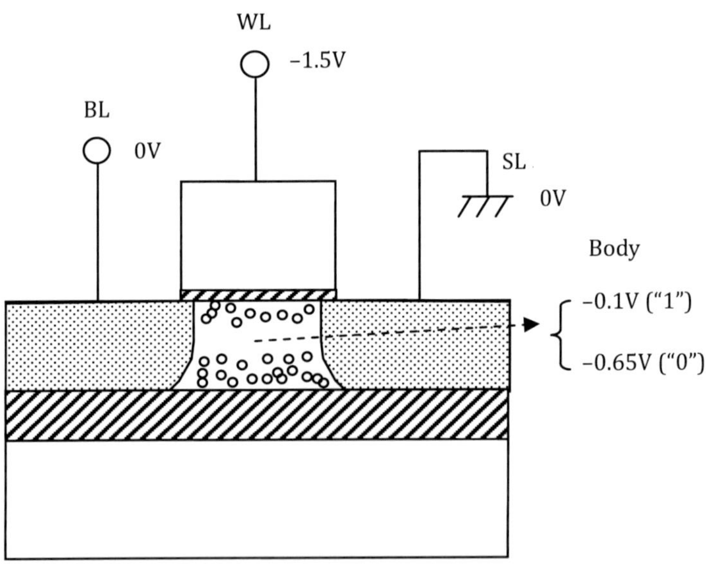

Figure 4.17 Bias conditions while data are held in the floating body.

[a] It is worth noting that the time scale of Fig. 4.18 is logarithm of time. This scale makes us feel as if the data "0" state degrades sigtnificantly between 1 ns and 10 ns in this graph. In reality, however, the holes are equally entering into the body and the "0" read current increases linearly with time.

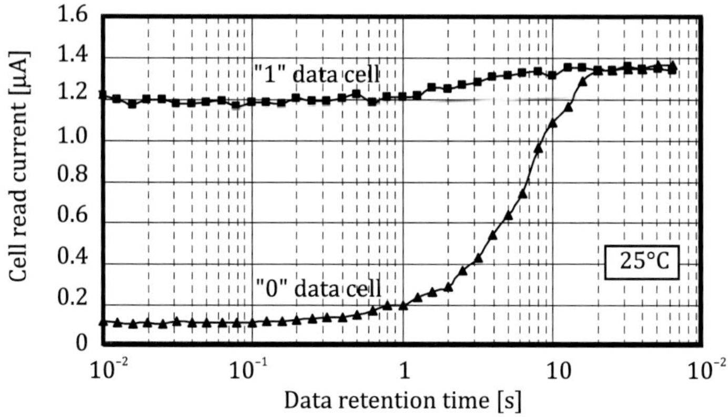

Figure 4.18 Retention time measurement results at room temperature.

0.9 × 10^{13} cm^{-2}, 1.2 × 10^{13} cm^{-2}, 1.5 × 10^{13} cm^{-2}, and 1.8 × 10^{13} cm^{-2}. By increasing the Boron dose from 0.9 × 10^{13} cm^{-2} to 1.5 × 10^{13} cm^{-2}, the retention time distributions become improved. This can be explained by the signal increase expressed in Eq. 2.4, for the Boron concentration N_A increases with the dose. However, if we increase it to 1.8 × 10^{13} cm^{-2}, the retention time distribution become worse than that for 1.5 × 10^{13} cm^{-2}. This setback will be attributed to the increase of the strength of electrical field in the depletion region which becomes thinner by increasing the boron concentration. This degradation of the data retention time is very similar to that occurred in one-transistor and one-capacitor (1T-1C) DRAM where the retention time becomes short with the thermionic field emission current increased[8,9] due

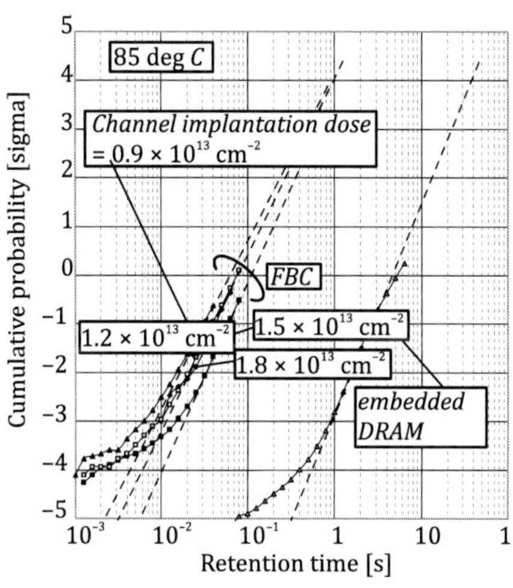

Figure 4.19 Retention time distribution measurement results of PD-SOI FBC at 85°C for various channel implantation doses in comparison with a typical embedded DRAM retention time distribution.

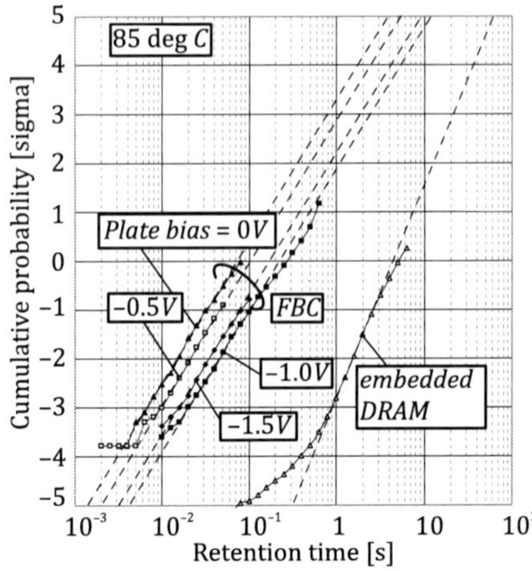

Figure 4.20 Retention time distribution measurement results of FD-SOI FBC at 85°C for various channel implantation doses in comparison with a typical embedded DRAM retention time distribution.

to enhanced electrical field strength. In the case of FBC, it also understood that the effect of increasing the thermionic field emission current is larger than the effect of increasing the signal, making the retention time shorter. Therefore, there is an optimum dose of channel impurity atoms which makes the retention time the longest.

Figure 4.20, on the other hand, shows retention time cumulative distributions for cells within an array built on FD-SOI with silicon thickness t_{Si} = 55 nm for various plate bias voltages.[10] The signal of FD-SOI FBC dependent on the substrate bias or plate bias as was shown in Fig. 4.4.

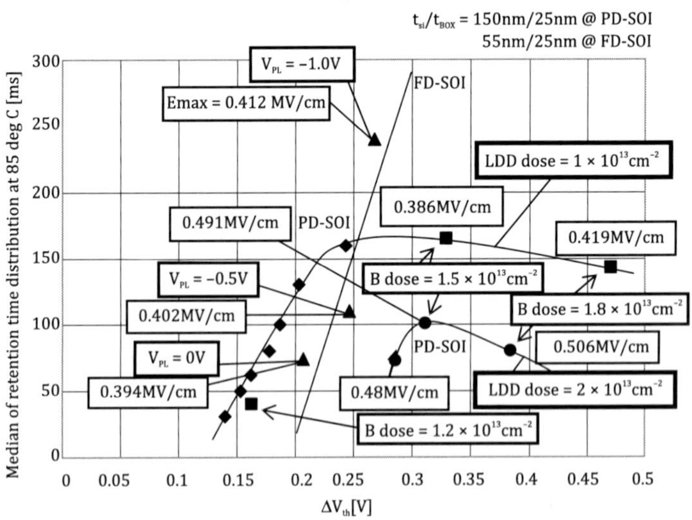

Figure 4.21 The relations between the medians of retention time distributions and the medians of threshold voltage differences for PD-SOI FBC and FD-SOI FBC.

Figure 4.20 indicates that increasing the absolute value of the plate bias from 0 V to −1.0 V improves the retention time distribution. This can be explained by the increased signal due to increasing the absolute value of the plate bias voltage. However, if we increases the plate bias to −1.5 V, the distribution becomes worse that that for −1.0 V. This setback is attributed to the increase in the electrical field strength in the body by increasing the absolute value of the plate bias voltage.

Figure 4.21 summarizes the mean retention time of both PD FBC and FD FBC as a function of the signal. In the case of PD FBC, the increase of the channel dose enhances the signal, which elongates the retention time. Too much channel dose, however, degrades the retention time due to gate-induced drain leakage (GIDL) current by the strong electric field. As a result, the relationship between the signal and the retention time is a bell shape. The best retention time is 160 ms. Two bell shapes, depending on the LDD phosphorus dose, are also plotted in the figure. The increase of the phosphorus

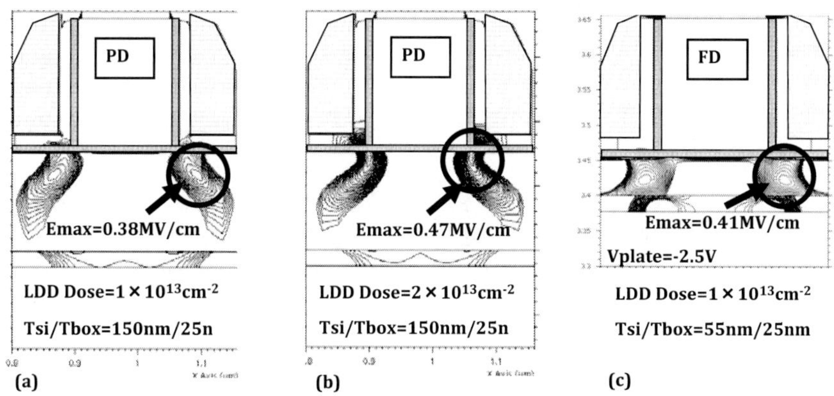

Figure 4.22 The counter plots of the electrical field. (a) PD FBC with LDD dose of 1×10^{13} cm^{-2}, (b) PD FBC with LDD dose of 2×10^{13} cm^{-2}, (c) FD FBC with LDD dose of 1×10^{13} cm^{-2}. The channel dose of the three structures is 1.2×10^{13} cm^{-2}. Applied bias is that of "0" hold; $V_{BL} = 0$ V, $V_{WL} = -1.5$ V, $V_{PL} = -2.5$ V (FD FBC).

dose makes the enhancement of the electrical field at the gate edge, which makes the degradation of the retention time. Figure 4.22 shows the contour plots of the electrical field at "0" hold. The electrical fields of PD FBC, which have LDD doses of 1×10^{13} cm^{-2} and 2×10^{13} cm^{-2}, are compared in Fig. 4.22a,b. The maximum electrical fields of them are 0.38 MV/cm and 0.47 MV/cm, respectively. This difference makes the two kinds of the bell shape in Fig. 4.21. The simulated electrical field strength in each device is also added in the figure.

In the case of FD FBC, the situation has drastically changed, which is also shown in Fig. 4.21. Besides the electrical field, the negative plate bias makes large improvement of the retention time. Deeper plate bias boosts up the retention time longer than the peak of PD FBC. Important to note is that the electric filed is larger than PD FBC of same LDD dose. The difference of the electrical field of PD FBC and FD FBC can be compared by Fig. 4.22a,c. The maximum electrical filed of them are 0.38 MV/cm and 0.41 MV/cm, respectively. We suppose that the deeper plate bias reduces the leakage current from the body/BOX interface because the accumulated layer will be formed there. We also suppose that the reduction of the silicon thickness contributes to the reduction of the junction leakage current because the depth of the metallurgical junction of FD FBC is one-third of that of PD FBC.

The cell capacitance C_s of the conventional 1T-1C DRAM is 20 ~ 40 fF, while the body total capacitance for the FBC with gate length L_g = 150 nm, channel width W = 165 nm, silicon thickness t_{Si} = 50 nm, gate dielectric thickness t_{ox} = 6 nm, BOX thickness t_{BOX} = 25 nm is about 0.1 ~ 0.2 fF. Therefore, the capacitance to hold data for FBC is more than two orders of magnitude smaller than that for the conventional 1T-1C DRAM. As for the leak current, the conventional 1T-1C DRAM cell has components mainly from the p-n junction leak, the GIDL, and the subthreshold leakage. FBC has basically the same components except for the subthreshold leakage component. The p-n junction cross section of FBC will be smaller than that of 1T-1C DRAM, because FBC is built in a thin film silicon on the BOX. However, GIDL of FBC will be larger than that of 1T-1C DRAM, because FBC holds data with WL −1.5 ~ −2.5 V which is lower than WL voltage of 1T-1C DRAM of 0 ~ −0.5 V. Anyway, the decreased capacitance to hold data has a dominant effect on the retention time, making it more than two orders of magnitude shorter than that of the conventional 1T-1C DRAM cell as is shown in Figs. 4.19 and 4.20 where a typical retention time distribution of embedded DRAM is compared with those of FBC.

It is considered that the status quo of the data retention time of FBC is long enough for embedded memory applications especially for high performance LSIs like second or third level cache memories of high performance microprocessors. In fact, since the standby current of these high performance LSIs are dominated by the logic transistors' subthreshold current, the refresh power of the FBC cache memory will be negligible. And refresh busy rate RBR which is defined by

$$\mathrm{RBR} = \frac{N_{\mathrm{ref}} \cdot t_{\mathrm{RC}}}{t_{\mathrm{ret}}}, \tag{4.1}$$

with refres cycle, Nref, refresh cycle time t_{RC}, and the with refresh cycle N_{ref}, refresh cycle time t_{RC}, and the retention time spec of the memory t_{ret} to characterize the memory's usability can be suppressed low enough, because the cycle time t_{RC} will be very small for the high-speed design. However, for stand-alone memory applications, the current value of the retention time of FBC is too short to satisfy the current DRAM specification.

This is also the case for memories embedded into low power LSIs which are used in mobile devices. For the applications, two orders of magnitude or more improvement is required. Recently, a long retention time of FBC was reported by improving the process to optimize the FBC's structure.[11] Systematic and detailed studies in this filed are expected to be made in the future. The FBC-specific refresh which makes both the data retention power and the RBR low enough even with a relatively poor retention time will be another direction for solving this problem. The refresh method, autonomous refresh, will be explained in detail in Chapter 7.

4.4 DISTURB

Since FBC is a single transistor memory cell whose data storage device is merged with the select device, several kinds of data disturbs are inevitable when the storage device is directly subjected to WL and BL voltage applications. They are categorized into three kinds of disturbs. These are the WL disturb, the BL disturb, and the bipolar disturb, respectively. The WL disturb is related to the charge pumping phenomenon.[12,13] This type of disturb will be discussed in detail in Chapter 6 in relation to the design of S/A circuits. In this section, we will show the degradation of retention time distributions under the influence of the BL disturbs. We will also show the cell-to-cell leakage caused by the bipolar disturb, and the countermeasure for this phenomenon.

4.4.1 BL Disturb

The mechanisms of BL "1" disturb and BL "0" disturb were explained in section 2.4. The "0" data cell at hold may be degraded when it is subjected to high BL voltage by which the data "1" is written to another cell which shares BL with the cell at hold. This is called BL "1" disturb caused by GIDL and can be mitigated by increasing the WL hold level V_{WLL}. On the other hand, the "1" data cell at hold may be degraded when it is subjected to low BL voltage by which the data "0" is written to another cell which shares BL with the cell at hold. This is due to p-n junction forward current and can be mitigated by decreasing the WL hold level V_{WLL}. Both BL disturbs, thus, are in the relationship of trade-off with respect to the WL hold voltage V_{WLL}.

Figure 4.23 is the retention time distribution of the data "0" at 25°C under the influence of BL "1" disturb for various duty ratio which is the time during which the cells are subjected to high BL voltage of 2.4 V divided by the each measured retention time. The data for the duty ratio 0% is the same as the pure retention time distribution we have been showing in the previous section which doesn't include BL disturb influence.

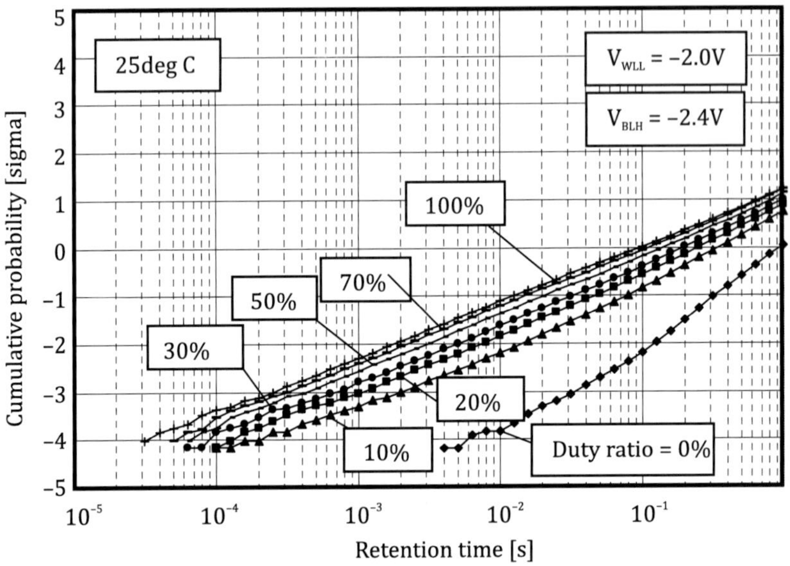

Figure 4.23 The retention time distributions of the data "0" under the influence of BL "1" disturb for various duty ratio of BL stress at V_{BLH} = 2.4 V at room temperature.

On the other extreme, the data for the duty ratio 100% corresponds to the retention time distribution which is measured with BL at 2.4 V all the time for the measurement. This graph clearly indicates that the retention time distribution becomes degraded as the duty ratio increases. This is because the number of holes flown into the body due to GILD increases as the time increases during which BL is raised high. Furthermore, the distributions exhibit almost straight lines for larger duty ratios, indicating the retention times for all cells distribute according to a normal distribution. However, as the ratio decreases, the distribution tends to be split into two parts of straight lines, i.e., a main body distribution line and a tail distribution line. And the tail distribution line seems to have the same gradient as that for large duty ratios, while the main body distribution line has another gradient which is steeper. This can be explained as follows. When the duty ratio is large the distribution is dominated by retention time which is determined by GIDL. As the ratio decreases, GILD influence decreases and the p-n junction leakage influence becomes larger instead. Therefore, another normal distribution having a different average and a different standard deviation regulates the main body of the distribution.

The same data were taken at 85°C as is shown in Fig. 4.24. This graph has basically the same tendency as Fig. 4.23, i.e., the distribution can be better approximated by a straight line as the duty ratio increases. However, it is worth noting that the tail lines at 85°C especially for higher duty ratio do

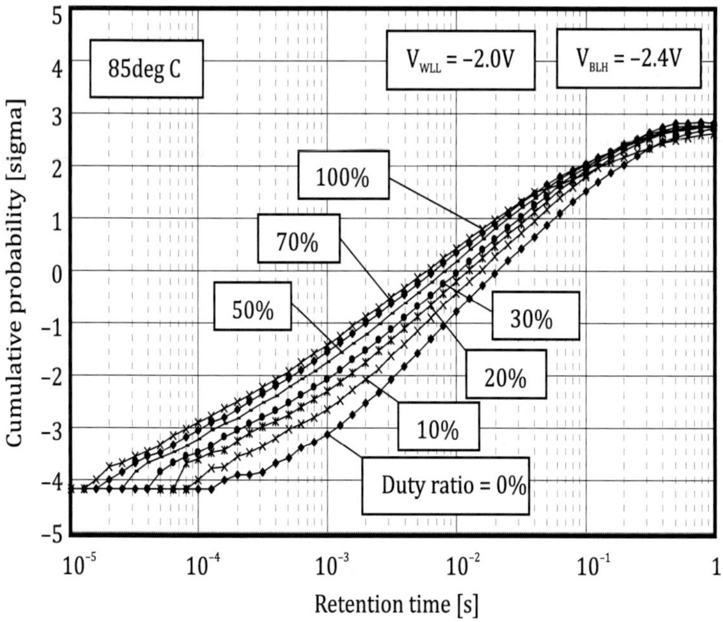

Figure 4.24 The retention time distributions of the data "0" under the influence of BL "1" disturb for various duty ratio of BL stress at V_{BLH} = 2.4 Vat 85°C.

not change significantly from those at 25°C on the contrary to the main body lines which are much more degraded from 25°C to 85°C. This tendency can be explained by the fact that GIDL which is the band-to-band tunneling current is basically independent of temperature on the contrary to the p-n junction current which becomes larger as temperature increases.

What is the duty ratio of the BL "1" disturb in actual memory operations? The duty ratio depends on the data access pattern which the software the device uses determines and it is not predicted in general. However, the worst case can be easily estimated and the estimation will be meaningful in setting a guideline for the immunity against the BL "1" disturb. If we consider embedded memory applications with high-speed random cycles. For example, for a memory system having the cycle time is t_{RC} = 3.5 ns with write duration t_W = 0.5 ns, the duty ratio for BL high for the worst case where other cells on the same BL as a cell in question which holds the data "0" are written the data "1" continuously for the time interval between the time when the cell in question is written the data "0" and the time when it is refreshed is to be estimated as DR $\leq t_w/t_{RC}$ = 14.3%. In Fig. 4.24, the retention time for a cell at -4.0σ for DR = 20% is 70 μs. On the other hand, for DR = 0%, the retention time for a cell at -4.0σ is 250 μs. Therefore, roughly, the retention time will be shortened to 1/3 due to the worst case BL "1" disturb.

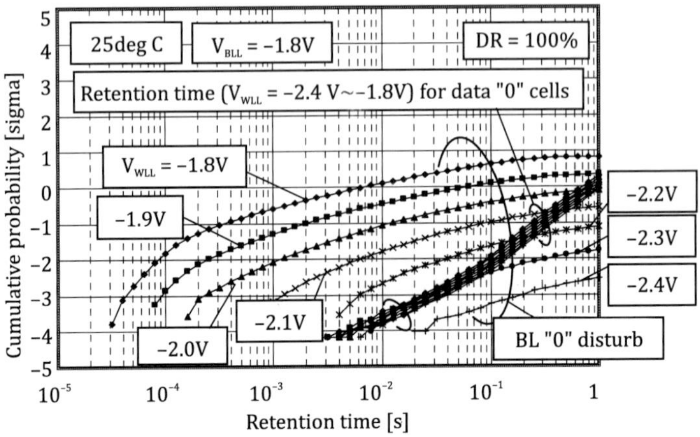

Figure 4.25 The retention time distributions of the data "1" under the influence of BL "0" disturb for various WL hold voltage V_{WLL} with V_{BLL} = −1.8 V at 25°C.

To check the influence of BL "0" disturb on retention time distributions, we measured retention time distributions of "1" data under the influence of BL "0" disturb. Figures 4.25 and 4.26 are the results of the retention time distribution of data "1" cells at 25°C and 85°C, respectively, with BL biased at −1.8 V throughout the test corresponding to DR = 100%. The WL hold voltage is changed from −1.8 V to −2.4 V. The retention time distribution for the data "0" cells for each temperature are superposed for reference. The retention time distributions under the influence of BL "0" disturb are improved at higher temperature. It can be observed that the distribution for the data "1" retention time under the influence of 100% duty BL "0" disturb at 25°C for $V_{WLL} \leq -2.0$ V is better than the distribution of data "0" retention time at 85°C. Though it is necessary to check the data at −10°C, the influence of the data "0" disturb can be avoided by lowering the WL hold voltage V_{WLL} at realistically worst condition DR ≤ t_w/t_{RC} = 14.3%.

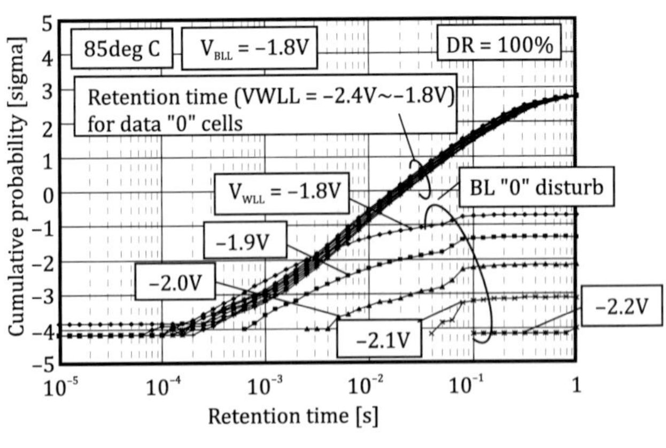

Figure 4.26 The retention time distributions of the data "1" under the influence of BL "0" disturb for various WL hold voltage V_{WLL} with V_{BLL} = −1.8 V at 85°C.

4.4.2 Bipolar Disturb

Figure 4.27 shows the array configuration of FBC. There are two kinds of cell-to-cell leakage caused by bipolar disturb. The memory cell with solid line is the disturbed cell. When data "1" is written to the neighboring cell sharing n^+ diffused layer connected to SL, the change of data "0" to data "1" may occur. We have defined this type of cell-to-cell leakage as "1" bipolar disturb. When data "0" is written to the neighboring cell sharing n^+ diffused layer connected to BL, the change of data "0" to data "1" may also occur. We have defined this type of cell-to-cell leakage as "0" bipolar disturb.

The bias condition of "1" bipolar disturb is schematically shown in Fig. 4.28a. Data "0" is written to the disturbed cell. The holes are swept out from the body region of the disturbed cell. When data "1" is written to the neighboring cell, holes are generated by the impact ionization. The generated holes accumulate in the body region of the neighboring cell. Once the body region is full, the generated holes overflow to the common n^+ diffused layer between the "1" written cell and the disturbed cell. And some of the holes diffuse into the n^+ layer and collect in the body of the disturbed cell. The collected holes in the body change the data of the disturbed cell from data "0" to data "1." This cell-to-cell leakage can be regarded

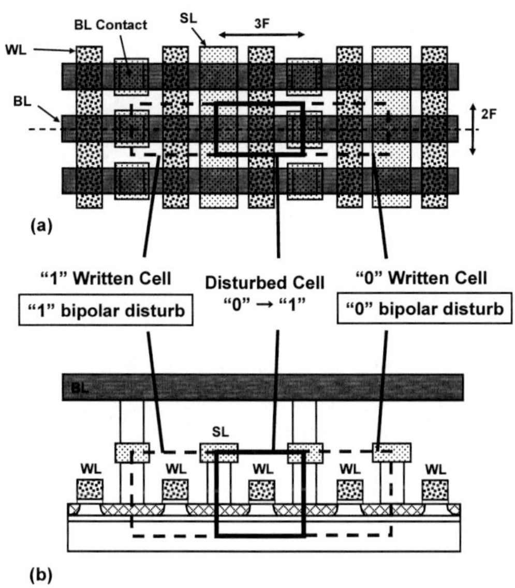

Figure 4.27 The array configuration of FBC. (a) Layout of FBC array. (b) Cross-section of FBC array. The memory cell with solid line is the disturbed cell.

as the operation of parasitic p-n-p bipolar transistor. The body of the "1" written cell, the n^+ diffused layer, and the body of the disturbed cell play the roles of emitter, base, and collector, respectively. Therefore, we have called this cell-to-cell leakage bipolar disturb. Holes flowing to the n⁺ diffused layer can be regarded as the emitter current, I_E. The base current, I_B, is due to the recombination inside the n⁺ diffused layer. Holes flowing to the body of the disturbed cell can be regarded as the collector current, I_C. Unlike a conventional bipolar transistor, the current gain h_{FE} ($=\Delta I_C/\Delta I_B$) is extremely small, about the order of 10^{-6} or less.

The bias condition of "0" bipolar disturb is schematically shown in Fig. 4.28b. When data "0" is written to the neighboring cell, holes stored in the body are extracted by the forward bias to the p-n junction consisting of body and n⁺ layer connected to BL. Some of the

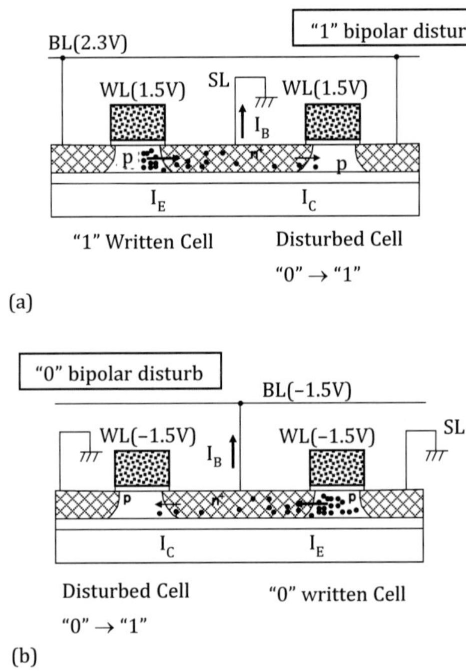

Figure 4.28 The bias condition of bipolar disturb. (a) The bias condition of "1" bipolar disturb. (b) The bias condition of "0" bipolar disturb. The solid circle represents hole.

holes extracted from the neighboring cell diffuse to the body of the disturbed cell. The collected holes in the body also change the data of the disturbed cell from data "0" to data "1."

This cell-to-cell leakage due to bipolar disturb is a serious obstacle to the reduction of the memory cell size. If this leakage cannot be controlled, the isolation between the neighboring cells along the BL direction is necessary, which leads to the enlargement of the cell size along BL direction from $3F$ to

5F. As a result, the memory cell size must increase from $6F^2$ to $10F^2$, which corresponds to a 67% enlargement.

To reduce cell-to-cell leakage attributable to bipolar disturb, the key point is how to control the hole diffusion to the neighboring cell in n^+ layer. We have verified that the combination of silicide and thin silicon film on the source and the drain region can reduce the diffusion of holes in n^+ layer. Figure 4.29 compares the FBC structure with and without silicide on the source and drain region. Figure 4.29a shows a cross-sectional view of the fabricated FBC structure of the 128 Mbit FBRAM.[3] The thickness of silicon at the body region is 55 nm. The other device and process parameters have already been published.[5] Figure 4.29b shows a schematic view of FBC without silicide. Figure 4.29c shows a schematic view of FBC with silicide. Co silicide was formed on the gate, source, and drain region. It should be noted that the thickness of the silicon layer with silicide is thinner than that without silicide because formation of Co silicide consumes the silicon at the source and the drain.

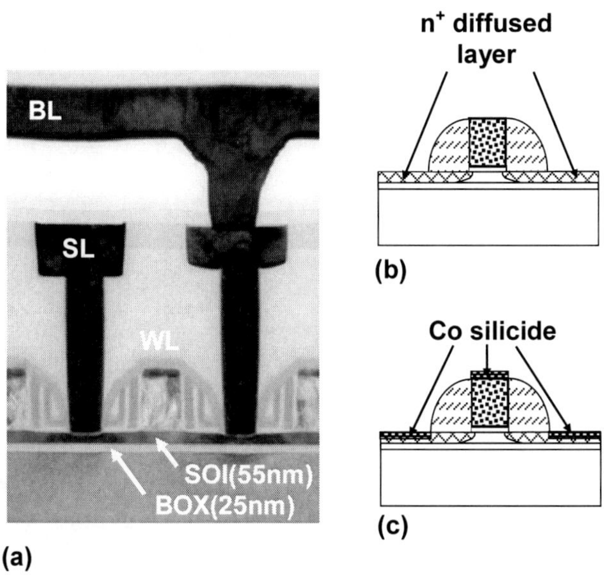

Figure 4.29 (a) Cross-sectional micrograph of the fabricated memory cell array in the BL direction. (b) The schematic view of FBC without silicide. (c) The schematic view of FBC with silicide.

Figure 4.30 shows the measurement result of the change of the stored data by "1" bipolar disturb in the case of FBC structure without silicide. The measurement data focuses on a particular memory cell in an array. The vertical line is the cell current in the read operation. The cell current of data "1" is larger than that of data "0" because the threshold voltage of data "1" is lower than that of data "0." The horizontal line is the disturb time, that is, the time of writing data "1" to the neighboring cell. It is clearly shown that the change of the data "0" to data "1" occurred within 2×10^{-8} owing to "1" bipolar

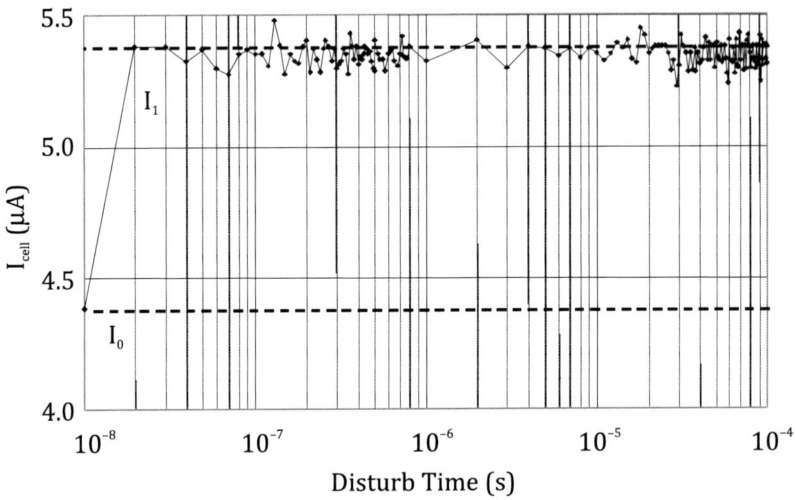

Figure 4.30 The measurement results of the change of the stored data by "1" bipolar disturb in the case of FBC structure without silicide. The measurement temperature was 25°C.

disturb. In this case, since the ability of the FBC structure to endure the bipolar disturb is slight, a countermeasure is necessary.

Figure 4.30 shows the measurement results in the case of FBC structure with silicide. The result of "1" bipolar disturb is plotted in Fig. 4.31a. The change of the data occurs after 1×10^{-3} s of "1" bipolar disturb. The result of "0" bipolar disturb is also plotted in Fig. 4.31b. Unlike "1" bipolar disturb, "0" bipolar disturb is dominated by the number of disturbances. This is because the number of holes of "0" disturb at one disturbance is limited to the number of holes stored in "1" body. The change of the data occurs after a disturb count of 10^4. Comparison of the "1" bipolar disturb of Fig. 4.30 and that of Fig. 4.31a reveals that 10^5 times improvement has been achieved by introducing the silicide layer on the source and drain region.

In the previous sentence, we clarified that the cell-to-cell leakage caused by bipolar disturb can be reduced by introducing silicide formed on the source and drain region. However, to guarantee that the failure rate due to bipolar disturb has been reduced to a level such that the production chip is unaffected, we must control the fail bit caused by bipolar disturb beneath -4σ distribution. Accordingly, we have investigated the dominant process factor for the fail bit distribution using the 128 Mbit FBRAM.

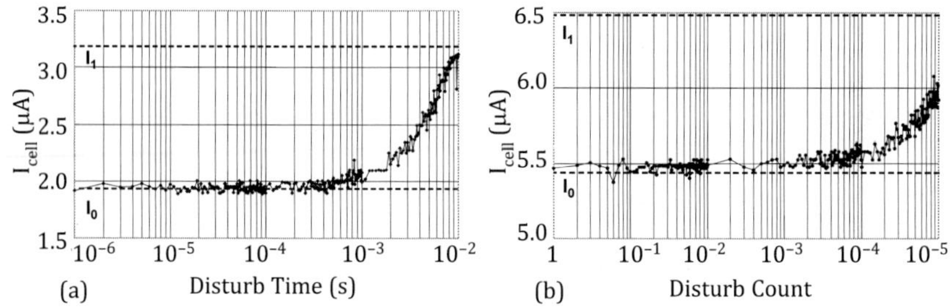

Figure 4.31 The measurement results of the change of the stored data in the case of FBC structure with silicide. (a) "1" bipolar disturb. (b) "0" bipolar disturb.

Figure 4.32 shows the cumulative distributions of fail bit caused by "1" bipolar disturb as a function of silicon thickness of the channel region under the gate. The silicon thickness under the silicide is thinner than these values.

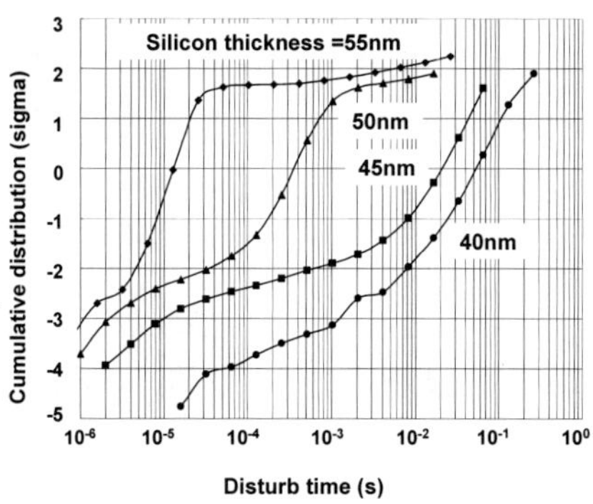

Figure 4.32 The cumulative distribution of fail bit caused by "1" bipolar disturb. The silicon thickness is that of the channel region. The measurement temperature was 25°C.

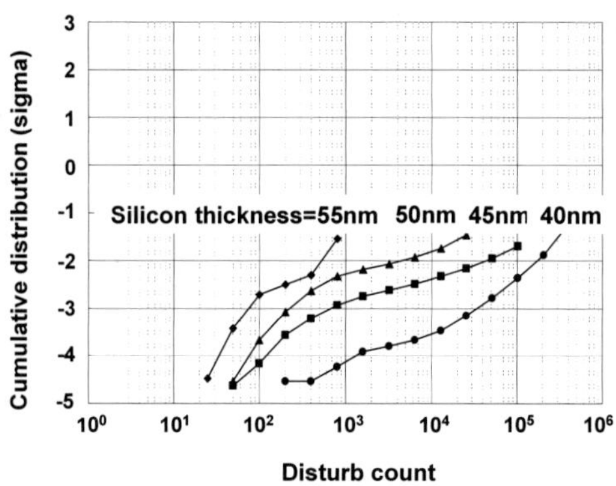

Figure 4.33 The cumulative distribution of fail bit caused by "0" bipolar disturb.

For example, as shown in a subsequent sentence, the silicon thickness under the silicide was 7.5 nm when the silicon thickness of the channel region under the gate was set to 45 nm. Reducing the silicon thickness form 55 nm to 40 nm drastically decreases the fail bit. For both the average cell region (0σ) and the tail cell region ($<-2\sigma$), disturb time is improved more than three orders. Figure 4.33 also shows the cumulative distribution of fail bit caused by "0" bipolar disturb. Similarly for "1" disturb, reducing the silicon thickness drastically decreases the fail bit. Assuming the redundancy system and the operation of the 128 Mbit FBRAM, we think the distribution of 40 nm silicon thickness is acceptable. The improvement of the disturb time indicates that the combination of silicide and thin silicon film on the source and drain region is indispensable.

To clarify the failure mechanism, physical analysis of the fail bit has been performed. The silicon thickness of channel region of the sample is set to 45 nm. The disturb time of the fail bit is 2 μs, which is almost the minimum time

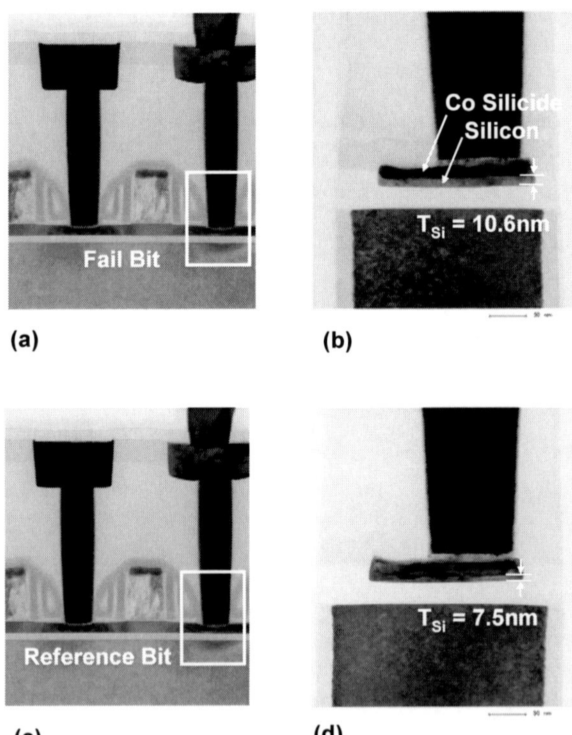

Figure 4.34 Failure analysis of "1" bipolar disturb. (a) The cross-sectional view of fail bit along BL direction. (b) The enlarged cross-section at n^+ diffused layer of the fail bit along WL. (c) and (d) show the cross-sectional view of the reference bit of the same direction.

in the tail distribution. The disturb time of the reference bit is 10 ms, which is almost the average value of the main distribution. Figure 4.34 shows the comparison of the cross-sectional TEM photograph of the fail bit and that of the reference bit. The cross-sectional view of the fail bit along BL direction is shown in Fig. 4.34a. The enlarged cross-section at n^+ diffused layer of the fail bit along WL is shown in Fig. 4.34b. The silicon thickness under Co silicide is 10.6 nm. Figure 4.34c,d show the cross-sectional view of the reference bit of the same direction. The silicon thickness under Co silicide is 7.5 nm, which is thinner than that of the fail bit.

From these experimental results, it has been clarified that the dominant factor of the cell-to-cell leakage by bipolar disturb is the remaining silicon thickness under the Co silicide. To reduce the failure rate of bipolar disturb in the 128 Mbit FBRAM, it is necessary to control the remaining silicon thickness after silicide formation.

To clarify the mechanism of the reduction of the cell-to-cell leakage in the FBC structure, the bipolar transistor characteristics have been investigated by the device simulation. Figure 4.35 shows the schematic view of the simulated structure and the bias condition. The physical dimensions are extracted from the FBC structure. The bias condition is that of "1" bipolar disturb. Two kinds of FBC structure have been investigated. One is FBC without silicide, in which Co silicide is not formed on n+ diffused layer.

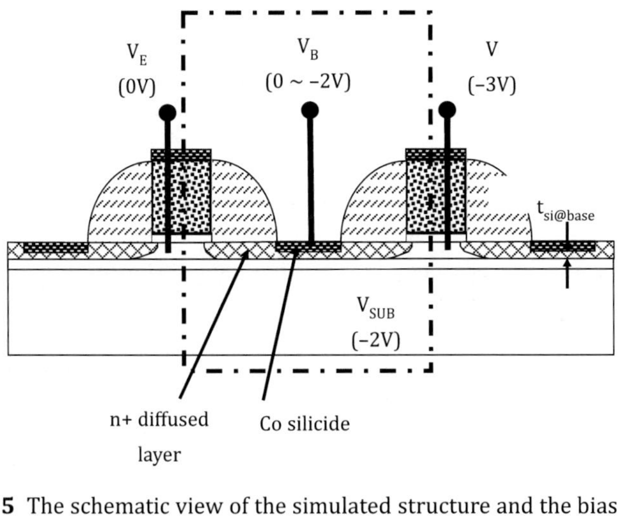

Figure 4.35 The schematic view of the simulated structure and the bias condition.

The other is FBC with silicide, in which Co silicide is formed there. The parameter of the simulation is the silicon thickness under Co silicide, $t_{si@base}$. In the case of FBC without silicide, $t_{si@base}$ is the silicon thickness at n^+ diffused layer.

Figure 4.36 compares the measurement result of the FBC structure and the simulation results. The FBC structure is with silicide. The silicon thickness of the channel region under the gate for the experimental results was 45 nm. Based on the cross-sectional TEM photograph of Fig. 4.34d, the silicon thickness under the silicide, $t_{si@base}$, was 7.5 nm. With regard to the base current, the simulation results are the same whether the silicon thickness and the structure have been changed. On the contrary, the collector current changed depending on the device structure and the silicon thickness. In the case of FBC structure without silicide, the collector current is almost two orders smaller than the base current. It should be noted that introduction of FBC structure with silicide drastically reduced the collector current. Moreover, the magnitude of the

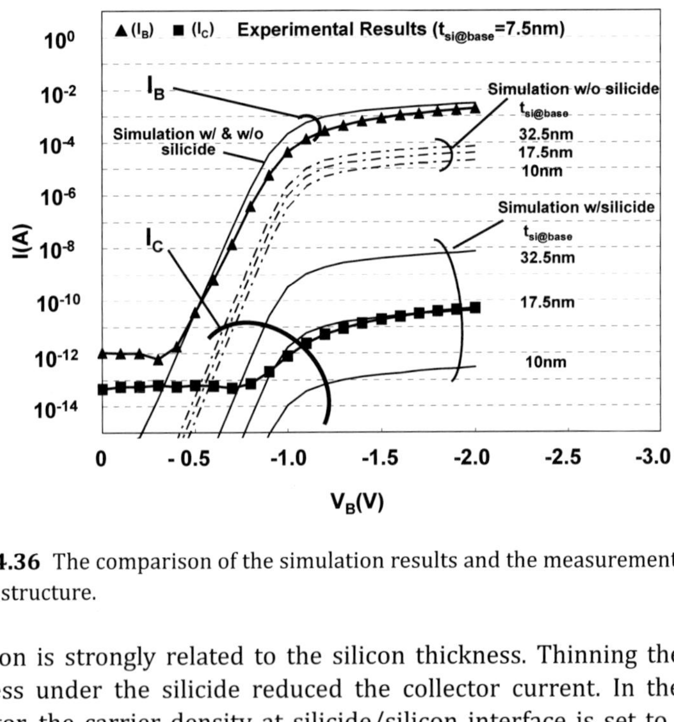

Figure 4.36 The comparison of the simulation results and the measurement result of the FBC structure.

reduction is strongly related to the silicon thickness. Thinning the silicon thickness under the silicide reduced the collector current. In the device simulator, the carrier density at silicide/silicon interface is set to thermal equilibrium. Therefore, the excess holes are captured at the interface. The change of the collector current depending on the silicon thickness of FBC with silicide indicates that the diffusion of holes inside n^+ region is restricted by the capture of holes at the silicide/silicon interface when silicon thickness is reduced. There is some discrepancy of silicon thickness between the experimental results and the simulation results. We suppose that may be due to the difference of structural parameters, notably the gradual change of the silicon thickness along the BL direction, or the electrical parameters, notably the carrier concentration of the base region. However, we think the mechanism of the reduction of the collector current has been clarified by these investigations.

From these simulation results, it can be concluded that the reduction of the cell-to-cell leakage current in the FBC structure derives from the combination of the silicide and the thin silicon film formed on the source and drain region.

References

1. T. Shino, T. Higashi, N. Kusunoki, K. Fujita, T. Ohsawa, N. Aoki, H. Tanimoto, Y. Minami, T. Yamada, M. Morikado, H. Nakajima, K. Inoh, T. Hamamoto and A. Nitayama, "Fully-depleted FBC (floating body cell) with enlarged signal window and excellent logic process compatibility," *IEDM Tech. Dig.*, pp. 281–284, 2004.
2. Y. Minami, T. Shino, A. Sakamoto, T. Higashi, N. Kusunoki, K. Fujita, K. Hatsuda, T. Ohsawa, N. Aoki, H. Tanimoto, M. Morikado, H. Nakajima, K. Inoh, T. Hamamoto and A. Nitayama, "A floating body cell (FBC) fully compatible with 90 nm CMOS technology (CMOS4) for 128 Mb SOI DRAM," *IEDM Tech. Dig.*, pp. 317–320, 2005.
3. T. Ohsawa, K. Fujita, K. Hatsuda, T. Higashi, M. Morikado, Y. Minami, T. Shino, H. Nakajima, K. Inoh, T. Hamamoto and S. Watanabe, "An 18.5ns 128Mb SOI DRAM with a floating body cell," *ISSCC Dig. Tech. Papers*, pp. 458–459, Feb. 2005.
4. T. Ohsawa, K. Fujita, K. Hatsuda, T. Higashi, T. Shino, Y. Minami, H. Nakajima, M. Morikado, K. Inoh, T. Hamamoto, S. Watanabe, S. Fujii and T. Furuyama, "Design of a 128-Mb SOI DRAM using the floating body cell (FBC)," *IEEE J. Solid-State Circuits*, vol. 41, no. 1, pp. 135–145, Jan. 2006.
5. T. Shino, T. Ohsawa, T. Higashi, K. Fujita, N. Kusunoki, Y. Minami, M. Morikado, H. Nakajima, K. Inoh, T. Hamamoto and A. Nitayama, "Operation voltage dependence of memory cell characteristics in fully depleted floating-body cell," *IEEE Trans. Electron Devices*, vol. ED-52, no. 10, pp. 2220–2226, Oct. 2005.
6. F. Matsuoka, T. Ohsawa, T. Higashi, H. Furuhashi, K. Hatsuda, K. Fujita, R. Fukuda, N. Ikumi, T. Shino, Y. Minami, H. Nakajima, T. Hamamoto, A. Nitayama and Y. Watanabe, "FBC's potential of $6F^2$ single cell operation in multi-Gbit memories confirmed by a newly developed method for measuring signal sense margin," *IEDM Tech. Dig.*, pp. 39–42, Dec. 2007.
7. S. Ogura, P. J. Tsang, W. W. Walker, D. L. Critchlow and J. F. Shepard, "Design and characteristics od the lightly doped drain-source (LDD) insulated gate field-effect transistor," *IEEE Trans. Electron Devices*, vol. ED-27, no. 8, pp. 1359–1367, Aug. 1980.

8. T. Hamamoto, S. Sugiura and S. Sawada, "On the retrention time distribution of dynamic random access memory (DRAM)," *IEEE Trans. Electron Devices*, vol. 45, no. 6, pp. 1300–1309, June 1998.
9. G. Vincent, A. Chantre and D. Bois, "Electric field effect on the thermal emission of traps in semiconductor junctions," *J. Appl. Phys.*, vol. 50, pp. 5484–5487, 1979.
10. T. Hamamoto, Y. Minami, T. Shino, N. Kusunoki, H. Nakajima, M. Morikado, T. Yamada, K. Inoh, A. Sakamoto, T. Higashi, K. Fujita, K. Hatsuda, T. Ohsawa and A. Nitayama, "A floating-body cell fully compatible with 90-nm CMOS technology node for a 128-Mb SOI DRAM and its scalability," *IEEE Trans. Electron Devices*, vol. 54, no. 3, pp. 563–571, March 2007.
11. T.-S. Jang, J.-S. Kim, S.-M. Hwang, Y.-H. Oh, K.-M. Rho, S.-J. Chung, S.-O. Chung, J.-G. Oh, S. Bhardwaj, J. Kwon, D. Kim, M. Nagoga, Y.-T. Kim, S.-Y. Cha, S.-C. Moon, S.-W. Chung, S.-J. Hong and S.-W. Park, "Highly scalable Z-RAM with remarkably long data retention for DRAM application," *2009 Symposium on VLSI Technology Digest of Technical Papers*, pp. 234–235, June 2009.
12. G. Groeseneken, H. E. Maes, N. Beltran and R. F. De Keersmaecker, "A reliable approach to charge-pumping measurements in MOS transistors," *IEEE Trans. Electron Devices*, vol. ED-31, no. 1, pp. 42–53, Jan. 1984.
13. S. Okhonin, M. Nagoga and P. Fazan, "Principles of transient charge pumping on partially depleted SOI MOSFETs," *IEEE Electron Device Letters*, vol. 23, no. 5, pp. 279–281, May 2002.

Chapter 5

Scaling of Floating Body Cell

Downscaling of the memory cell of the conventional dynamic random access memory (DRAM) has become more and more difficult because the storage capacitance of the memory cell must be kept constant. The stacked capacitor cell with metal/high-ε/metal capacitor has been used down to 50 nm generation. However, the development of sub-40 nm technologies has become tremendously difficult, and manufacturable solutions are not known. This is called the red brick wall of DRAM technologies.[1] To overcome these difficulties, several kinds of capacitor-less DRAM cell have been proposed and developed.[2–18] Floating body cell (FBC) is a promising candidate from the view point of the simplest structure and its scalability. Based on the fabrication and measurement results shown in Chapter 4, the scalability of FBC down to 32 nm technology node will be discussed.

There are three issues concerning the scaling of FBC. The first one is a large signal. The threshold voltage difference of 0.4 V is necessary. The second one is a long retention time of "0" hold. As is shown in the previous section, the retention time can be controlled by the electrical field of p-n junction. The electrical field strength can be kept constant even in scaled FBCs. The third one is the chip functionality of the single-cell operation. The signal-to-noise ratio of 18 is also necessary. The scalability of fully depleted FBC will be investigated by the device simulation.

5.1 SCALING OF CONSTANT SIGNAL

To achieve the adequate signal in the scaled floating body cell (FBC), the designs of partially depleted (PD) FBC and fully depleted (FD) FBC are different. Figure 5.1 compares the difference of the scalability between PD and FD FBC. ΔV_b represents the difference of the body potential between "1" read and "0" read.

Floating Body Cell: A Novel Capacitor-less DRAM Cell
By Takashi Ohsawa and Takeshi Hamamoto

www.panstanford.com

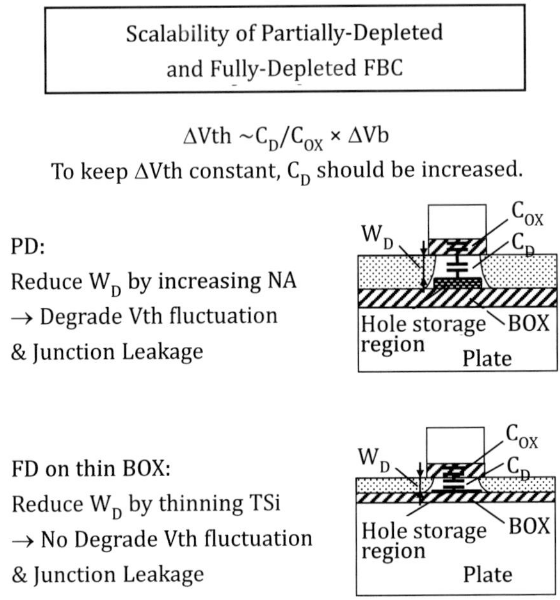

Figure 5.1 Comparison of scalability of PD and FD FBC.

When the FBC is scaled down, ΔV_b must be reduced because the operation voltage reduces. To keep the ΔV_{th} constant, the capacitance of depletion layer (C_D) should be increased, which is realized by reducing the depletion layer width (W_D). In the case of PD FBC, the doping density of the channel region must increase for reducing W_D. However, the degradation of the V_{th} fluctuation and junction leakage must occur. In the case of FD FBC, silicon thickness must reduce to reduce W_D. It is not necessary to increase the doping density of the channel region. Therefore, it is expected that no degradation of the V_{th} fluctuation and junction leakage will occur. We have selected FD FBC from these advantages.

The signal ΔV_{th} of FBC built on partially depleted silicon-on-insulator (PD-SOI) is expressed in Eq. 2.4 as was explained in section 2.1. This expression is written again as

$$\Delta V_{\mathrm{th}} = \frac{\sqrt{2\varepsilon_{\mathrm{Si}} q N_{\mathrm{A}}}}{C_{\mathrm{ox}}}\left(\sqrt{2\phi_{\mathrm{B}} - V_{\mathrm{b0}}} - \sqrt{2\phi_{\mathrm{B}} - V_{\mathrm{b1}}}\right). \tag{5.1}$$

As the cell size shrinks, t_{ox} must become smaller to suppress the short channel effect and accordingly the operating voltages must become smaller, making the voltage difference between V_{b1} and V_{b0} smaller. The changes reduce ΔV_{th} calculated by Eq. 5.1. To maintain ΔV_{th} constant against the changes, it is necessary to make the impurity concentration of the body N_A increase in accordance with the rule

$$N_A \propto \frac{1}{t_{ox}^2\left(\sqrt{2\phi_B - V_{b0}} - \sqrt{2\phi_B - V_{b1}}\right)^2}. \tag{5.2}$$

Though this scaling scenario certainly keeps the signal ΔV_{th} constant, it will be clear that the increase of the doping concentration of the body by following Eq. 5.2 inevitably increases the threshold voltage variation due to the random dopant fluctuation, which makes the signal-to-noise ratio (SNR) too small to satisfy the condition which was derived in section 3.3 for acquiring an enough memory chip yield.[19] Therefore, it will be assumed that the scaling of FBC is studied based upon the fully depleted silicon-on-insulator (FD-SOI) FBC in which the concentration of the body impurity is determined independent from the channel length.

The signal ΔV_{th} of FBC built on FD-SOI is expressed in Eq. 2.19 as was explained in section 2.5. This expression is written again as

$$\Delta V_{th} = \frac{C_{Si}}{C_{ox}}(V_{b1} - V_{b0}) = \frac{C_{Si}}{C_{ox}}\Delta V_b = \frac{\varepsilon_{Si}}{\varepsilon_{ox}}\frac{t_{ox}}{t_{Si}}\Delta V_b, \tag{5.3}$$

where ΔV_b stands for the difference in the body voltage between the data "1" cell and the data "0" cell both at read, i.e., $V_{b1} - V_{b0}$. This voltage difference is certainly decreases as cell shrinks, because the operating voltage of bit line (BL) and word line (WL) must be decreased as the device shrinks to avoid the reliability problems of dielectric breakdown and the increase of power. Especially, V_{b0} will decrease definitively by decreasing the absolute value of the low level of BL V_{BLL} for writing the data "0" according to Eq. 2.15 in section 2.3. Therefore, the ratio t_{ox}/t_{Si} should be increased with generation to compensate the decrease in ΔV_b for the signal to be kept constant. This corresponds to the rule that the scaling of t_{ox} needs to be milder than the scaling of t_{Si}. Here, the scaling law of the FD-SOI FBC for the constant signal is expressed as follows. If the scaling coefficient of silicon thickness is written by κ_{Si}, the silicon thickness of generation n, $t_{Si}^{(n)}$, will be expressed from the silicon thickness of the previous generation, $t_{Si}^{(n-1)}$, as

$$t_{Si}^{(n)} = \kappa_{Si} \times t_{Si}^{(n-1)}. \tag{5.4}$$

If the scaling coefficient of gate dielectric thickness is written by κ_{ox}, the oxide thickness of generation n, $t_{ox}^{(n)}$, will be expressed from the oxide thickness of the previous generation, $t_{ox}^{(n-1)}$, as

$$t_{ox}^{(n)} = \kappa_{ox} \times t_{ox}^{(n-1)}. \tag{5.5}$$

And if the scaling coefficient of voltage is written by κ_V, the body voltage difference of generation n, $\Delta V_b^{(n)}$, will be expressed from the body voltage difference of the previous generation, $\Delta V_b^{(n-1)}$, as

$$\Delta V_b^{(n)} = \kappa_V \times \Delta V_b^{(n-1)}. \tag{5.6}$$

By substituting Eqs. 5.4–5.6 into Eq. 5.3 for generation n, we have

$$\Delta V_{th}^{(n)} = \frac{\kappa_{ox} \cdot \kappa_V}{\kappa_{Si}} \Delta V_{th}^{(n)}, \tag{5.7}$$

where $\Delta V_{th}^{(n)}$ stands for the threshold voltage difference between the data "1" cell and the data "0" cell at read for the FBC of n-th generation. By setting $\Delta V_{th}^{(n)} = \Delta V_{th}^{(n-1)}$,we have

$$\kappa_{ox} = \frac{\kappa_{Si}}{\kappa_V}. \tag{5.8}$$

This equation indicates that the scaling coefficient of the gate oxide thickness must be larger than that of the silicon thickness by the inverse of the scaling coefficient of the operating voltage. This situation is illustrated in Fig. 5.2 where the silicon thickness is necessary to be shrunk more aggressively than the gate oxide thickness to maintain the signal of FD-SOI FBC constant throughout generations by compensating the decrease of the signal due to the operational voltage shrink as generation proceeds. This is a simple scaling law derived from the theoretical equation of the threshold voltage of an FD-SOI MOSFET. In the next section, we will verify the theoretical law by using TCAD device simulation.

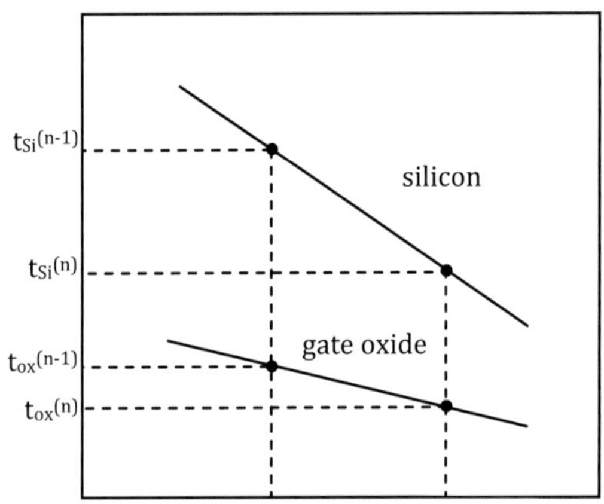

Figure 5.2 The scaling law of FD-SOI FBC which maintains the signal constant.

5.2 VERIFICATION BY TCAD SIMULATION

Table 5.1 shows the device parameters and the characteristics obtained in TCAD process and device simulators for FBCs from 90 nm through 32 nm technology nodes which are scaled according to the scaling law of constant signal, i.e., Eq. 5.8.[20] The structures and the impurity concentration profiles for the cells were formed by using a process simulator. Their characteristics were analyzed by using a device simulator. The gate length shrinks to 70% from the previous generation on average, resulting in 50% shrink for the cell size per generation on average. The silicon thickness and the buried oxide (BOX) thickness are also shrunk to 70% on average. On the other hand, the averaged scaling coefficient of the gate oxide thickness is 90% which is larger than the other scaling coefficients of the film thicknesses. Although the scaling coefficients of the operating voltages are different among them, they are within 80 ~ 90%. And the BL low voltage to write the data "0" V_{BLL}, the voltage which directly affects the body voltage difference ΔV_b between the data "1" cell and the data "0" cell at read, shrinks to 80% from the previous generation on average. Therefore, $\kappa_{ox} = 0.9$, $\kappa_{Si} = 0.7$ and $\kappa_V = 0.8$. These coefficients roughly satisfy the scaling

Table 5.1 An example of the scaling of FD-SOI FBC with the signal ΔV_{th} kept constant. κ stands for the scaling coefficient.

Generation	90 nm	65 nm	45 nm	32 nm	κ (on average)
Cell size [μm^2]	0.1372	0.0068	0.0336	0.02035	50%
L_g [nm]	150	110	75	55	70%
t_{Si} [nm]	42	32	21	16	70%
t_{BOX} [nm]	25	17	12.5	9	70%
t_{ox} [nm]	6.0	5.6	5.2	4.8	90%
N_A [cm^{-3}]	3.0×10^{17}	2.0×10^{17}	2.0×10^{17}	1.0×10^{17}	–
V_{WLH} [V]	1.5	1.35	1.2	1.0	85%
V_{WLL} [V]	-2.3	-2.0	-1.7	-1.45	85%
V_{BLH} [V]	2.2	1.9	1.7	1.5	90%
V_{BLL} [V]	-1.5	-1.2	-0.9	-0.7	80%
V_{PL} [V]	-3.0	-2.7	-2.4	-2.1	90%
ΔV_{th} [V]	0.397	0.4	0.397	0.368	100%
N_1	392	241	109	56	50%
N_0	25	8	1	0	30%
ΔN_{hole}	367	233	108	56	50%
E_{max} [MV/cm]	0.72	0.7	0.7	0.74	100%

law Eq. 5.8. The body doping concentration N_A is almost constant throughout generations. Figures 5.3–5.5 illustrate the transition of the parameters from 90 nm technology node to 32 nm technology node indicated in Table 5.1. Figure 5.6 shows the FBC performance trend for the signal ΔV_{th}, the hole number difference between the data "1" cell and the data "0" cell ΔN_{hole} and the maximum electrical field strength within the body at the "0" data hold state E_{max}. The cross sections of the cells for the four generations obtained by the TCAD process simulations, their operational waveforms, and the cell read characteristics obtained by the TCAD device simulations are shown in Figs. 5.7–5.9. The signal or the threshold voltage difference between the data "1" cell and the data "0" cell ΔV_{th} is shown to be almost constant 0.37 ~ 0.4 V as was predicted by the theory. The maximum strength of the electric field in the body at data hold is also kept constant 0.7 ~ 0.74 MV/cm, indicating that the leakage current per unit p-n junction area is kept constant.[21] Since the junction area shrinks to 50% per generation, the leakage current per cell is estimated to decreases to 50% per generation. The difference in the number of holes between the data "1" cell and the data "0" cell ΔN_{hole} decreases to 50% per generation.

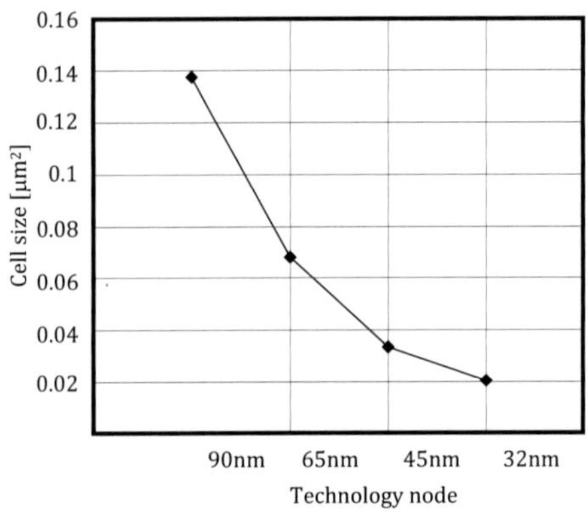

Figure 5.3 The cell size shrink trend in the scaling law of constant signal.

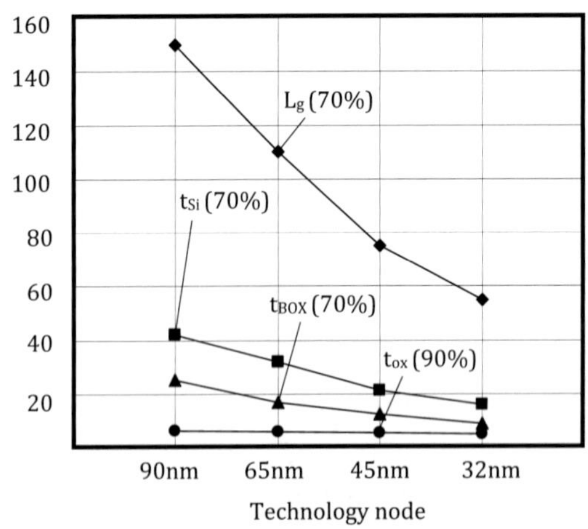

Figure 5.4 The shrink trend of the gate length and three film thicknesses in the scaling law of constant signal.

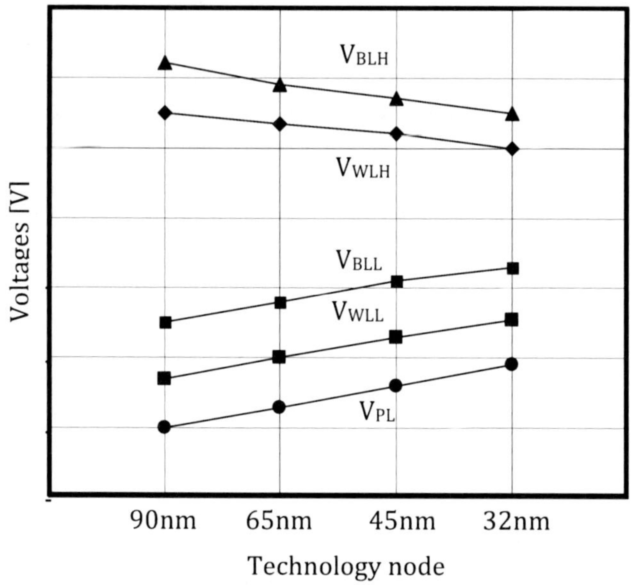

Figure 5.5 The shrink trend of the operating voltages in the scaling law of constant signal.

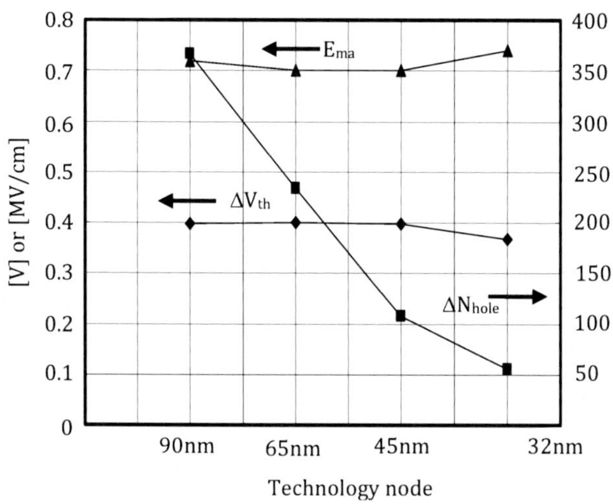

Figure 5.6 The FBC performance trend in the scaling law of constant signal.

The data retention time may be kept constant, because the reduction factor of ΔN_{hole} is the same as that of the leakage current per cell. The reduction of the number of holes accumulated in the "1" data cell with generation may raise a concern about the increase of the threshold voltage fluctuation due to the fluctuation of the number of holes accumulated in the data "1" cell's body. The true scaling law should maintain the SNR constant instead of keeping only the signal amount constant. In the next section, the scaling law of constant SNR will be examined for the FD-SOI FBC.

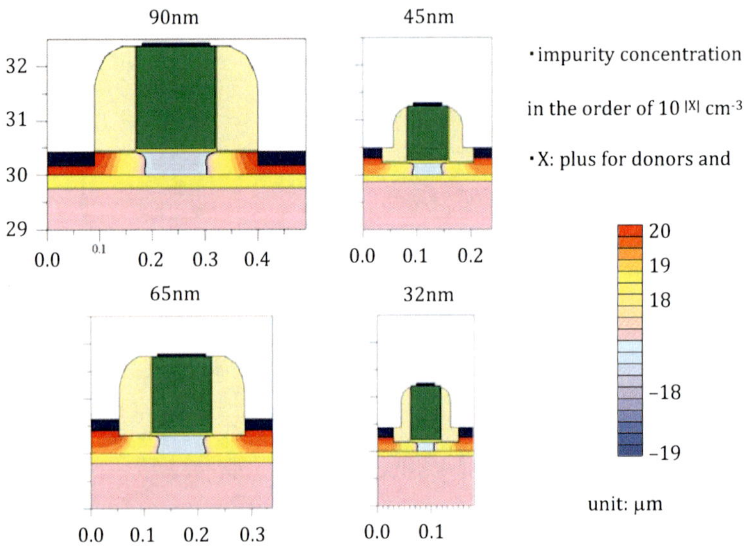

Figure 5.7 The cross section of the FBCs from 90 nm to 32 nm technology nodes obtained by the TCAD process simulations.

5.3 SCALING OF CONSTANT SNR

Though the scaling law which maintains the signal constant was discussed in the previous section, the parameter which should be constant of generation is not the signal but the chip yield As was analyzed in section 3.3, the chip yield is determined by $(\mathrm{SNR})_{V_{\mathrm{th}}} = \Delta V_{\mathrm{th}}/\sigma_{V_{\mathrm{th}}}$. Therefore, the scaling law which maintains $(\mathrm{SNR})_{V_{\mathrm{th}}}$ constant is necessary to be pursued. As a beginning, the FBCs which follow the scaling law of constant signal obtained in the previous section will be examined whether the $(\mathrm{SNR})_{V_{\mathrm{th}}}$ is maintained or not.

The threshold voltages of the FBC which are observed when reading the data "1" cell and the data "0" cell are considered to fluctuate in a cell array statistically due to the following reasons:

1. The intrinsic threshold voltage fluctuates because the number of the dopant atoms both in the body and in the source and the drain varies and the position of the dopant atoms varies for cells in an array. Here, by the intrinsic threshold voltage we mean the threshold voltage of a MOSFET whose body is fixed at a constant voltage.

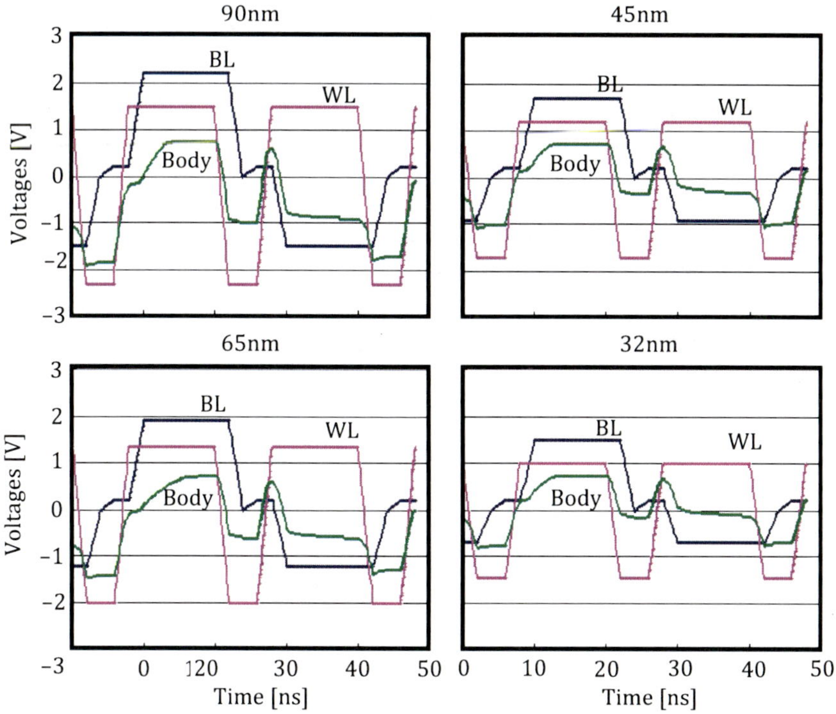

Figure 5.8 The operational waveforms of BL and WL applied to the FBCs and the body voltages obtained by the TCAD device simulations.

2. The intrinsic threshold voltage fluctuates due to the variation of the gate length for cells in an array (short channel effect).
3. The intrinsic threshold voltage fluctuates due to the variation of the channel width of cells in an array (narrow channel effect).
4. The intrinsic threshold voltage fluctuates due to the variation of the oxide thickness of cells in an array.
5. The intrinsic threshold voltage fluctuates due to the variation of the silicon film thickness of cells in an array.
6. The intrinsic threshold voltage fluctuates due to the variation of the BOX thickness of cells in an array.
7. The "1" cell body voltage at write fluctuates due to the impact ionization current variation and the p-n diode current variation which are originated from the variation of the number and the position of the dopant atoms both in the body and in the source and the drain in an array.

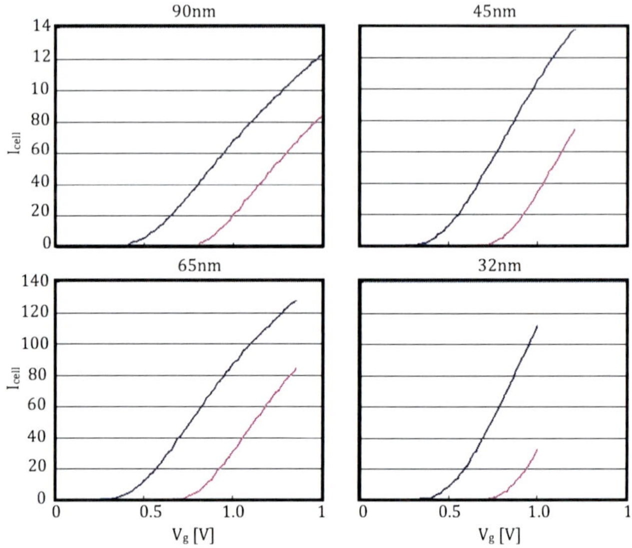

Figure 5.9 The cell read characteristics of the FBCs obtained by the TCAD device simulations.

8. The "0" cell body voltage at write fluctuates due to the p-n diode current variation which is originated from the variation of the number and the position of the dopant atoms both in the body and in the source and the drain in an array.
9. The body voltage at read fluctuates due to the variations of the device structures. Some of them are C_{ox}, C_{Si}, C_{BOX}, C_{db}, C_{sb}. The variations affect the processes of capacitive coupling illustrated in Fig. 3.1 in section 3.1.
10. The body voltage at write is also affected by the FBC's structural fluctuation.
11. The body voltage at write and read fluctuates due to the variations of the operating voltages such as V_{WLHW}, V_{WLL}, V_{WLHR}, V_{BLH}, V_{BLL}, V_{BLR}, and V_{PL}.
12. The body voltage fluctuates due to the variation of the parameters which affect the data retention characteristics such as the body capacitances and the cell leak current when retaining data.
13. The coefficient of the body effect fluctuates due to the variation of the device structure for cells in an array.
14. The coefficient of the body effect fluctuates due to the variation of the number and the position of the dopant atoms in the body for cells in an array.

Figure 5.10 illustrates the process in which the threshold voltages of the data "1" cell and the data "0" cell at read V_{th1}, V_{th0} are determined along with the sources of fluctuations which affect the variations in the threshold voltages within the process. The reasons (1)–(6) are for fluctuating the intrinsic threshold voltage. The reasons (7)–(12) are for fluctuating the body voltage. The reasons (13) and (14) are for fluctuating the coefficient of the body effect, i.e., the extent to which the threshold voltage responds to the body voltage. Among the reasons, we exclude the effect of data retention (12). This corresponds to considering only the variations of the signal which is read just after write. Though the variation of cell leak current is another important theme to be studied, we separate it from the fresh signal fluctuation which is to be considered here. And we also exclude the variations of the operating voltages (11), because the voltages can be generated and applied to all cells very accurately with a proper design, though systematic variations are inevitable to some extent. Furthermore, we neglect the device structure variations (2)–(6), (9), (10), and (13), i.e., all cells in an array are assumed to have exactly the same gate length, channel width, gate oxide thickness, BOX thickness, and silicon film thickness. In this manner, we include all effects which originate solely from the random

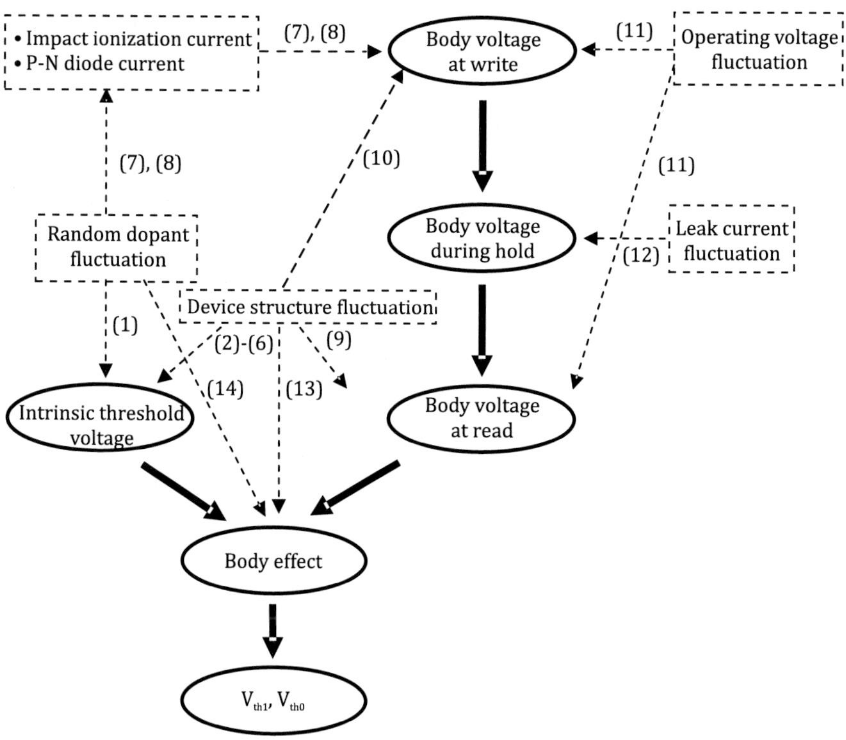

Figure 5.10 The process to determine the threshold voltages of the data "1" cell and the data "0" cell and the fluctuations which affect the variations of the threshold voltages in the process.

dopant fluctuation (1), (7), (8), and (14). The intrinsic threshold voltage variation due to the random dopant fluctuation has been studied for a long time and well characterized.[19,22,23] However, the predicted and measured standard deviation of the threshold voltage fluctuation due to the random dopant fluctuation is rather small compared with what has been observed at read in the standard deviation of the threshold voltage fluctuation for the data "1" cell and the data "0" cell that is about 20 ~ 30 mV for L_g = 150 nm with T_{OX} = 6 nm. The predicted number of the standard deviation for the MOSFET intrinsic threshold voltage variation with the same parameters is less than 20 mV.[24]

Since the experiments are performed with the MOSFET body fixed at a constant voltage, the discrepancy may be attributed to the body voltage variation which is superposed to the intrinsic threshold voltage variation. Now, we concentrate on the effects of the intrinsic threshold voltage variation, the body voltage variation at write, and body effect variation all originated from the random dopant fluctuation of donors and acceptors.

The Monte Carlo simulations were performed in the TCAD device simulation to predict the FBC's threshold voltage fluctuation at read for the data "1" and the data "0." The Monte Carlo simulations were performed within the framework of the conventional device simulations by corresponding a donor or an acceptor concentration that is assumed to vary according to a Poisson distribution to the averaged one at each control volume.

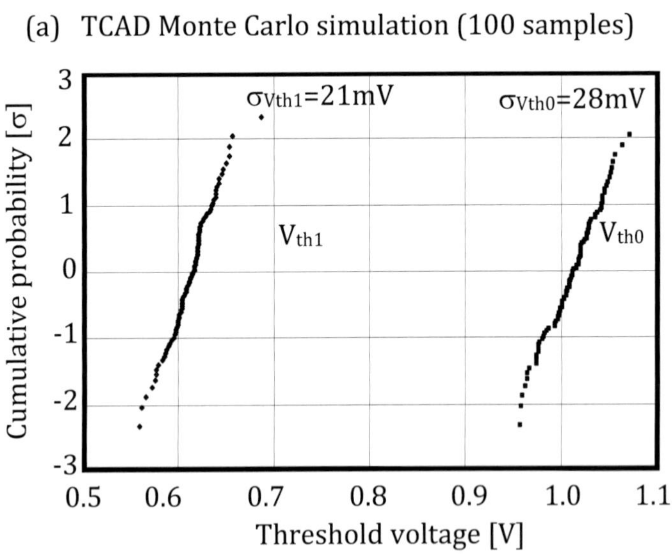

(b) HW measurement (128Mbit FBRAM)

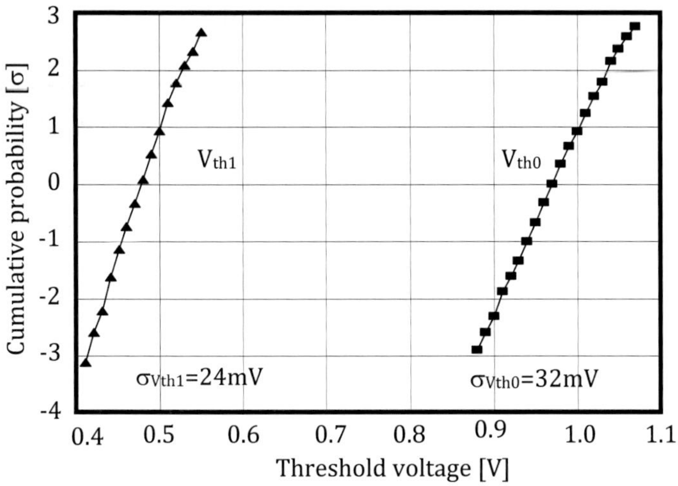

Figure 5.11 The fluctuations of the data "1" cells and the data "0" cells for the FBCs with L_g = 150 nm, W = 140 nm, t_{ox} = 6 nm, t_{Si} = 55 nm, and t_{BOX} = 25 nm. (a) TCAD Monte Carlo simulation results, (b) Hardware measurement results.

Figure 5.11 shows the Monte Carlo simulation results (100 samples) for the threshold voltage fluctuations for the data "1" FBCs and the data "0" FBCs with L_g = 150 nm, W = 140 nm, t_{ox} = 6 nm, t_{Si} = 55 nm, and t_{Box} = 25 nm along with the measurement results of the corresponding fluctuations taken from a cell array of the 128 Mbit floating body RAM prototype sample

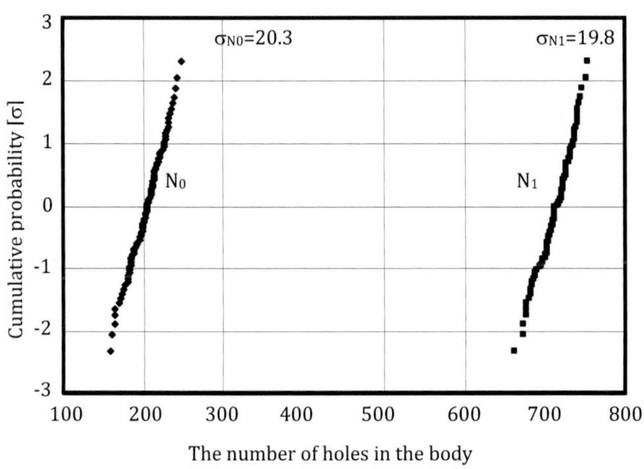

Figure 5.12 The fluctuations of the numbers of holes accumulated in the bodies of the "1" data cell and the "0" data cell obtained by the TCAD Monte Carlo simulations.

consisting of FBCs with the same device parameters.[25] The averaged values for the both distributions which were simulated are larger than those which were measured, especially for V_{th1}. However, the standard deviations for the simulated distributions are very close to those which were measured, although the simulated ones are 3 ~ 4 mV smaller than those were measured. The comparisons indicate that the assumption we made in the Monte Carlo simulations for predicting the FBC's threshold voltage fluctuation at read was realistic to some extent. In other words, the threshold voltage variations of FBCs at read are mainly explained by the random dopant fluctuations of donors and acceptors. Figure 5.12 shows the fluctuations of the number of holes accumulated in the body in the above Monte Carlo simulations. The results show that the averaged number of holes accumulated in the "1" cell body N_1 is about 720, while that in the "0" cell body N_0 is about 200. The difference is about 500 or a little bit larger. The standard deviation of N_1 is 19.8, while that of N_0 is 20.3. And it is observed that the both distributions are nearly the normal ones. The fluctuations of the numbers of holes accumulated in the "1" cell and the "0" cell may cause the fluctuations of the body voltages which lead to a part of the fluctuations of the threshold voltages observed at reading the cells. A theoretical approach also verifies the above simulation results as follows. From Eq. 2.16 in section 2.5, the threshold voltage of the FD-SOI FBC can be written as

$$V_{th} = V_{th}^{(i)} - \frac{q \cdot C_{Si}}{C_{ox} \cdot C_b} N_{hole}, \tag{5.9}$$

where $V_{th}^{(i)}$ stands for the intrinsic threshold voltage, C_b is the body total capacitance, and N_{hole} is the number of the holes accumulated in the body. This equation gives the expression for the standard deviation of the threshold voltage as

$$\sigma_{V_{th}^2} = \sigma_{V_{th}^{(i)2}} + \left(\frac{q \cdot C_{Si}}{C_{ox} \cdot C_b} \right)^2 \sigma_{N_{hole}}^2. \tag{5.10}$$

The first term is the contribution from the fluctuation of the intrinsic threshold voltage and the second term is that from the fluctuation of the number of holes accumulated in the body. From Eq. 5.10, we have

$$\sigma_{N_{hole}} = \left(\frac{C_{ox} \cdot C_b}{q \cdot C_{si}} \right) \sqrt{\sigma_{V_{th}}^2 - \sigma_{V_{th}^{(i)2}}}. \tag{5.11}$$

If we take the "1" cell as an example, $\sigma_{V_{th}} = 0.021$ V from the result of the Monte Carlo simulation. The standard deviation of the intrinsic threshold voltage of the MOSFET in this size can be estimated to be about $\sigma_{V_{th}^{(i)}} = 0.016$ V.[24]

If we use the silicon film thickness at the channel and the oxide thickness of the FBC as and

$$C_{ox}/C_{Si} = (\varepsilon_{ox}/\varepsilon_{Si})\cdot(t_{Si}/t_{ox}) = (3.9/11.9)\cdot(50\text{ nm}/6\text{ nm}) = 2.73$$

and the body total capacitance as 0.1 fF, thc standard deviation of the number of holes accumulated in the "1" cell's body will be

$$\sigma_{N_1} = 2.73 \times \frac{0.1\times10^{-15}}{1.6\times10^{-19}}\sqrt{0.021^2 - 0.016^2} = 23.2. \tag{5.12}$$

This number is roughly equal to the simulated standard deviation of the hole number fluctuation for the "1" cell. The fluctuation of the number of holes accumulated in the FBC's body thus consists of a large part of the threshold voltage variation of the FBC through the intrinsic threshold voltage and the number of holes or the body voltage.

It is worth studying to clarify the reason why the body voltage of FBC at write varies due to the random dopant fluctuation. It seems to be certain that the random dopant fluctuation gives rise to a variation of the impact ionization current and a variation of the p n diode current and that the variations lead to the fluctuation of the write level of the body voltage both

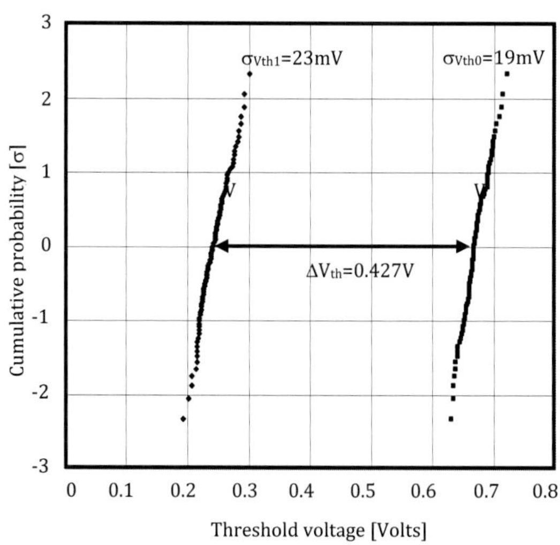

Figure 5.13 The fluctuations of the threshold voltages of the "1" data cell and the "0" data cell obtained by the TCAD Monte Carlo simulations for the FBC with L_g = 55 nm, W = 50 nm, t_{ox} = 4.8 nm, t_{Si} = 16 nm, t_{BOX} = 9 nm. The acceptor concentration N_A is 5.0×10^{16} cm^{-3}.

for the data "1" and for the data "0." However, microscopic pictures for the variation of the current characteristics need to be investigated as future important works.

Figure 5.13 exhibits the TCAD Monte Carlo simulation results (100 samples) for the threshold voltage variations for the data "1" cell and the data "0" cell for the FBC with L_g = 55 nm, W = 50 nm, t_{ox} = 4.8 nm, t_{Si} = 16 nm, and t_{Box} = 9 nm.[6] In this simulation, the body impurity concentration N_A is set to be at a very low level: $N_A = 5.0 \times 10^{16}$ cm^{-3}. The standard deviations of the threshold voltages of the data "1" cell and the data "0" cell are $\sigma_{V_{th1}}$ = 23 mV and $\sigma_{V_{th0}}$ = 19 mV, respectively. It is worth noting that the both extents to which the threshold voltages fluctuate are comparable or rather small compared with those at the generation of L_g = 150 nm. Figure 5.14 shows the standard deviations of the threshold voltage flunctuations $\sigma_{V_{th1}}$ and $\sigma_{V_{th0}}$ plotted as

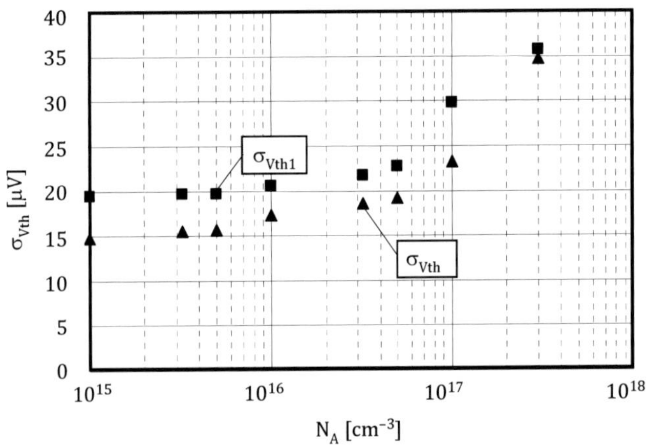

Figure 5.14 The dependence of the standard deviations of the threshold voltages of the data "1" cell and the data "0" cell on the acceptor concentration. The data are obtained by the TCAD Monte Carlo simulations for the FBC with L_g = 55 nm, W = 50 nm, t_{ox} = 4.8 nm, t_{Si} = 16 nm, t_{BOX} = 9 nm.

a function of the acceptor concentration N_A in the body of FBC.[25] As is shown in this graph, the standard deviations of the both fluctuations of the threshold voltages decrease as the acceptor concentration of the body decreases. This is a big advantage of the FD-SOI FBC over the PD-SOI counterpart.

Table 5.2 shows the scaling of device parameters and their device characteristics obtained by the TCAD simulations, which keeps the SNR with respect to threshold voltage $(\mathrm{SNR})_{V_{th}}$ constant.[25] All the parameters except for the acceptor concentrations N_A at the generations of 45 nm and 32 nm are the same as those in Table 5.1 which is the scaling law to keep the signal constant. The acceptor concentrations N_A at the generations of 45 nm and 32 nm are optimized to smaller values than those in Table 5.1. As is seen in this table, the SNR with respect to threshold voltage $(\mathrm{SNR})_{V_{th}}$ is maintained to be 17.9 ~ 20.5, indicating that it almost exceeds the critical value 18 which is necessary to make a Gbit class memory functional as was discussed in section 3.3. Since this scaling scenario only refers to the maintenance of $(\mathrm{SNR})_{V_{th}}$, it is certain that parasitic resistances in the source and the drain must be lowered enough for

Table 5.2 An example of the scaling of FD-SOI FBC with $(\mathrm{SNR})_{V_{th}} = \Delta V_{th}/\sigma_{V_{th}}$ kept constant.

	Generation	90 nm	65 nm	45 nm	32 nm
Structure	Cell size [mm^2]	0.1372	0.068	0.0336	0.02035
	t_{Si} [nm]	42	32	21	16
	N_A [cm^{-3}]	3.0×10^{17}	2.0×10^{17}	1.0×10^{17}	5.0×10^{16}
	t_{BOX} [nm]	25	17	12.5	9
	L [nm]	150	110	75	55
	W [nm]	140	100	70	50
	t_{ox} [nm]	6.0	5.6	5.2	4.8
Bias	V_{wlh} [V]	1.5	1.35	1.2	1.0
	V_{wll} [V]	−2.3	−2.0	−1.7	−1.45
	V_{blh} [V]	2.2	1.9	1.7	1.5
	V_{bll} [V]	−1.5	−1.2	−0.9	−0.7
	V_{pl} [V]	−3.0	−2.7	−2.4	−2.1
Results	V_{th1} [V]	0.310	0.235	0.214	0.241
	V_{th0} [V]	0.716	0.660	0.665	0.668
	ΔV_{th} [V]	0.406	0.425	0.451	0.427
	E_{max} [MV/cm]	0.72	0.70	0.70	0.74
	SNR	18.3	17.9	17.9	20.5

the SNR with respect to cell read current $(\mathrm{SNR})_{I\mathrm{cell}}$ to exceed the critical value 18.

References

1. International Technology Roadmap for Semiconductors (ITRS), San Jose, CA, 2008.
2. S. Okhonin, M. Nagoga, J. M. Sallese and P. Fazan, "A capacitor-less SOI 1T-DRAM concept," *Proc. IEEE Int. SOI Conf.*, pp. 153–154, 2001.

3. T. Ohsawa, K. Fujita, T. Higashi, Y. Iwata, T. Kajiyama, Y. Asao and K. Sunouchi, "Memory design using one-transistor gain cell on SOI," *Proc. IEEE Int. Solid-State Circuits Conf.*, pp. 152–153, 2002.
4. K. Inoh, T. Shino, H. Yamada, H. Nakajima, Y. Minami, T. Yamada, T. Ohsawa, T. Higashi, K. Fujita, T. Ikehashi, T. Kajiyama, Y. Fukuzumi, T. Hamamoto and H. Ishiuchi, "FBC (floating body cell) for embedded DRAM on SOI," *VLSI Symp. Tech. Dig.*, pp. 63–64, 2003.
5. A. Villaret, R. Ranica, P. Mazoyer, P. Candelier, F. Jacquet, S. Cristoloveanu and T.Skotnicki, "Triple-well nMOSFET evaluated as a capacitor-less DRAM cell for nanoscale low-cost & high density applications," *Proc. VLSI Nanoworkshop*, pp. 40–41, 2003.
6. E. Yoshida and T. Tanaka, "A design of a capacitorless 1T-DRAM cell using gate-induced drain leakage (GIDL) current for low-power and high-speed embedded memory," *IEDM Tech. Dig.*, pp. 913–916, 2003.
7. R. Ranica, A. Villaret, P. Malinge, P. Mazoyer, D. Lenoble, P. Candelier, F. Jacquet, P. Masson, R. Bouchakour, R. Fournel, J. P. Schoellkopf and T. Skotnicki, "A one transistor cell on bulk substrate (1T-Bulk) for low cost and high density eDRAM," *VLSI Symp. Tech. Dig.*, pp. 128–129, 2004.
8. T. Shino, T. Higashi, K. Fujita, T. Ohsawa, Y. Minami, T. Yamada, M. Morikado, H. Nakajima, K. Inoh, T. Hamamoto and A. Nitayama, "Highly scalable FBC (floating body cell) with 25nm BOX structure for embedded DRAM applications," *VLSI Symp. Tech. Dig.*, pp. 132–133, 2004.
9. T. Shino, T. Higashi, N. Kusunoki, K. Fujita, T. Ohsawa, N. Aoki, H. Tanimoto, Y. Minami, T. Yamada, M. Morikado, H. Nakajima, K. Inoh, T. Hamamoto and A. Nitayama, "Fully-depleted FBC (floating body cell) with enlarged signal window and excellent logic process compatibility," *IEDM Tech. Dig.*, pp. 281–284, 2004.
10. T. Ohsawa, K. Fujita, K. Hatsuda, T. Higashi, M. Morikado, Y. Minami, T. Shino, H. Nakajima, K. Inoh, T. Hamamoto and S. Watanabe, "An 18.5ns 128Mb SOI DRAM with a floating body cell," *Proc. IEEE Int. Solid-State Circuits Conf.*, pp. 458–459, 2005.
11. K. Hatsuda, K. Fujita and T. Ohsawa, "A 333MHz random cycle DRAM using the floating body cell," *Proc. IEEE Custom Integrated Circuits Conf.*, pp. 259–262, 2005.
12. T. Shino, T. Ohsawa, T. Higashi, K. Fujita, N. Kusunoki, Y. Minami, M. Morikado, H. Nakajima, K. Inoh, T. Hamamoto and A. Nitayama, "Operation voltage dependence of memory cell characteristics in fully depleted floating-body cell," *IEEE Trans. Electron Devices*, vol. ED-52, no. 10, pp. 2220–2226, Oct. 2005.
13. Y. Minami, T. Shino, A. Sakamoto, T. Higashi, N. Kusunoki, K. Fujita, K. Hatsuda, T. Ohsawa, N. Aoki, H. Tanimoto, M. Morikado, H. Nakajima, K. Inoh, T. Hamamoto and A. Nitayama, "A floating body cell (FBC) fully compatible with 90nm CMOS technology (CMOS4) for 128Mb SOI DRAM," *IEDM Tech. Dig.*, pp. 317–320, 2005.

14. T. Ohsawa, T. Higashi, K. Fujita, K. Hatsuda, N. Ikumi, T. Shino, H. Nakajima, Y. Minami, N. Kusunoki, A. Sakamoto, J. Nishimura, T. Hamamoto and S. Fujii, "A 128Mb floating body RAM (FBRAM) on SOI with multi-averaging scheme of dummy cell," *2006 Symp. on VLSI Circuits*, pp. 224–225.

15. T. Shino, N. Kusunoki, T. Higashi, T. Ohsawa, K. Fujita, K. Hatsuda, N. Ikumi, F. Matsuoka, Y. Kajitani, R. Fukuda, Y. Watanabe, Y. Minami, A. Sakamoto, J. Nishimura, H. Nakajima, M. Morikado, K. Inoh, T. Hamamoto and A. Nitayama, "Floating body RAM technology and its scalability to 32nm node and beyond," *2006 IEDM Tech. Dig.*, pp. 569–572.

16. T. Hamamoto, Y. Minami, T. Shino, N. Kusunoki, H. Nakajima, M. Morikado, T. Yamada, K. Inoh, A. Sakamoto, T. Higashi, K. Fujita, K. Hatsuda, T. Ohsawa and A. Nitayama, "A floating-body cell fully compatible with 90-nm CMOS technology node for a 128-Mb SOI DRAM and its scalability," *IEEE Trans. ED*, vol. 54, no. 3, pp. 563–571, 2007.

17. F. Matsuoka, T. Ohsawa, T. Higashi, H. Furuhashi, K. Hatsuda, K. Fujita, R. Fukuda, N. Ikumi, T. Shino, Y. Minami, H. Nakajima, T. Hamamoto, A. Nitayama and Y. Watanabe, "FBC's potential of $6F^2$ single cell operation in multi-Gbit memories confirmed by a newly developed method for measuring signal sense margin," *2007 IEDM Tech. Dig.*, pp. 39–42.

18. S. Okhonin, M. Nagoga, E. Carman, R. Beffa and E. Faraoni, "New generation of Z-RAM," *2007 IEDM Tech. Dig.*, pp. 925–928.

19. T. Mizuno, J. Okamura and A. Toriumi, "Experimental study of threshold voltage fluctuation due to statistical variation of channel dopamt number in MOSFET's," *IEEE Trans. Electron Devices*, vol. 41, no. 11, pp. 2216–2221, Nov. 1994.

20. T. Shino, T. Ohsawa, T. Hamamoto and A. Nitayama, "Overview and future challenges of floating body RAM technologies," *Ext. Abs. Int. Conf. Solid State Devices Materials*, pp. 226–227, Sept. 2007.

21. T. Hamamoto, S. Sugiura and S. Sawada, "On the retention time distribution of dynamic random access memory (DRAM)," *IEEE Trans. Electron Devices*, vol. 45, no. 6, pp. 1300–1309, June 1998.

22. R. W. Keyes, "Physical limits in digital electronics," *Proc. IEEE*, vol. 63, no. 5, pp. 740–767, May 1975.

23. D. J. Frank, R. H. Dennard, E. Nowak, P. M. Solomon, Y. Taur and H.-S. P. Wong, "Device scaling limits of Si MOSFETs and their application dependencies," *Proc. IEEE*, vol. 89, no.3, pp. 259288, March 2001.

24. T. Mizuno, "New channel engineering for sub-100nm MOS device considering both carrier velocity overshoot and statistical performance fluctuations," *IEEE Trans. Electron Devices*, vol. 47, no. 4, pp. 756–761, April 2000.

25. H. Furuhashi, T. Shino, T. Ohsawa, F. Matsuoka, T. Higashi, Y. Minami, H. Nakajima, K. Fujita, R. Fukuda, T. Hamamoto and A. Nitayama, "Scaling scenario of floating body cell (FBC) suppressing Vth variation due to random dopant fluctuation," *IEEE Int. SOI Conf.*, pp. 33–34, Oct. 2008.

Chapter 6

Cell Array Architecture and Sense Amplifier Design

Design of sense amplifier circuit (S/A) will be presented which is specific and appropriate to floating body RAM. At first, imperfection of non-destructivity for reading data from floating body cell (FBC) is discussed, and it is shown that placing a S/A for each bit line pair is necessary. Examples of S/A circuits for the twin cell and the single cell schemes are explained with special emphasis on dummy cell design for the latter scheme. Measurement results of memory chip functionality and retention time are analyzed to show optimum design of dummy cell systems. An open bit line cell array architecture for FBC which is immune against noise between bit lines is analyzed to show a big advantage over the one-transistor and one-capacitor dynamic random access memory cell.

6.1 WORD LINE DISTURB DUE TO CHARGE PUMPING AND SENSE AMPLIFIER CIRCUIT ARRANGEMENT

Strictly speaking, floating body cell (FBC) is a destructive read-out cell, as was touched on at the end of section 2.1, though it is much more immune against data destruction at read than the conventional one-transistor and one-capacitor (1T-1C) dynamic random access memory (DRAM) cell. It loses a few holes at the largest in a word line (WL) activation cycle at read due to the charge pumping phenomenon.[1,2] Since the difference between the number of holes accumulated in the body of the "1" cell and that of the "0" cell for L_g = 32, is more than 50 as was shown in Chapter 5, the read operation itself is not largely affected by the loss of holes by the charge pumping. FBC, in this sense, can be categorized as a quasi non-destructive read-out cell.[3,4] However, repetitive read operations between a refresh interval will inevitably degrade the data "1" state leading to a fail. Figure 6.1 shows "1" cell data degradation in WL pumping cycles which is compared

Floating Body Cell: A Novel Capacitor-less DRAM Cell
By Takashi Ohsawa and Takeshi Hamamoto

www.panstanford.com

with data retention characteristics. The gate length is L_g = 150 nm and the channel width is W = 165 nm for the cell measured in Fig. 6.1. Since the difference between the number of holes accumulated in the body of the "1" cell and that of the "0" cell for L_g = 150 nm is about 500 as was shown in Chapter 5 and the averaged number of interface states in the cell is 2.85 as will be shown later, it is reasonable that the cell read current of the data "1" cell decreases and approaches to that of the data "0" cell when WL is being pumped more than 100 times. This WL disturb forces sense amplifier circuit (S/A) to be placed per a bit line (BL) pair just like the conventional 1T-1C DRAM as is shown in Fig. 6.2 where (a) indicates the twin cell scheme and (b) indicates the single cell scheme. In the twin cell scheme, one memory bit consists of a pair of memory cells where different data

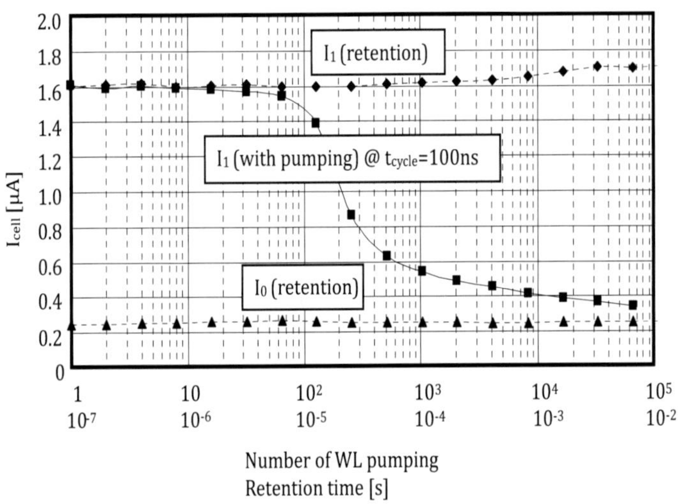

Figure 6.1 Phenomenon of "1" cell data degradation in WL pumping cycles which is compared with data retention characteristics. L_g = 150 nm, W = 165 nm.

are written. The pair cells are two adjacent cells on a WL in Fig. 6.2a. Therefore, one cell acts as a reference cell of the other, making the sensing system simple with the dummy cells unnecessary at the expense of doubled cell size. In addition to the simplicity, the twin cell scheme has almost the signal doubled, leading to higher yield and shorter access time. On the other hand, single cell scheme has advantages in small cell size. However, it needs dummy cells to provide reference current for S/A to distinguish the data state "1" from the data state "0" and vice versa with signal almost a half of the twin cell scheme. The adoption of the cell scheme will depend on the memory application. In general, the single cell scheme is suitable for high-density memories with relatively slow access time, while the twin cell scheme is suitable for high speed embedded memories in the field where SRAM is used traditionally.

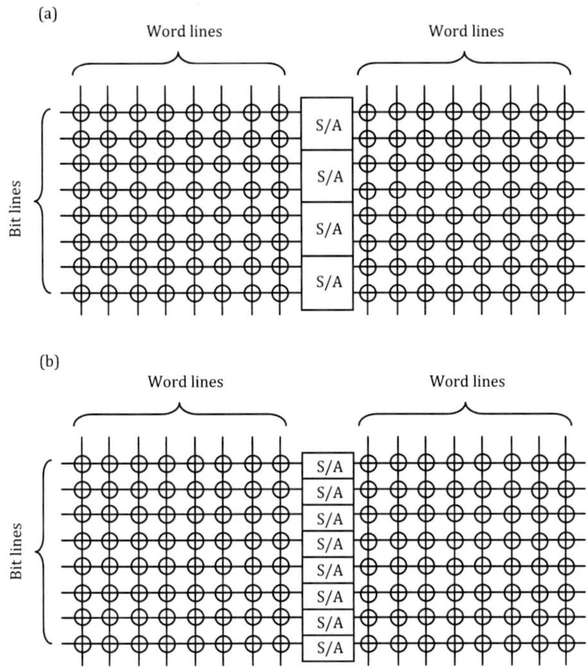

Figure 6.2 S/A placement for (a) twin cell scheme and (b) single cell scheme.

The above guide line is applied to FBC which reads data in MOS current. If FBC is designed to read data in bipolar current[5] as was explained in 2.6.4.2, it is not necessary to place a S/A per each BL pair, because the holes lost by the

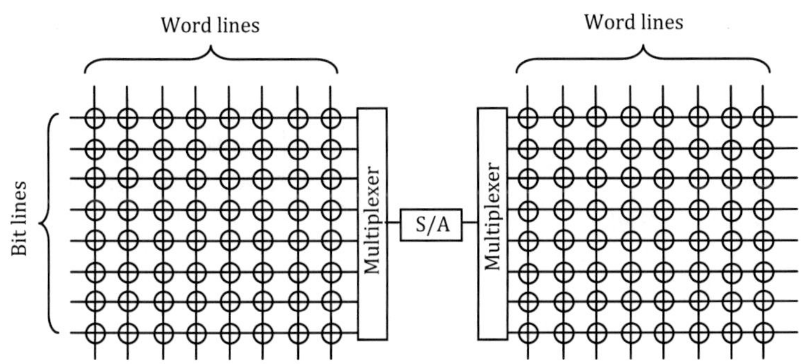

Figure 6.3 An example of S/A placement possible for bipolar read FBC in a single cell scheme. This scheme drastically reduces S/A area. However, the number of cells to be accessed in a page mode is reduced.

charge pumping can be replenished by impact ionization of collector current during read. Each FBC itself acts as its own S/A, as it were. In this case, a S/A is just placed per-multiple pairs of BLs which can be switched to be connected to the S/A in multiplexer as is shown in Fig. 6.3 for the single cell scheme as an example where one S/A is placed per eight BL pairs. This scheme drastically reduces the S/A area in a memory chip. Since a large area is occupied by the S/A area in the scheme shown in Fig. 6.2, this design contributes largely to reducing the chip size. However, it must be noted that the number of cell to be accessed in a page mode is reduced as many.

In the following sections, we will explain S/A circuits separately for the twin cell scheme and for the single cell scheme using dummy cells. The FBC in MOS read is assumed. The methods of write are assumed to be by the MOS impact ionization for "1" and by the BL minus p-n forward bias for "0" whi ch are shown in Tables 2.1 and 2.2, respectively.

6.2 S/A IN TWIN CELL SCHEME

To read data from FBC, its current flowing between the drain and the source in linear region must be detected. Therefore, on the contrary to the conventional 1T-1C DRAM, a voltage V_{BLR} which is larger than that of source line (SL) (0 V) and lower than $V_{WLHR} - V_{th}$ is necessary to be applied to BL at read, where V_{WLHR} and V_{th} are WL voltage at read and the threshold voltage of the cell. Usually, V_{BLR} is 0.2 V ~ 0.5 V. Therefore, the S/A needs a current load circuit in addition to the dynamic flip-flop circuit which is commonly used in the S/A of the conventional 1T-1C DRAM.

Figure 6.4 exemplifies a S/A which is used for memories in the twin cell scheme.[6] This is a S/A shared among two pairs of BLs in both sides of the S/A. The twin cell scheme is also called two-cell-per-bit (2cpb) scheme and restores data of opposite polarity into the two cells at the intersection of either BLL_j, $/BLL_j$, and WLL_i or BLR_j, $/BLR_j$, and WLR_i. This scheme has an advantage in chip yield and access time over the single cell scheme which will be discussed in section 6.3 and is also called one-cell-per-bit (1cpb), because the signal will be almost doubled. Only $i - th$ row and $j - th$ column in the both sides are shown in Fig. 6.4. Actually, cells are two-dimensionally arranged for $i = 0 \sim 511$ and $j = 0 \sim 4095$ for example, making 4 Mbit cell array (2 Mbit memory capacity). And a memory chip consists of multiple units of the array. A pair of sense nodes SN_i and $/SN_i$ which are isolated from two BL pairs by NMOS and CMOS transfer gates are connected to an n-type dynamic flop-flop and a p-type dynamic flip-flop which are to be activated by/SAN and SAP, respectively. And there is an n-type MOSFET to short circuit SN_i to $/SN_i$ which is gated by SHORT. Although there seems to be no current load circuit, the p-type flip-flop plays the role of current load as well as the data latch as will be understood in the explanation of the operation of the circuit. The merger of the two different functions into a single flip-flop is possible due to the high symmetry between SN_i and $/SN_i$ for the twin cell scheme and largely contributes to reduction of chip size. As will be explained in section 6.3, this symmetry will be broken for the single cell scheme and the current load circuit cannot be merged with the data latch circuit because of the necessity of dummy cells. This is one of the merits of adopting the twin cell scheme. The read data or the write data are transferred to or from a data line pair DQ and /DQ through a pair of n-type transfer gates which are gated by the *j – th* column select line CSL_j, just like the S/A in the conventional 1T-1C DRAM. A pair of CMOS transfer gates which are connected to the opposite side of a pair of sense nodes SN_i and $/SN_i$ are used for restoring data or writing data to the cells. The reason the connection is interchanged is that the read data which is amplified and latched at the sense nodes has an opposite polarity to what is stored in the cell in the case where cell current is loaded from the p-type MOSFETs. The equalizers which force BLs to 0 V in the pre-charge periods are also required, though they are not shown in Fig. 6.4. Extra cells along a WL or several WLs can play the role of the equalizers which set all BLs to V. Since FBC flows

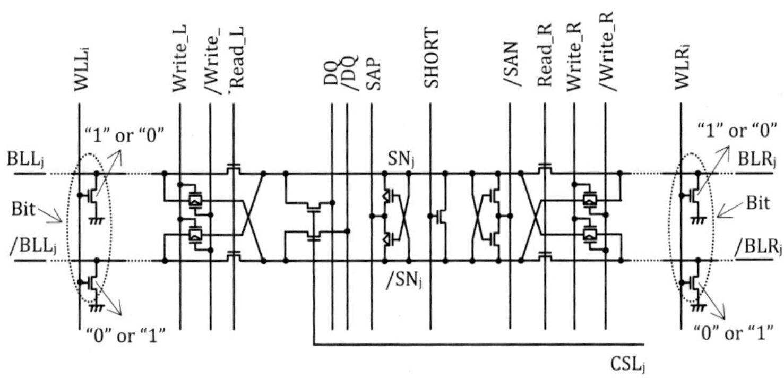

Figure 6.4 Sense amplifier circuit for the twin cell scheme.

current during write, the channel widths of the CMOS flip-flop and those of CMOS transfer gates must be large enough to bias BL at the voltages as close as possible to the power voltages of SAP during "1" write and /SAN during "0" write. When the circuits are built on PD-SOI substrate, the body contacts needs to be tied to fixed voltages especially for the dynamic flip-flop latches to avoid unfavorable threshold voltage off-set to be introduced between the cross coupled transistor pairs.

Figure 6.5 illustrates the operational waveforms of the S/A for the twin cell scheme at read. Prior to read operation (the state is called the "pre-charge" period), all word lines WLL_i, WLR_i are at their negative hold level

V_{WLL}, all bit lines BLL_j, $/BLL_j$, BLR_j, $/BLR_j$ are at 0 V, the voltage of gates of n-type MOSFET transfer gates Write _L, Write_R, Read_L, Read_R are at the positive BL "1" write level V_{BLH}, the voltage of gates of p-type MOSFET transfer gates /Write_L, Write_R are at the negative BL "0" write level V_{BLL}, the activation signal of the p-type flip-flop SAP is at V_{BLL}, the activation signal of the n-type flip-flop /SAN is at V_{BLH}, the equalizing signal between the sense nodes SHORT is at V_{BLH}, the column select line CSL_j is at V_{BLL}, the data line pair DQ, /DQ are at V_{BLH}. Since all WLs are at the hold level, data are held and all BLs are set at 0 V by BL pre-charge transistors which are not shown in Fig. 6.4 but made by the cell transistors whose gates are controlled by the BL pre-charge signal which are raised at the positive WL high voltage in write V_{WLHW}. Since all transfer gates between BLs and the sense nodes are

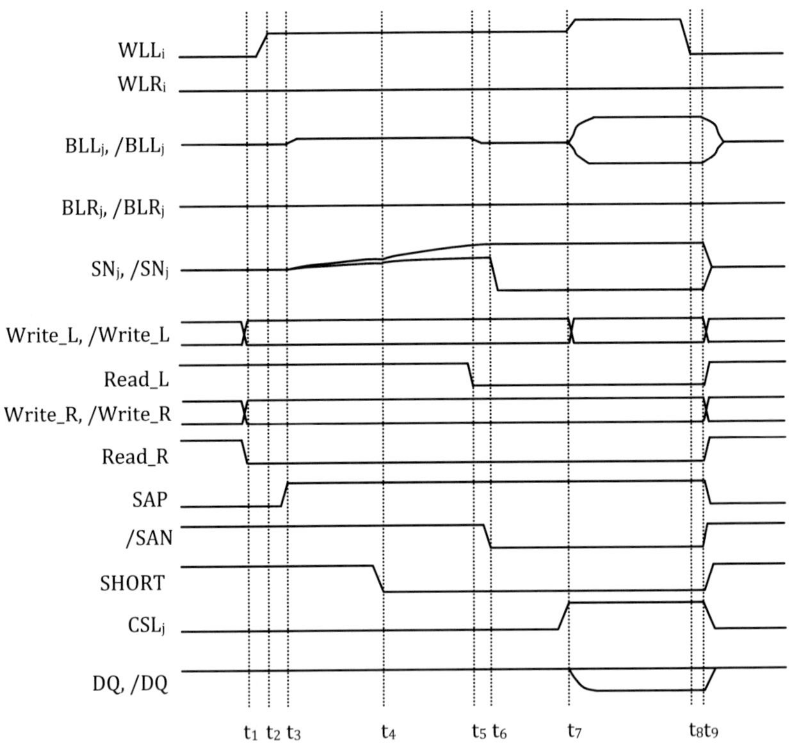

Figure 6.5 Operational waveforms of the sense amplifier circuit at read for the twin cell scheme.

turned on, the sense nodes SN_j and $/SN_j$ are also at 0 V. The n-type MOSFET transfer gate gated by SHORT which is at V_{BLH} makes the voltages of all pairs of sense nodes are highly equalized. The dynamic flip-flop is not activated. All the column select lines CSL_j are at low level to disconnect between the sense nodes and the data lines. At time t_2, a word line WLL_i is selected to be raised to the world line read level V_{WLHR}. Earlier, at time t_1, corresponding to the input row address which select the word line WLL_i, all write transfer gates are turned off with Wriute_L and Write_R lowered to V_{BLL} and /Write_L and /Write_R raised to V_{BLH}. The read transfer gates in the right array are also turned off with Read_R lowered to V_{BLL}, remaining only the read transfer gates in the left array turned on with Read_L at V_{BLH}. At time t_3 which is just after WLL_i rises, the activation signal of the p-type flip-flop circuit SAP rises from 0 V to V_{BLH}. This triggers the current loading to the cells selected by WLL_i for developing the signal between SN_j and $/SN_j$. However, since the signal SHORT is high to short circuit SN_j to $/SN_j$, the gain of the p-type flip-flop positive feedback is suppressed to a low level. Therefore, the voltage difference between the sense nodes SN_j and $/SN_j$ is amplified very slowly. When the voltage difference is amplified to a certain extent, the signal SHORT is lowered to V_{BLL} at time t_4, making the feedback gain increased so that the signal is amplified rapidly. The reason why t_4 cannot be equal to t_3 is that the large feedback gain at initial stage of amplification might induce malfunction of S/A due to the threshold voltage mismatch of the p-type MOSFET pair which works stronger than the signal development direction due to the data difference between the cell transistors pair. Then at t_5, the sense nodes are isolated from the left BLs by lowering Read_L to V_{BLL}, followed by the activation of the n-type flip-flop by lowering /SAN to V_{BLL} at time t_6 to latch the signal in the S/A. At time t_7, a column select line CSL_j is selected to read the latched signal to the data lines. At this time, the read and amplified data are restored to the cells by raising the word line WLL_j, from the read level V_{WLHR} to the write level V_{WLHW} and by turning on the write transfer gates of the left array with Write_L raised to V_{BLH} and /Write_L lowered to V_{BLL}. Finally, at time t_8 and t_9, the word line WLL_i and the bit lines BLL_j,/ BLL_j are reset to their pre-charge levels. It should be noted that t_8 and t_9 can be equal on the contrary to the operation of the 1T-1C DRAM. This contributes to shortening the cycle time.

The reason why the WL level is raised to V_{WLHW} at the end of the read cycle is that the cell loses several holes due to the charge pumping as was explained in section 6.1 and that the hole must be replenished by impact ionization. The refresh cycle is the same as Fig. 6.5 except for the operation of the column select line which is not activated in the refresh cycle with the data lines kept at V_{BLH}. On the contrary to the 1T-1C DRAM, the read cycle and refresh cycle can be very short because the data restore is just for a few holes which are lost by the charge pumping in a WL cycle. Therefore, though FBC is not a perfect non-destructive read-out cell, its feature of quasi

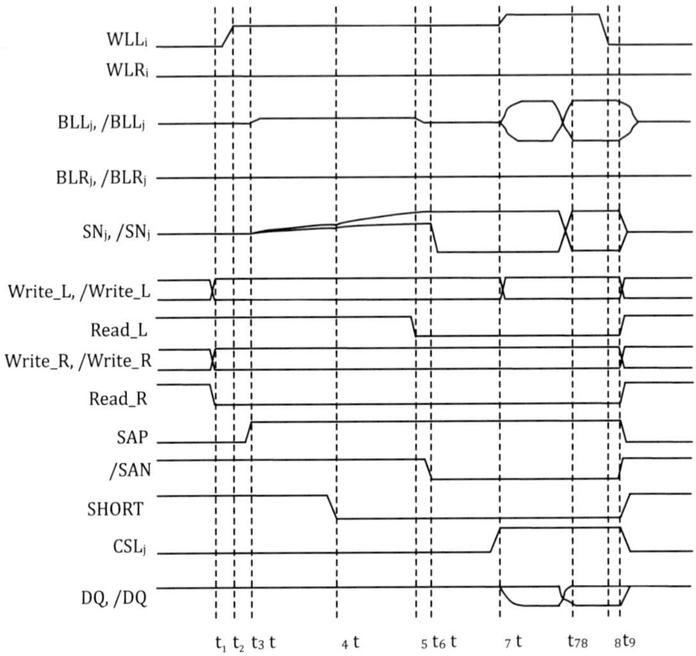

Figure 6.6 Operational waveforms of the sense amplifier circuit at write for the twin cell scheme.

non-destructive read-out is fully enjoyed in shortening the read and the refresh cycle for the floating body RAM (FBRAM). The actual performance of the S/A for the twin cell scheme will be simulated in detail in section 7.3.

Figure 6.6 illustrates the operational waveforms of the S/A for the twin cell scheme at write. Until time t_{78}, the operation in Fig. 6.6 is the same as Fig. 6.5. When the write data is input to the data line pair at time t_{78}, they changes the state of the flip-flop through the n-type transfer gates selected by the column select line CSL_j. Then the changed data is written to the cell through the CMOS transfer gates.

6.3 S/A IN SINGLE CELL SCHEME

Figure 6.7 exemplifies a S/A circuit which is used for memories in the single cell scheme.[3,4,7] Though this S/A also corresponds to *j – th* column which exists at the center of *j – th* BL pair like Fig. 6.4, the S/A in Fig. 6.7 is for the open BL architecture unlike Fig. 6.4. In this scheme, a reference current I_{ref} which is between the "1" cell read current I_{cell} and the "0" cell read current I_{cell0} to distinguish between the data "1" and the data "0" is necessary to be supplied to the S/A. In Fig. 6.7, a word line WLL_i in the left array is selected and accordingly the dummy WL in the right array DWLR is activated. If a WL in the right array WLR_i is selected, the dummy WL in the left array DWLL is activated, though they are not shown in Fig. 6.7. Furthermore, in Fig, 6.7, the dummy cell is represented by a single cell which is written the data "1/2." However, this expression is rather for convenience. Actual implementation of dummy cells into the sensing system will be explained in detail in the next section. A current load circuit of p-type MOSFET current mirror is placed at the center of the S/A apart form the data latch flip-flop circuit in contrast with the S/A of the twin cell scheme. The signals CM and /CM change the configuration of the current mirror circuit depending on which sense node SN_j or $/SN_j$ connects to the dummy cell for generating the reference current I_{ref} through the n-type transfer gate which is gated by Read_L or Read_R. If a word line WLL_i in the left array is selected with the dummy word line DWLR in the right array activated, I_{ref} flows on $/SN_j$ through the transfer gate which is gated by Read_R. In this case, CM and /CM are set at V_{BLH} and V_{BLL}, respectively, making the common gate nodes of the current mirror circuit connected to $/SN_j$. The CMOS transfer gates gated by Write_L, /Write_L and Write_R, /Write_R are used in writing or refreshing the cells.

Figure 6.8 illustrates the operational waveforms of the S/A for the single cell scheme at read. In the pre-charge period, all word lines WLL_i, WLR_i are at their negative hold level V_{WLL}, both dummy word lines DWLR, DWLL are also at V_{WLL}, all bit lines BLL_j, BLR_j are at 0 V, the voltage of gates of n-type MOSFET transfer gates Write_L, Write_R, Read_L, Read_R are at the positive BL "1" write level V_{BLH}, the voltage of gates of p-type MOSFET transfer gates /Write_L, /Write_R are at the negative BL "0" write level V_{BLL}, the activation signal of the p-type flip-flop SAP is at V_{BLL}, the activation signal of the n-type

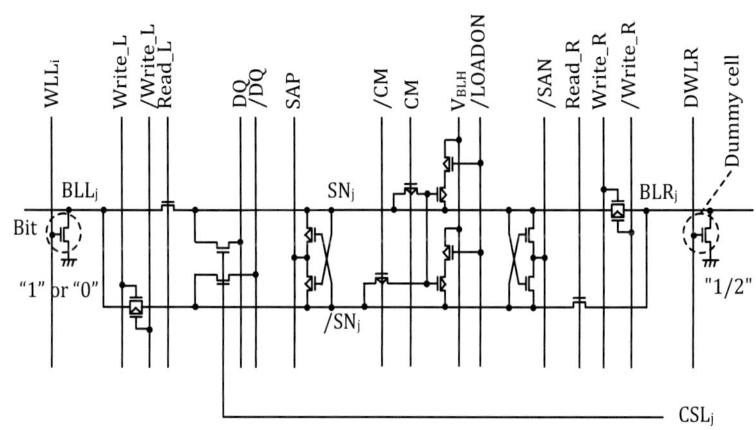

Figure 6.7 Sense amplifier circuit for the single cell scheme.

flip-flop /SAN is at V_{BLH}, the current mirror configuration signals CM, /CM, are at V_{BLL}, the activation signal of the current load circuit /LOADON is at V_{BLH}, the column select line CSL_j is at V_{BLL}, the data line pair DQ, /DQ are at V_{BLH}. Since all WLs and dummy WLs are at the hold level, data are held and all BLs are set at 0 V by BL pre-charge transistors which are not shown in Fig. 6.7 but made by the extra cell transistors whose gates are controlled by the BL pre-charge signal which are raised at the positive WL high voltage in write V_{WLHW}. Since all transfer gates between BLs and the sense nodes are turned on, the sense nodes SN_j and $/SN_j$ are also at 0 V. The dynamic flip-flop is not activated. The current load circuit is not activated, either. All the column select lines CSL_j are at low level to disconnect between the sense nodes and the data lines. Prior to a WL activation of WLL_i at time t_1, all write CMOS transfer gates are turned off with Write_L and Write_R lowered to V_{BLL} and /Write_L and /Write_R raised to V_{BLH}. The read n-type MOSFET transfer gates remain turned on. At the same time, the current mirror configuration signal CM rises to V_{BLH} in accordance with the selected dummy cell which

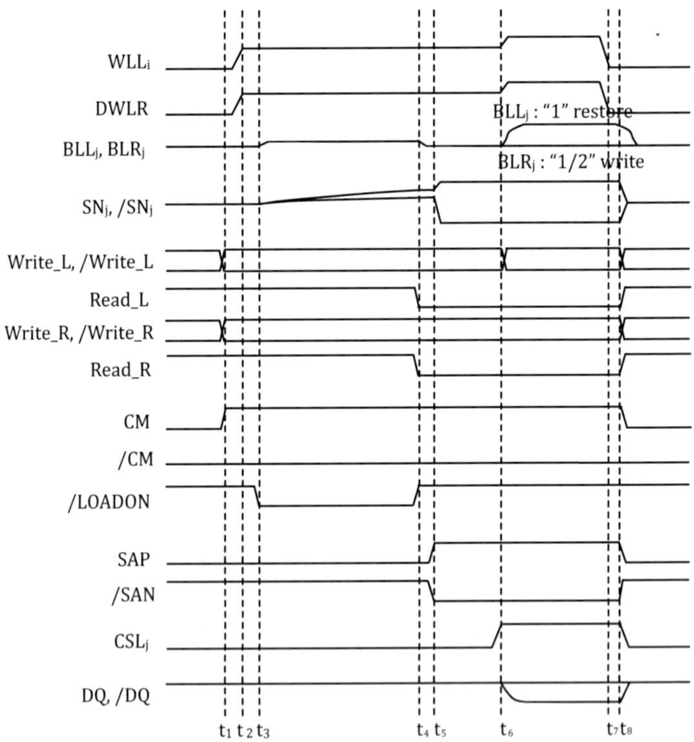

Figure 6.8 Operational waveforms of the sense amplifier circuit at read for the single cell scheme.

cell, which connects the common gate nodes of the current mirror circuit to $/SN_j$. At time t_3 immediately after the WL activation, the current load circuit is activated with /LONDON lowered to V_{BLL}. The signal is being developed between the sense nodes pair SN_j and $/SN_j$. When the signal is amplified to more than the threshold imbalance between the pair transistors in the dynamic flip-flop, the activation signals of the flip-flop SAP and /SAN transit to V_{BLH} and V_{BLL}, respectively, to latch the signal at the sense nodes at time t_5. At time t_4 immediately before the latch, the sense nodes are isolated from BLs by lowering Read_L and Read_R. At the same time, the current load circuit is also inactivated by raising /LOADON. At time t_6, a column select line CSL_j is selected to read the latched signal to the data lines. At this time, the read and amplified data are restored to the cells by raising the word line WLL_i from the read level V_{WLHR} to the write level V_{WLHW} and by turning on the write transfer gates of the left array with Write_L raised to V_{BLH} and /Write_L lowered to V_{BLL}. The dummy cells are also written the data "1/2" simultaneously. The actual method of dummy cell writing will be explained later in the next section. Finally, at time t_7 and t_8, the word line WLL_i and the bit lines BLL_j, $/BLL_j$ are reset to their pre-charge levels. The actual performance of the S/A for the single cell scheme will be simulated in detail in section 7.4.

Figure 6.9 illustrates the operational waveforms of the S/A for the single cell scheme at write. Until time t_{67}, the operation in Fig. 6.9 is the same as Fig. 6.8. When the write data is input to the data line pair at time t_{67}, they changes the state of the flip-flop through the n-type transfer gates selected by the column select line CSL_j. Then the changed data is written to the cell through the CMOS transfer gates.

6.4 DESIGN OF DUMMY CELL SYSTEM IN SINGLE CELL SCHEME

Dummy cell implementation is one of the most critical issues in designing the S/A in the single cell scheme. z shows a dummy cell implementation method where a pair of FBCs on adjacent BLs with opposite data written are short circuited to generate the reference current which is the average of the "0" cell read current and the "1" cell read current $I_{ref} = (I_{cell0} + I_{cell1})/2$.[3,4,8] Although this one-pair averaging method is simple and generates I_{ref} whose value changes depending on device parameters and temperature in the same manner as those of memory cells, the reference current suffers from some statistical fluctuations among dummy cells in a cell array. Figure 6.11 illustrates this situation where the reference current standard deviation due to the dummy cell fluctuation is $\sigma/\sqrt{2}$ with σ being the standard deviation of the cell read current fluctuations. This fluctuation decreases the yield of chip functionality with the cell read current fluctuations combined, although we derived the criteria for the signal-to-noise ratio of cells under the assumption of this one-pair averaging dummy cell system. Here, to increase the chip yield, we consider a method of averaging many pairs of the "1" cell and the "0" cell for generating I_{ref} as is shown in Fig. 6.12 where the read currents of multiple dummy cells along a DWL are averaged by short circuiting adjacent BLs in the unselected array both in parallel and in series with the signals PAVRG and SAVRG, respectively.[3,4,8] The standard deviation

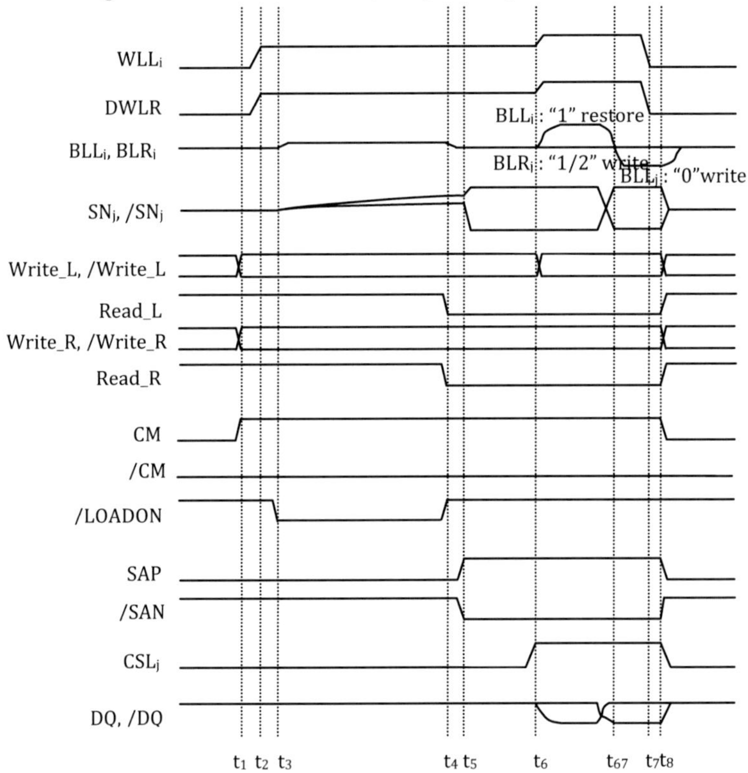

Figure 6.9 Operational waveforms of the sense amp. circuit at write for the single cell scheme.

of the dummy cell current distribution in an array is ideally expected to be $\sigma\sqrt{N2}$ as is shown in Fig. 6.13 where σ is the standard deviation of the cell read current distribution and N is the number of averaged dummy cell pairs. This multi-averaging method of dummy cell is thus expected to contribute greatly to generating a very accurate reference current in sensing data in the S/A for decreasing the probability of data sensing error. The serial average by the signal SAVRG has an ability to force each BL to "1/2" because it is direct short cut between "1" and "0." However, it suffers from the relatively high channel resistance of the averaging MOSFETs, making it difficult to average all cells spreading over a long distance in the direct dirction.

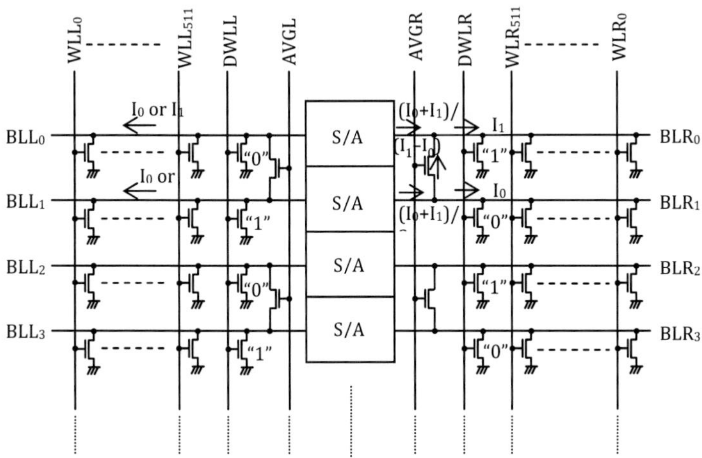

Figure 6.10 One-pair averaging method for implementing dummy cells.

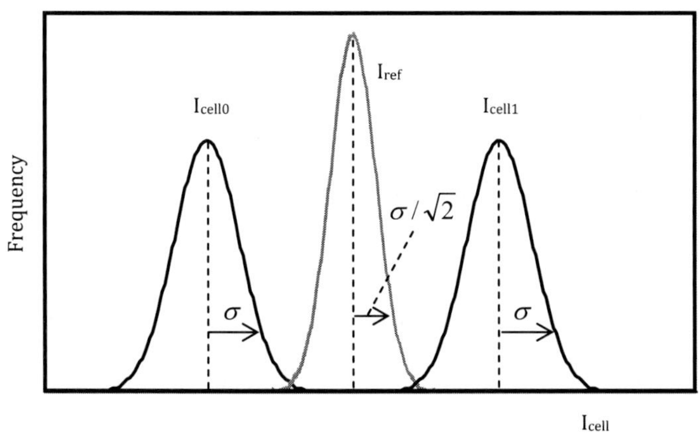

Figure 6.11 Fluctuations of cell read currents and dummy cell read current in a cell array in the one-pair averaging method.

In contrast, the parallel average by the signal PAVRG is to connect each BL to a low resistance wire, making it easy to propagate averaged state over a long distance. However, since it is connection between "1" or "0" and "1/2,"

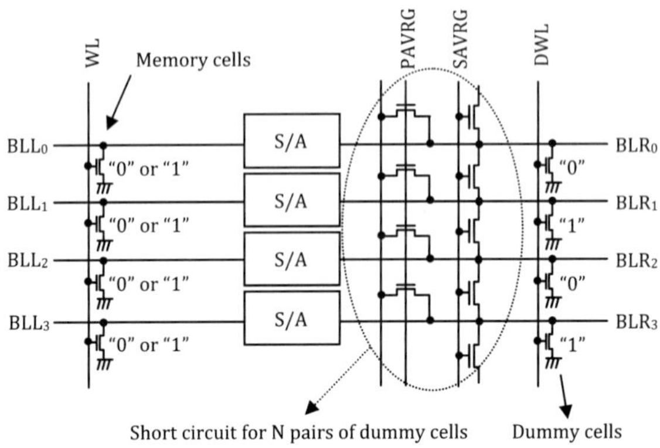

Figure 6.12 Multi-pair averaging method for implementing dummy cells which is illustrated only for the case where a WL in the left array is selected.

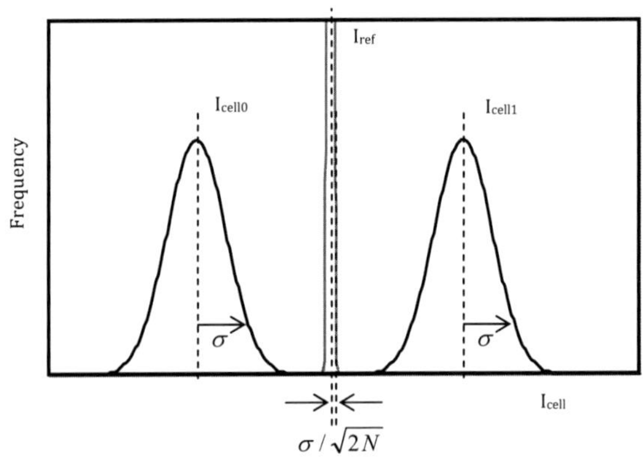

Figure 6.13 Fluctuations of cell read currents and dummy cell read current in a cell array in the multi-pair averaging method, where N is the number of averaged dummy cell pairs.

it lacks the ability to force each BL to "1/2." To compensate the weak point of each method, the hybrid method which averages the dummy cells in series along with in parallel is recommended for designing an effective dummy cell system. Figure 6.14 shows the Monte Carlo simulation results of the signal developments on the sense nodes from 10 ns when /LOADON is lowered to trigger load current which compare (a) one-pair dummy cell averaging and (b) 128-pair dummy cell averaging where 1000 simulation results with normally distributed cell threshold voltages ($\Delta V_{th} = \langle V_{th0} \rangle - \langle V_{th1} \rangle = 0.4$ V, $\sigma_{V_{th0}} = \sigma_{V_{th1}} = 30$ mV) are superposed in each case. In the simulations, the transistors' variations in the p-type MOSFET current mirror pair and the n-type MOSFET transfer gate pair are also included (both pairs were assumed to be normally distributed with the same standard deviations of 12 mV and 4 mV for their absolute value and their imbalance, respectively). In this graph,

SN_0, SN_1, and $SN_{1/2}$ stand for the sense nodes corresponding to the data "0" cell, data "1" cell and the dummy cell, respectively. The fluctuation of the reference voltages on $SN_{1/2}$ is much smaller and the signal voltages of the data "1" and the data "0," respectively, on SN_1 and SN_0 are amplified further for the case (b) than the case (a). Both are due to the tighter control of the reference voltage distribution by adopting the multi-averaging method. Figure 6.15 shows the cross-section of the signal and the reference voltages developed in 10 ns from the /LOADON activation where the vertical axis stands for a standard deviation of the cell Array Architecture and Sense Amplifier Design cumulative distributions. It should be noted that $SN_{1/2}$ is separated into two groups: one is the sense nodes connected to the "0" dummy cell $SN_{1/2(0)}$ and

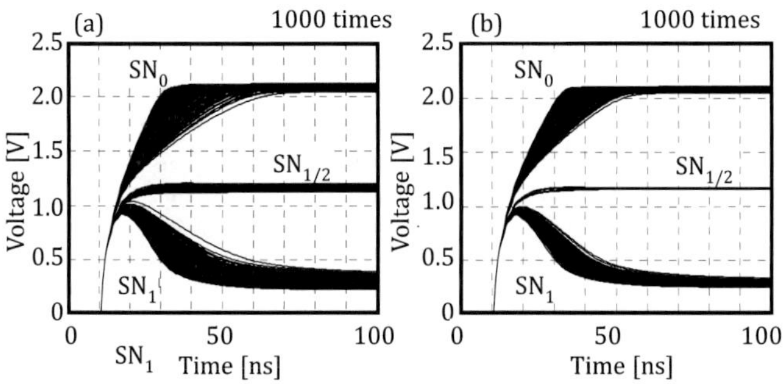

Figure 6.14 Monte Carlo simulation results of signal developments for (a) one-pair averaging of dummy cells and (b) 128-pair averaging of dummy cells.

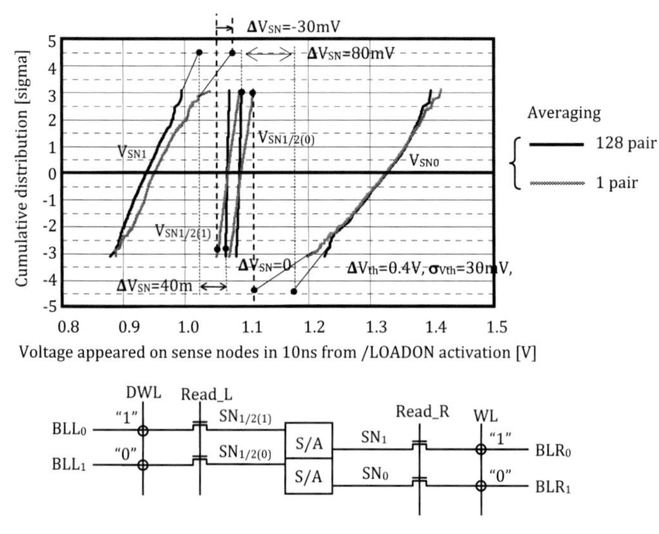

Figure 6.15 Comparison of sense margins between the on-pair dummy cell averaging and the 128-paircdummy cell averaging.

the other is those connected to the "1" dummy cells $SN_{1/2(1)}$. If we assume that the redundancy system can repair eight random fails in a 2 Mbit array and that the number of dummy cells in the array is 1024, we need to extrapolate the cumulative distributions to ±4.5σ for the signal voltages and ±3.0σ for reference voltages to evaluate the worst cases for the signal development. For the case (a), the voltage difference at the worst point is negative (−30 mV) and therefore correct sensing cannot be expected. For the case (b), on the other hand, the voltage difference is estimated to be 40 mV which is regarded to be larger enough for correct sensing by considering the dynamic flip-flop's threshold voltage variations. The scheme of the multi-averaging of dummy cells is thus shown to be effective for realizing a high-density memory using FBC.

6.5 EVALUATION OF S/A IN FBRAM PROTOTYPE SAMPLE

In this section, hardware evaluation results of S/A designed for the 128 Mbit FBRAM prototype sample (PTS) according to sections 6.3 and 6.4 are disclosed.[3,4] The FBRAM uses the FBC whose size is 0.33 μm × 0.515 μm = 0.17 μm^2 with device parameters: gate length L_g = 150 nm, channel width W = 165 nm, gate oxide thickness t_{ox} = 6 nm, silicon thickness t_{Si} = 42.5 nm, buried oxide thickness t_{Box} = 25 nm. The chip size measures 7.6 nm × 8.5 mm = 64.6 mm^2. The peripheral circuits are also designed by MOSFETs on SOI with their bodies tied to the corresponding power voltages. The typical channel length is 450 nm and 400 nm for n-type MOSFETs and p-type MOSFETs, respectively. The other device parameters for the MOSFETs of the peripheral circuits are the same as FBC. Figure 6.16 shows the chip photograph and Fig. 6.17 shows the cross-sectional transmission electron microscopy (TEM) image of FBC. The features of the PTS are summarized in Table 6.1.

Figure 6.18 shows the functional test results of a 16 Mbit array in 128 Mbit FBRAM PTS. The bias voltage of plate (substrate) V_{PL} is changed as (a) −3.0 V, (b) −2.5 V, (c) −2.0 V, (d) −1.5 V, and (e) −1.0 V. The horizontal axis is WL voltage at read V_{WLHR} and the vertical axis is the number of fails in the 16 Mbit array. There are five groups of data points labeled as 2 AV, 256 PAV, 256 SAV, 256 HAV, and 2 cpb which stand for one-pair average, 128-pair parallel average, 128-pair parallel average, 128-pair serial average, 128-pair hybrid average (parallel plus serial), and twin cell (2cpb) mode, respectively. Other measurement conditions are summarized in the box at the bottom right of Fig. 6.18. The silicon film in which the cells are placed is 42.5 nm thick and the cells can be regarded as those on fully depleted silicon-on-insulator. Therefore, the fail bit count increases as the absolute value of V_{PL} decreases, because it becomes difficult for holes to be accumulated in the body unless the absolute value of V_{PL} is large enough. It can be observed that the number of fails decreases less than 10 bit in a wide range of V_{WLHR}

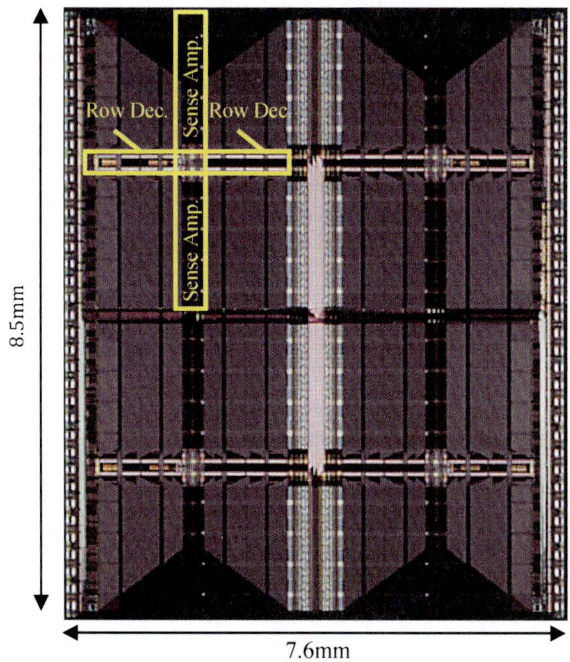

Figure 6.16 Chip photograph of 128 Mbit FBRAM prototype sample.

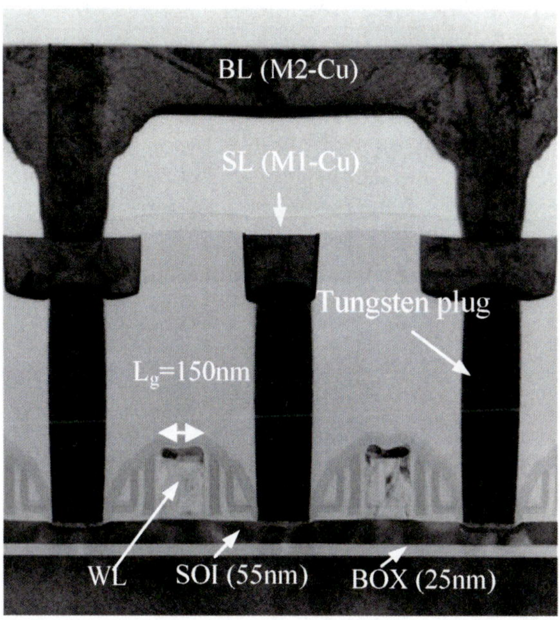

Figure 6.17 TEM cross-sectional view of FBC used in the 128 Mbit FBRAM prototype sample.

Table 6.1 Features of 128 Mbit FBRAM prototype sample

Bit organization	8 M word × 16 bit
Cell structure	$t_{ox}/t_{Si}/t_{BOX}$ = 6 nm/55 nm/25 nm, L_g = 150 nm
Cell size	0.33 μm × 0.515 μm = 0.17 μm^2
Peripheral	Bodied tied SOI with t_{ox} = 6 nm, L_{gn} = 450 nm, L_{gp} = 400 nm
Random access	18.5 ns (normal mode), 25.7 ns (VSRAM mode)
Refresh cycle	4 K
Redundancy	Eight redundant LWLs/1 Mbit and 16 redundant LBLs/2 Mbit
Special modes	Fast page mode, VSRAM mode
Chip size	7.6 mm × 8.5 mm = 64.6 mm^2

for $V_{PL} \leq -2$ V. Though the twin cell scheme is very stable with the fail count 10 or less for a very wide range of V_{WLHR} and for all V_{PL}s, the performance of the single cell scheme is comparable for $V_{PL} \leq -2$ V and $V_{WLHR} \geq 1.0$ V. The serial average method seems to exhibits the best performance which is almost comparable with the hybrid average method. The parallel average looks the worst and even worse than the one-pair average.

If we measure the data retention time distribution for various dummy cell methods, the differences among the methods are more clearly manifested. Figure 6.19 is the measurement results of data retention time distributions for the FBRAM PTS taken by changing the dummy cell averaging methods including the twin cell scheme (2 cpb) for reference. All data are at the temperature of 85°C. The simple one-pair averaging (2 AV) exhibits the worst characteristics, i.e., the shortest retention time distribution. The performance of the hybrid averaging (HAV) is the best among the single cell schemes, although it falls short of the twin cell scheme (2 cpb). It can be observed that the retention time of the single cell scheme is about two to three times shorter than that of the twin cell scheme for the averaged cells or for the -4σ cells if the hybrid averaging method is adopted.

6.6 QUASI-SHIELDED OPEN BL SCHEME

As was discussed in section 1.2, it has become inevitable for the cell array architecture of the conventional 1T-1C DRAM to migrate from the traditional folded BL architecture to an open BL one when the cell size shrinks from

$8F^2$ to $6F^2$ and $4F^2$. The problem which arises in this migration is that the effective method of BL twist cannot be applied to the open BL architecture for reducing the noise between BLs which will become more severe as the cell size shrinks. As was discussed in section 2.1, on the contrary, FBC is a gain cell which has an ability in itself to draw current from BL or SL.[9,10] This means that the signal of FBC develops in the balance between the cell current and the load current which is supplied from S/A. Therefore, if the

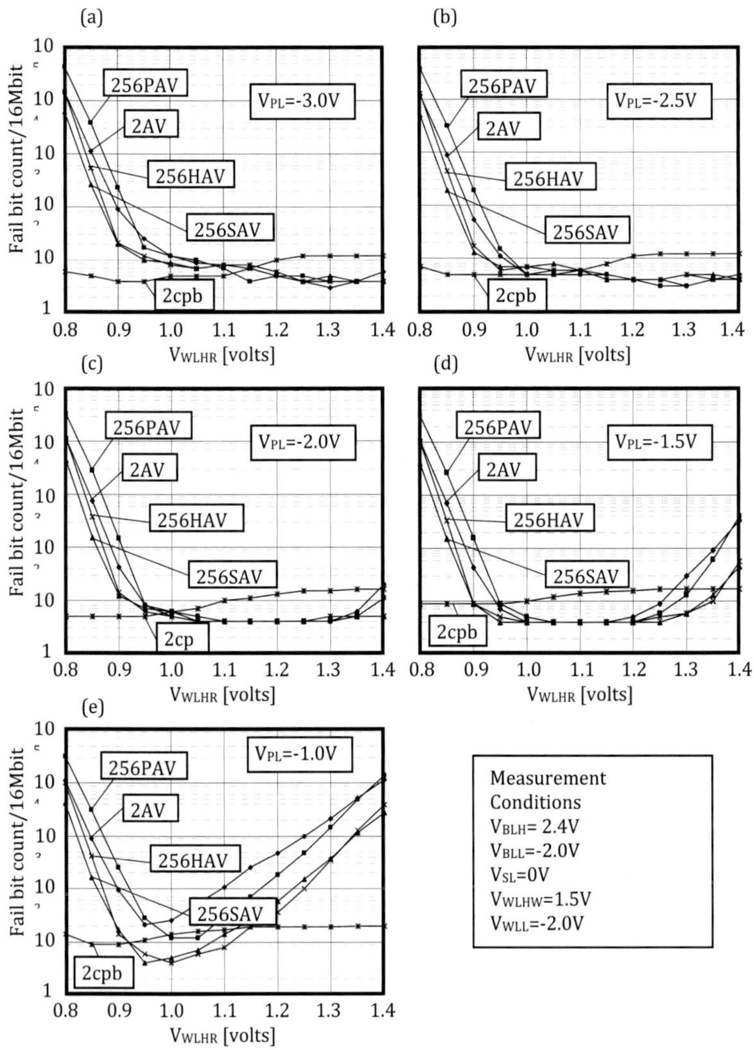

Figure 6.18 Functional test results of a 16 Mbit array in 128 Mbit FBRAM prototype sample. The plate bias V_{PL} is changed from (a) –3.0 V to (e) –1.0 V.

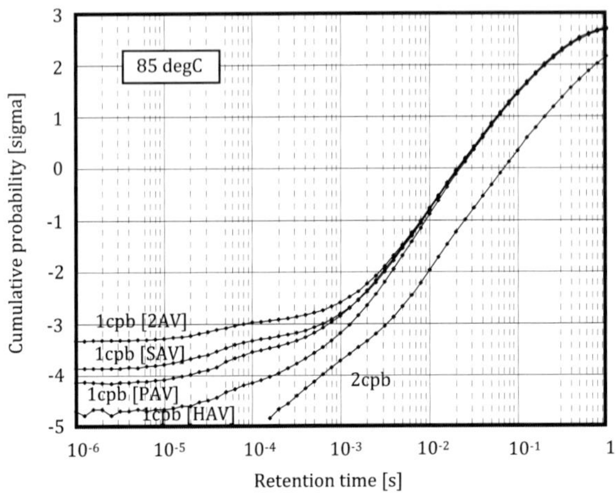

Figure 6.19 Measurement results of retention time distributions of 128 Mbit FBRAM prototype sample for various averaging methods of dummy cells. That of the twin cell scheme is added for reference.

signal is read in a long enough time, the signal amount is not affected by the noise between BLs, because it is determined by the balanced DC level. Actually, it is true that the noise affects the signal development speed, leading to the access time degradation, because the capacitive coupling between BLs slows down the signal development. However, the gain cell can have inherently a higher immunity against the noise between BLs than the conventional 1T-1C DRAM cell whose signal is determined by the charge sharing between the cell capacitance and the BL capacitance. Furthermore, since FBC is a quasi non-destructive read-out cell,[3,4] the start of current loading can be waited for a certain period after a WL rising without degrading the signal. By making use of this feature, the signal development phase of the odd BL group can be delayed from that of the even BL group, or vice versa, to shield each other's inter-BL noise by biasing a fixed voltage to the every other quiescent BLs. FBC, like this, can cancel the inter-BL noise to a large extent in contrast to the conventional 1T-1C DRAM cell even in the open BL architecture, providing a big advantage over the 1T-1C DRAM in the future generations. And it is further advantageous for FBC that the data restore operation for a BL group which started the sense operation with a delay (referred to "background bit lines") can be completed earlier than that of another BL group which started sensing earlier (foreground BLs) again due to the feature of the quasi non-destructive read-out by assigning the cells connected to the foreground BLs to be read or written and those connected to the background BLs not to be selected to be accessed but just restored with a very few holes for the data "1" which are lost by charge pumping, leading to no access penalty by adopting this interchanging sensing method. This constitutes a superior array design specific to FBC that cancels inter-BL noises by shielding the capacitance coupling effect in sensing two interchanging BL groups in different periods without penalty of the cycle time degradation. In this section, the design of this method, the quasi-shielded open BL scheme, will be explained in detail.[11]

Figure 6.20 shows a selected cell, a dummy cell, and the S/A which will be simulated in this section and is the same as that shown in Fig. 6.7 except for the write/restore path of the CMOS transfer gates and the DQ line pair and the transfer gates to and from the pair. The p-type and the n-type dynamic latches are denoted by a box F/F. The short circuit path between the sense nodes SN /SN and is added in Fig. 6.20. And there are cell transistors which pre-charge both BLs to V_{SL} and are gated by the BL pre-charge signal PRCH. Figure 6.21 shows the cell array architecture for the quasi-shielded open BL scheme. Only three cell arrays are shown which are related to a WL activation in the array B. The BLs are in the open BL architecture and the sense amplifiers are arranged at the both end of each array connecting to every other BL.[a] The dummy cell averaging circuits are not shown in Fig. 6.20. Now we assume that a word line WL in the array B is activated. According to this activation, two dummy word lines DWL_A and DWL_C are also activated. This address input to this memory is assumed to be the address multiplex type[b] and the even bit lines BL_e denoted by the solid lines and the odd bit lines BL_0 denoted by the dotted lines are assumed to be assigned to different row addresses. This means that either the celle on the even BLs or on the odd BLs accessed to be read or to be written when a set of row address are input depending on the address assignment design. Therefore, data are read from or written to the cells only on the even BLs or on the odd BLs in the event of the subsequent input of column addresses. In Fig. 6.21, row addresses which select the cells on the even BLs are assumed to be input the memory. The simulated waveforms of the WL, BLs, and sense nodes are shown in Fig. 6.22 where BL_e "1 ," BL_e "0," BL-O "0" SN "1", SN "0",/SN "1", and SN "0" represent the even BL connected to the data "1" cell, the even BLs connected to the data "0" cell, the odd BLs connected to the data "1" cell, the odd BLs connected to the data "0" cell, the sense node connected to the data "1" cell, the sence node connected to the data "0" cell, the sense node connected to the dummy cell in reading the data "1" cell, and the sense node connected to the dummy cell in reading the data "0," respectively. In the phase 1 which is just after the WL activaction, the S/A s which are connected to even BLs are activated and the

[a] This S/A arrangement is commonly called the double-ended open bit line S/A.

[b] The address multiplex is commonly adopted in the DRAM to reduce the pin count of the chip package where a set of row addresses is input prior to a set of column addresses into the common address pins.

signal is developing between the BL pair, through the BL connected to the dummy cell is not shown in Fig. 6.22.

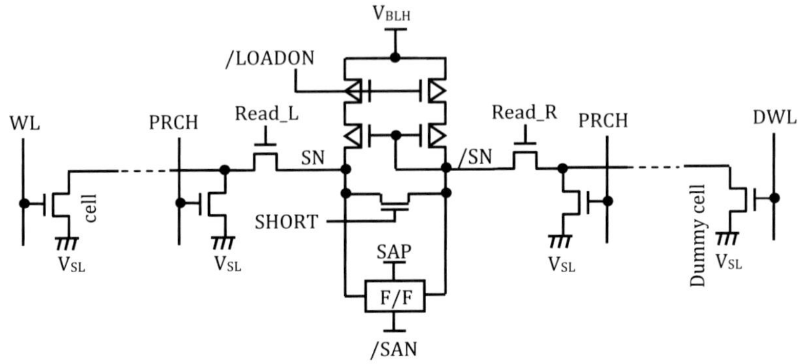

Figure 6.20 The sense amplifier circuit to be simulated in analyzing the quasi-shielded open bit line scheme.

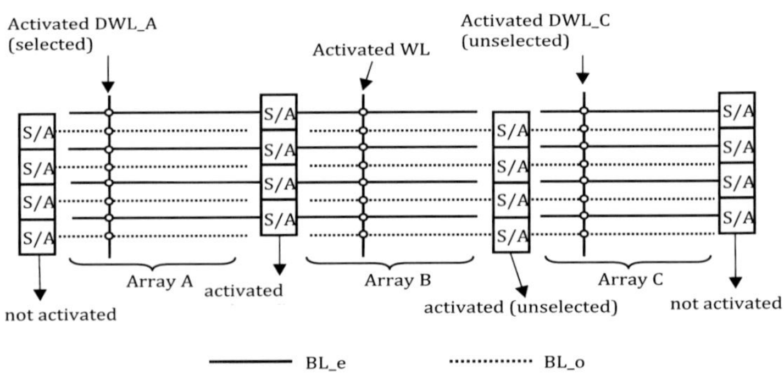

Figure 6.21 Three cell arrays in the quasi-shielded open bit line scheme. Actually, more cell arrays are arranged to the left and the right, though they are not shown here. The solid and the dotted bit lines belong to the even bit line group which are selected to be accessed and the odd bit line group which are not selected but just restored, respectively.

However, SN and /SN "1" or and /SN "0" are equalized by the n-type MOSFET gated by the signal SHORT to a certain point of time. After lowering the signal SHORT to isolate between a pair of sense nodes at 9 ns. the sense node pair starts to be sensed and signal develops suddenly. At 10 ns, the dynamic flip-flop is activated by the signal SAP and /SAN with Read_L and Read_R lowered and then the amplified signals are written to BLs through write paths which are not shown in Fig. 6.22. During this signal development in phase 1, odd BLs are kept pre-charged V_{SL} (0 V) by the n-type MOSFET gated by PRCH, making even BLs shielded to each other.

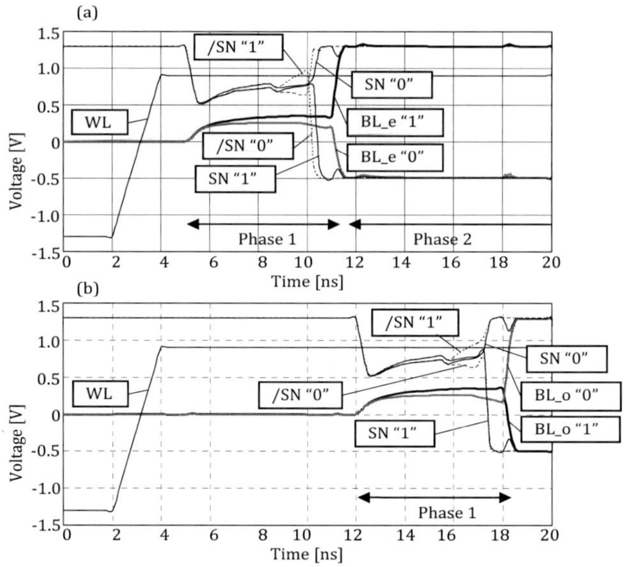

Figure 6.22 The simulated results (circuit simulation) of data sensing operation for the quasi-shielded open bit line scheme. (a) shows the selected even bit lines and the related sense nodes and (b) shows the unselected odd bit lines and the related sense nodes.

This shielding cancels the inter BL noise significantly. After latching the data developed on the even BLs, the S/As which are connected to odd BLs are activated. In this period of phase 2 for even BLs (phase 1 for the odd BLs), since even BLs are biased to V_{BLH} or V_{BLL}, the inter BL noise is again cancelled in developing the signal on the odd BLs.

Figure 6.23 illustrates the results of Monte Carlo simulation performed for signal development of "0" data cell with "1" data cells on the adjacent BLs with and without the quasi-shielded open BL scheme. In this simulation, it was assumed that both the n-type MOSFET pair gated by Read_L, Read_R, and the p-type MOSFET of current mirror load circuit fluctuate according to the normal distribution with the standard deviation 5 mV and that FBC's current fluctuates according to the normal distribution of Fig. 6.23 for RDT = 1 μs. The number of cells per BL is assumed to be 2048 with its capacitance 110 fF and the inter BL capacitance 89 fF. These values correspond to the cell of the BL half pitch (blhp) 35 nm. Temperature is assumed to be 85°C. This graph indicates the cumulative probability measured in standard deviation for the signal ΔV_{SN} appeared on the sense nodes when they are arranged in ascending order for the signal development time (SDT)[c] 4 ns, 4.5 ns, and 5 ns in the case with BLs shielded and 10 ns,

[c] SDT is measured from the time when WL is activated.

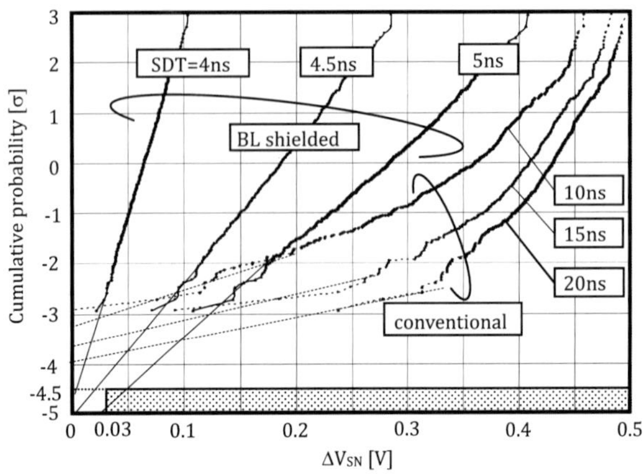

Figure 6.23 The results of Monte Carlo simulation for signal development of the "0" data cell with "1" data cells on the adjacent bit lines with and without the quasi-shielded open bit line scheme.

15 ns, and 20 ns in the case of without shielded. The "0" data cell signals are simulated with the "1" data cells on the adjacent BLs. In the case of without shield, as is shown in this graph, ΔV_{SN} is less than 30 mV for the worst cell corresponding to -4.5σ even for SDT = 20 ns.. In the case of with shield in the quasi-shielded open BL scheme, on the contrary, ΔV_{SN} grows to more than 30 mV for the worst cell corresponding to -4.5σ for SDT = 5 ns. These results clearly show the effectiveness of the quasi-shielded open BL scheme with respect to inter bit line noise canceling.

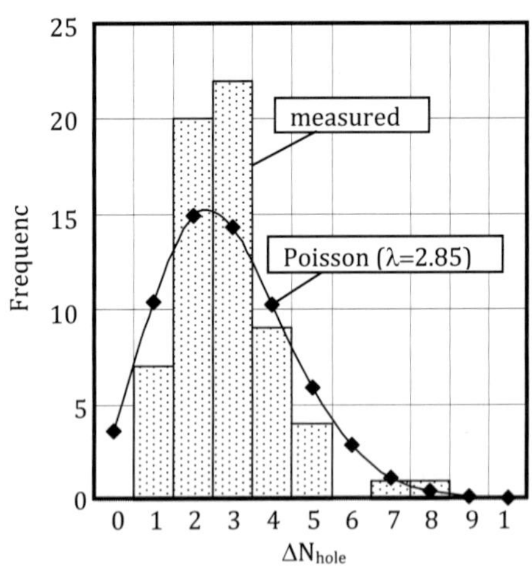

Figure 6.24 The measured histogram in the number of holes lost by a word line pumping for the cell with gate length L_g = 150 nm and channel width W = 180 nm. The Poisson distribution is added for its average λ = 2.85.

Next, let us consider a possible impact of this scheme on the cycle time degradation in delaying the start of the sensing for the every other BLs. Figure 6.24 shows the measured histogram in the number of holes lost by a WL pumping of FBC with gate length L_g = 150 nm and channel width W = 180 nm. This is due to charge pumping phenomenon and the measurement results provide the information about the number of the interface state at the silicon surface of the FBC. The method of determining the interface state by charge pumping for the MOSFET with its body floating without a body contact developed by S. Okhonin *et al.*[2] was used. In this graph, the Poisson distribution function with its average λ = 2.85 is added. Though the measured distribution seems steeper than the theoretical curve, the data are well fitted to the Poisson distribution.[d] The interface state density is estimated to 1.056×10^{10} cm^{-2}. If the density is assumed constant for future generations, the number of the interface states for the cell with gate length L_g = 35 nm, and channel width W = 35 nm, will be 0.129.

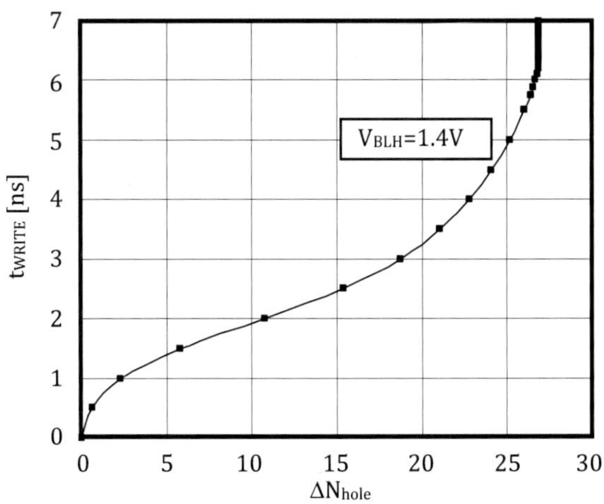

Figure 6.25 The time t_{WRITE} which is needed to write the full "1" level to an FBC with $L_g = W$ = 35 nm from the state where the number of holes is smaller than the full level (27 holes) by ΔN_{hole}.

This indicates that the largest number of holes which will be lost by a WL pumping will be 0.68 in the error rate of 0.25 ppm. Figure 6.25 shows the TCAD simulation results which calculate the time t_{WRITE} which is necessary to write the full "1" level to an FBC with gate length L_g = 35 nm and channel width W = 35 nm from the state where the number of holes is less than the full level (27 holes) by ΔN_{hole}. This graph shows that t_{WRITE} is 500ps at the longest to write the full "1" level for ΔN_{hole} = 0.68. We denote the t_{WRITE} as

[d] The discrepancy between the measured number of holes lost by charge pumping and the Poisson distribution will be discussed in section 7.5 of Chapter 7.

a replenish time t_{rep}. In this analysis, we assume t_{rep} = 0.8 ns with enough margin included. The full "1" write time t_{wf1}, on the other hand, can be read from Fig. 6.25 as 7 ns which corresponds to ΔN_{hole} = 27: t_{wf1} = 7 ns. And the time needed for phase 1 t_{ph1} is assumed to be 6 ns with 1 ns included as a margin: t_{ph1} = 6 ns. Under these assumptions, the inequality holds.

$$t_{wf1}\ (7\ \text{ns}) \geq t_{ph1}\ (6\ \text{ns}) + t_{rep}\ (0.8\ \text{ns}), \tag{6.1}$$

Therefore, BL_o which starts sensing behind by the time t_{ph1}= 6 ns can complete the replenishment of holes lost by charge pumping earlier than BL_e which must inject 27 holes to write "1" fully. The quasi-shielded open BL scheme, thus, can reduce the inter BL noise effectively without any impact on the cycle time degradation as is shown in Fig. 6.26 where circuit simulation results for BL voltages and sense node voltages similar to Fig. 6.22 are combined with TCAD simulation results for body voltages.

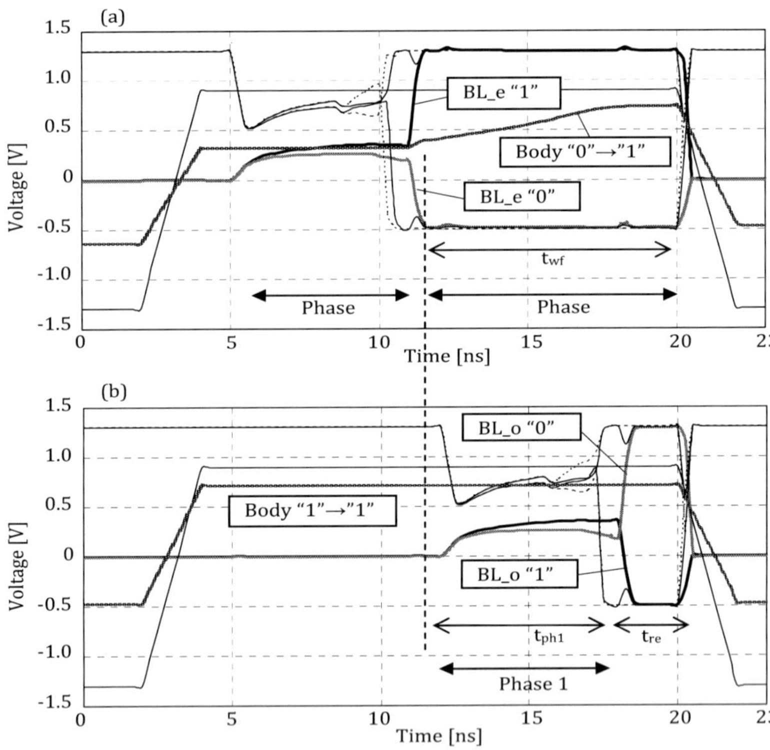

Figure 6.26 The simulated results (circuit and TCAD simulations combined) of data sensing and write/restore operations for the quasi-shielded open bit line scheme. (a) shows the selected even bit lines and the related sense nodes and (b) shows the unselected odd bit lines and the related sense nodes.

References

1. G. Groeseneken, H. E. Maes, N. Beltran and R. F. De Keersmaecker, "A reliable approach to charge-pumping measurements in MOS transistors," *IEEE Trans. Electron Devices*, vol. ED-31, no. 1, pp. 42–53, Jan. 1984.
2. S. Okhonin, M. Nagoga and P. Fazan, "Principles of transient charge pumping on partially depleted SOI MOSFETs," *IEEE Electron Device Lett.*, vol. 23, no. 5, pp. 279–281, May 2002.
3. T. Ohsawa, K. Fujita, K. Hatsuda, T. Higashi, M. Morikado, Y. Minami, T. Shino, H. Nakajima, K. Inoh, T. Hamamoto and S. Watanabe, "An 18.5ns 128Mb SOI DRAM with a floating body cell," *ISSCC Dig. Tech. Papers*, pp. 458–459, Feb. 2005.
4. T. Ohsawa, K. Fujita, K. Hatsuda, T. Higashi, T. Shino, Y. Minami, H. Nakajima, M. Morikado, K. Inoh, T. Hamamoto, S. Watanabe, S. Fujii and T. Furuyama, "Design of a 128-Mb SOI DRAM using the floating body cell (FBC)," *IEEE J. Solid-State Circuits*, vol. 41, no. 1, pp. 135–145, Jan. 2006.
5. S. Okhonin, M. Nagoga, E. Carman, R. Beffa and E. Faraoni, "New generation of Z-RAM," *IEDM Tech. Dig.*, pp. 925–928, Dec. 2007.
6. T. Ohsawa, "Semiconductor memory device," *U.S. Patent 7,502,270*, March 10, 2009.
7. T. Ohsawa, "Semiconductor storage device," *U.S. Patent 7,145,811*, Dec. 5, 2006.
8. T. Ohsawa, T. Higashi, K. Fujita, K. Hatsuda, N. Ikumi, T. Shino, H. Nakajima, Y. Minami, N. Kusunoki, A. Sakamoto, J. Nishimura, T. Hamamoto and S. Fujii, "A 128Mb floating body RAM (FBRAM) on SOI with multi-averaging scheme of dummy cell," *2006 Symposium on VLSI Circuits Digest of Technical Papers*, pp. 224–225, June 2006.
9. T. Ohsawa, K. Fujita, T. Higashi, Y. Iwata, T. Kajiyama, Y. Asao and K. Sunouchi, "Memory design using one-transistor gain cell on SOI," *ISSCC Dig. Tech. Papers*, pp. 152–153, Feb. 2002.
10. T. Ohsawa, K. Fujita, T. Higashi, Y. Iwata, T. Kajiyama, Y. Asao and K. Sunouchi, "Memory design using a one-transistor gain cell on SOI," *IEEE J. Solid-State Circuits*, vol. 37, no. 11, pp. 1510–1522, Nov. 2002.
11. K. Fujita, T. Ohsawa, R. Fukuda, F. Matsuoka, T. Higashi, T. Shino and Y. Watanabe, "Array architecture of floating body cell (FBC) with quasi-shielded open bit line scheme for sub-40nm node," *IEEE Int. SOI Conf.*, pp. 31–32, Oct. 2008.

Chapter 7

Design Practices of Floating Body Random Access Memory

Two different memory designs which make full use of the floating body cell (FBC) characteristics are presented in detail and assessed quantitatively in speed and macro size; one is a high-speed oriented embedded memory design using the twin cell scheme and the other is a high-density oriented stand-alone or embedded memory design using the single cell scheme. Autonomous refresh of FBC which drastically increases the refresh efficiency is discussed as one of the most important technologies for FBC to be used in a stand-alone memory and in a low-power embedded memory macro.

7.1 CHIP COST OF FLOATING BODY RANDOM ACCESS MEMORY ON SILICON-ON-INSULATOR

An aspect of floating body cell (FBC) technology which should be taken into account at the start of designing a floating body random access memory (FBRAM) is that the memory cell typically needs to be built on silicon-on-insulator (SOI) that does not provide a common and standard platform in the present-day semiconductor industries except for several high-performance fields in general-purpose microprocessors[1] and in broadband engine processors.[2] Though it is possible to build an FBC on bulk silicon substrate,[3] it is not usual because the increase in cell size due to the isolation region between adjacent cells along a bit line (BL) is unavoidable. A method has been proposed to transform selected regions of SOI substrate into bulk silicon substrate by removing the buried oxide (BOX) and growing epitaxial silicon layer.[4] To make BOX layer partially from a bulk silicon substrate, on the other hand, a method has been proposed to make use of silicon epitaxial growth on SiGe which is also grown epitaxially

Floating Body Cell: A Novel Capacitor-less DRAM Cell
By Takashi Ohsawa and Takeshi Hamamoto

www.panstanford.com

on silicon substrate. Afterwards, the SiGe layer is removed partially and the hollowed spaces are thermally oxidized to be filled with SiO_2 which forms BOX.[5] In general, however, the adoption of SOI wafers as a starting material is common for building FBC,[a] constituting one of factors that make the designers hesitate to adopt FBC as an embedded memory cell especially in the field of application-specific integrated circuits (ASIC) where many functional macros which are specifically optimized for transistors on bulk silicon have been standardized and strictly modeled as libraries. The high cost of SOI wafers is also considered to be a bottleneck that impedes common production of SOI chips in general, although the superiority in performance of transistors on SOI over that on bulk silicon has been widely recognized and verified experimentally for a long time. At the moment, therefore, only a limited number of product kinds are designed and fabricated on SOI that are required to have very high performances and to be produced in quite large volumes at the same time.

However, the chip cost of the FBRAM, whether it is embedded in an LSI or a stand-alone RAM, can be reduced to lower than the DRAM with stacked cell or the deep-trench cell. Now let us compare the chip cost of an LSI chip which embeds an FBRAM macro on SOI substrate with that of an LSI chip which embeds a one-transistor and one-capacitor (1T-1C) dynamic random access memory (DRAM) macro having the same number of bits on bulk silicon substrate. We denote the standard CMOS process cost, the additional process cost for making the 1T-1C DRAM cell (mainly for making the stacked or the deep-trench capacitor), and the additional process cost for making the FBRAM (mainly for making the FBC transistor which may be different from what is used in the logic LSI[b]) all per wafer as PC_{CMOS}, PC_{DRAM}, and PC_{FBRAM}, PC_{FBRAM}, respectively. And we denote the bulk silicon wafer cost and the SOI wafer cost as $WC_{bulk\ silicon}$ and WC_{SOI}, respectively. Furthermore, we denote the areas of logic LSI part in the both chips and embedded DRAM part in the latter chip as A_{logic} and A_{DRAM}, respectively. The area of FBRAM part in the former chip which has the same number of information bits as the DRAM part is written as A_{FBRAM}. Using these notations, the chip cost $CC_{LSI(DRAM)}$ of the LSI with the 1T-1C DRAM embedded is written as

$$CC_{LSI\,(DRAM)} = \frac{(PC_{CMOS} + PC_{DRAM} + WC_{bulk\ silicon})}{\text{the number of good chips per wafer}}$$
$$= \frac{(PC_{CMOS} + PC_{DRAM} + WC_{bulk\ silicon}) \times (A_{logic} + A_{DRAM})}{Y_{DRAM} \times Y_{logic} \times A_{wafer}}, \qquad (7.1)$$

[a] An FBC which uses a vertical transistor instead of a planar one can be built on bulk silicon substrate. This type of FBC will be discussed in Chapter 8.
[b] This addition, however, can be zero in many cases as was shown in Chapter 4.

where Y_{DRAM}, Y_{logic}, and A_{wafer} stand for the DRAM macro yield, the logic part yield, and the wafer area, respectively. The chip cost $\text{CC}_{\text{LSI(FBRAM)}}$ of the LSI with the FBRAM embedded is written as

$$\text{CC}_{\text{LSI (FBRAM)}} = \frac{(\text{PC}_{\text{CMOS}} + \text{PC}_{\text{FBRAM}} + \text{WC}_{\text{SOI}})}{\text{the number of good chips per wafer}}$$
$$= \frac{(\text{PC}_{\text{CMOS}} + \text{PC}_{\text{FBRAM}} + \text{WC}_{\text{SOI}}) \times (A_{\text{logic}} + A_{\text{FBRAM}})}{Y_{\text{FBRAM}} \times Y_{\text{logic}} \times A_{\text{wafer}}}, \tag{7.2}$$

where Y_{FBRAM} stands for the FBRAM macro yield. The ratio $\eta_{\text{FBRAM/DRAM}}$ of the cost of the LSI with the FBRAM embedded to that with the DRAM embedded is expressed a where the extra process cost of the

$$\eta_{\text{FBRAM/DRAM}} = \frac{\text{CC}_{\text{LSI (FBRAM)}}}{\text{CC}_{\text{LSI (DRAM)}}}$$
$$= \frac{(\text{PC}_{\text{CMOS}} + \text{PC}_{\text{FBRAM}} + \text{WC}_{\text{SOI}}) \times (A_{\text{logic}} + A_{\text{FBRAM}}) \times Y_{\text{DRAM}}}{(\text{PC}_{\text{CMOS}} + \text{PC}_{\text{DRAM}} + \text{WC}_{\text{bulk silicon}}) \times (A_{\text{logic}} + A_{\text{DRAM}}) \times Y_{\text{FBRAM}}}$$
$$= \frac{(1 + \lambda_{\text{FBRAM}} + \beta \times \mu_{\text{bulk silicon}})}{(1 + \lambda_{\text{DRAM}} + \mu_{\text{bulk silicon}})} \times [1 - \xi(1 - \theta)] \times \gamma, \tag{7.3}$$

FBRAM and that of the 1T-1C DRAM over the standard CMOS process are normalized by the standard CMOS process cost as $\lambda_{\text{FBRAM}} = \text{PC}_{\text{FBRAM}}/\text{PC}_{\text{CMOS}}$ and $\lambda_{\text{DRAM}} = \text{PC}_{\text{DRAM}}/\text{PC}_{\text{CMOS}}$, respectively, and the bulk silicon wafer cost is also normalized by the standard CMOS process cost per wafer as $\mu_{\text{bulk silicon}}/\text{WC}_{\text{bulk silicon}}/\text{PC}_{\text{CMOS}}$, the SOI wafer cost is assumed to be β times the cost of the bulk silicon wafer cost as $\beta = \text{WC}_{\text{SOI}}/\text{WC}_{\text{bulk silicon}}$, θ stands for the ratio of the FBRAM macro size to the 1T-1C DRAM macro size with the same number of memory cells which is assumed to be independent of the memory capacity as $\theta = A_{\text{FBRAM}}/A_{\text{DRAM}}$, and $\xi = A_{\text{DRAM}}/(A_{\text{DRAM}} + A_{\text{logic}})$ defines the ratio of embedded 1T-1C DRAM macro area to the total LSI chip area. The ratio of the DRAM macro yield divided by the FBRAM macro yield is denoted by $\gamma = Y_{\text{DRAM}}/Y_{\text{FBRAM}}$. The yields can be calculated using the negative binomial yield model[6] as

$$Y_{\text{DRAM}} = \frac{1}{\left(1 + \frac{A_{\text{DRAM}} \times D_0}{\alpha}\right)^{\alpha}}, \tag{7,4}$$

$$Y_{\text{FBRAM}} = \frac{1}{\left(1 + \frac{A_{\text{FBRAM}} \times D_0}{\alpha}\right)^{\alpha}}, \tag{7.5}$$

where D_0 and α are a defect density and a cluster parameter, respectively, and they both are assumed to have the same values for the 1T-1C DRAM and the FBRAM.

Figure 7.1 shows the calculated results of the cost ratio $\eta_{\text{FBRAM/DRAM}}$ as a function of ξ for $\lambda_{\text{FEBRAM}} = 0.05$, $\lambda_{\text{DRAM}} = 0.2$, $\mu_{\text{bulk silicon}} = 0.03$. The two parameters θ and β are changed: $\theta = 0.5$, 0.7, and 0.9; $\beta = 5$, 6, 7, 8, 9, and 10. For the calculations of the yields, we assumed that the cluster

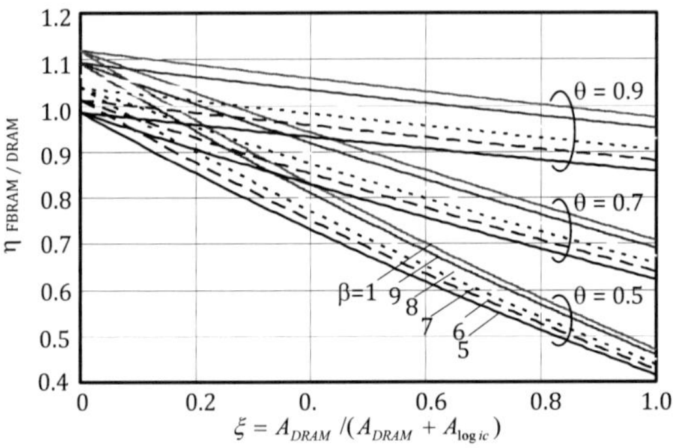

Figure 7.1 The chip cost ratio $\eta_{\text{FBRAM/DRAM}}$ of the LSI with the FBRAM embedded to that with the DRAM embedded having the same memory bits that is calculated for $\lambda_{\text{FBRAM}} = 0.05$, $\lambda_{\text{DRAM}} = 0.2$, and $\mu_{\text{bulk silicon}} = 0.03$. The macro size ratio θ of the FBRAM to the DRAM is changed as 0.5, 0.7, and 0.9. The cost of the SOI wafer is changed from 5 to 10 times that of the bulk silicon wafer ($\beta = 5 \sim 10$).

parameter $\alpha = 2$ and that the defect density $D_0 = 4000$ faults/m^2.[7] The total LSI chip area with the 1T-1C DRAM embedded is assumed to be 100 mm^2. Though an SOI wafer is much more expensive than a bulk silicon wafer, the chip of the LSI on SOI with the FBRAM embedded can be less expensive than the chip of the LSI on bulk silicon with the 1T-1C DRAM embedded due to the replacement of the extra process for making DRAM capacitor whose cost will be estimated roughly as high as 20% of the standard CMOS process cost to that for making FBC whose cost will be as low as 5% of the standard CMOS process cost. The chip gross per wafer increases and the memory yield enhances both by replacing the DRAM by the FBRAM whose area is smaller than the DRAM, contributing to further reduction of the chip cost. Since both effects increase as the memory area increases in an LSI, the larger cost reduction is achieved by replacing the DRAM by the FBRAM as the memory area increases in an LSI. An important conclusion from the above simple estimation teaches that the expensive SOI wafers don't necessarily lead to an expensive LSI if the wafer cost can be reduced to as low as 10 times the bulk silicon one or lower, because the smaller FBRAM can be implemented into the LSI with minimal process addition on the contrary to the DRAM. It is worth noting that the cost of the stand-alone FBCRAM on SOI is most effectively reduced from that of the 1T-1C DRAM as is shown at $\xi = 1.0$ in Fig. 7.1 corresponding to the cost reduction of the stand-alone memory.

7.2 DESIGN STRATEGY OF FBRAM

As was understood in the analyses throughout the previous chapters, FBC has the following advantages over the conventional 1T-1C DRAM cell: (1) a quasi non-destructive read-out cell without a necessity of a full restore in a read and a refresh cycle[8,9] and with high immunity against inter BL coupling noises when designed in the quasi-shielded open BL architecture,[10] (2) a gain cell with large signal even when scaled down,[11,12] (3) a small cell with $4F^2$ footprint ideally,[8,9] (4) a simple cell to be fabricated by a standard CMOS process plus a minor addition if any,[13] (5) a scalable cell with its signal-to-noise ratio (SNR) kept constant throughout generations.[14] On the other hand, it has several drawbacks: (6) a cell with shorter retention time than the 1T-1C DRAM cell,[15] (7) a cell subjected to a variety of disturbs such as BL disturb and word line (WL) disturb.[8,9,16]

With the above pros and cons of the FBRAM in mind, the following strategy can be laid out for designing memories by using FBC:

a. The twin cell or the two-cell-per-bit (TCPB) scheme will be applied to high-performance embedded memories. A relatively short retention time can afford to keep decent chip characteristics, since the high-performance logic part usually consumes a large stand-by current due to the low threshold voltages for MOSFETs it is designed by. And the short cycle time for refresh in the twin cell scheme can reduce the refresh busy rate down to a usable level even with a relatively short retention time spec as is calculated by Eq. 4.1. The cell size of $10F^2 \sim 12F^2$ ($2 \times 5F^2 \sim 2 \times 6F^2$) might be still cost effective over embedded DRAM and static random access memory (SRAM).

b. The single cell or the one-cell-per-bit (OCPB) scheme will be applied to low-power embedded memories with medium performance and medium density. The random cycle time in the range of tens of nanoseconds which will be achieved by the single cell scheme can satisfy the specifications of some memories in this category. However, data retention power and refresh busy rate must be a few milliamps per gigabit or lower and a few % at the largest, respectively. The retention time of FBC, therefore, must be improved about two orders of magnitudes or more from the status quo unless some FBC-specific refresh method is realized that improves the refresh efficiency drastically. The requirement for $6F^2 \sim 8F^2$ cell size can be satisfied by adopting the single cell scheme.

c. Single cell or OCPB scheme will be applied to stand-alone memories. This is especially an attractive application for the FBRAM, because the huge market is expected to be acquired that the conventional 1T-1C DRAM now enjoys. The cell size constraint is the most severe in this application, for $4F^2$ size will be required. Not only the planar type of cell but also some 3D cells such as FinFET or surrounding gate transistor (SGT) type cell must be taken into consideration in reducing cell size for this application.[c] The retention time specifications will be very severe for stand-alone memories. Data retention current of a few milliamps or less than one milliamp must be achieved per gigabit especially for the chips in low-power version. The retention time must be improved by two or three orders of magnitude from the value achieved by state-of-the-art FBC unless some FBC-specific refresh method is realized that improves the refresh efficiency drastically.

In this chapter, let us discuss the FBC's electrical performance as well as the macro size estimation in the twin cell scheme for the application (a), followed by the FBC's electrical performance in the

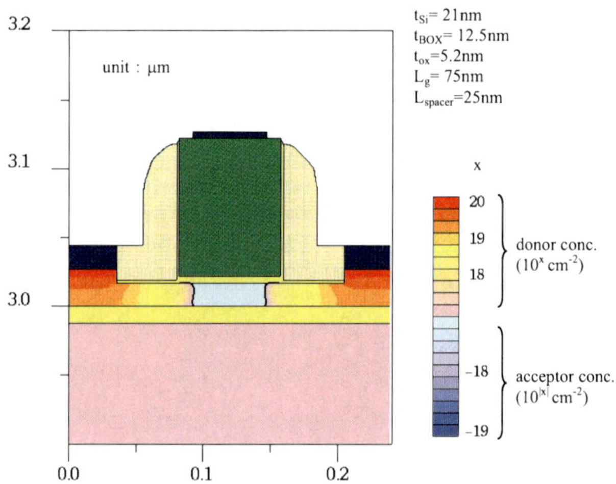

Figure 7.2 The cross-sectional view of the FBC which is simulated to estimate its access and random cycle time in the twin cell scheme. The impurity concentration profile is obtained as a result of the TCAD process simulation.

The retention time improvement for low-power applications are one big challenge yet to be conquered. In the last section in this chapter, a novel refresh method named autonomous refresh will be explained that makes use of the current anomaly of impact ionization and will increase the refresh efficiency drastically when applied successfully. This technology is still under investigation. However, if it is developed successfully, the retention time specifications for FBC in the low-power applications can be relaxed so much that the applications (b) and (c) come within the range of reality even with the retention time of state-of-the-art FBC.

[c] For the 3D cells, refer to Chapter 8.

7.3 TWIN-CELL-BASED HIGH-PERFORMANCE EMBEDDED MEMORY

In section 6.2, we considered the sense amp. circuit (S/A) that will be used in the FBRAM adopting the twin cell scheme for fast access and random cycle time. In this section, we estimate the random cycle time of the twin cell scheme FBRAM by performing circuit simulations combined with Monte Carlo simulations as a function of its macro size.

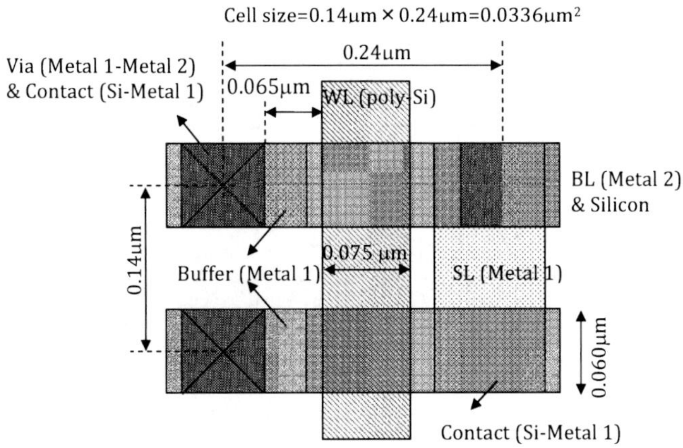

Figure 7.3 The plan view of the FBC which is simulated to estimate its access and random cycle time in the twin cell scheme. The macro sizes of embedded FBRAMs are also estimated on the basis of this cell layout.

As examples of applications, the secondary (L2) or the third-level (L3) caches in high-performance microprocessors are taken into account. The target value for the random cycle time t_{RC} is 3.5 or faster. The gate length of the cell is assumed to be L_g = 75 nm and the cell size is 0.0336 μm². The cross-sectional view obtained by the TCAD process simulation and the plan view of the cell is shown in Figs. 7.2 and 7.3, respectively. The write and read characteristics of this cell that is obtained by 2D TCAD device simulation are shown in Fig. 7.4. The WL voltage at write V_{WLHR}, the WL voltage at read V_{WLHR}, the WL voltage at hold V_{WLL}, the BL voltage at write "1" V_{BLH}, the BL voltage at write "0", V_{BLL} and the plate voltage V_{PL} are V_{WLHW} = 1.0 V, V_{WLHR} = 1.0 V, V_{WLL} = −1.5 V, V_{BLH} = 1.7 V, and V_{PL} = −2.4 V, respectively. The median of the threshold voltage difference between the "1" cell and the "0" cell is obtained to be ΔV_{th} = 0.377 V. The median of the cell read current of the data "1" cell and the data "0" cell are 92 μA/μm (I_1 = 5.52 μA/cell for the actual channel width W = 0.06 μm) and 25 μA/μm (I_0 = 1.5 μA/cell for the actual channel width W = 0.06 μm). The cell current which flows during the "1" write and the "0" write are 139 μA/μm (I_1 = 8.34 μA/

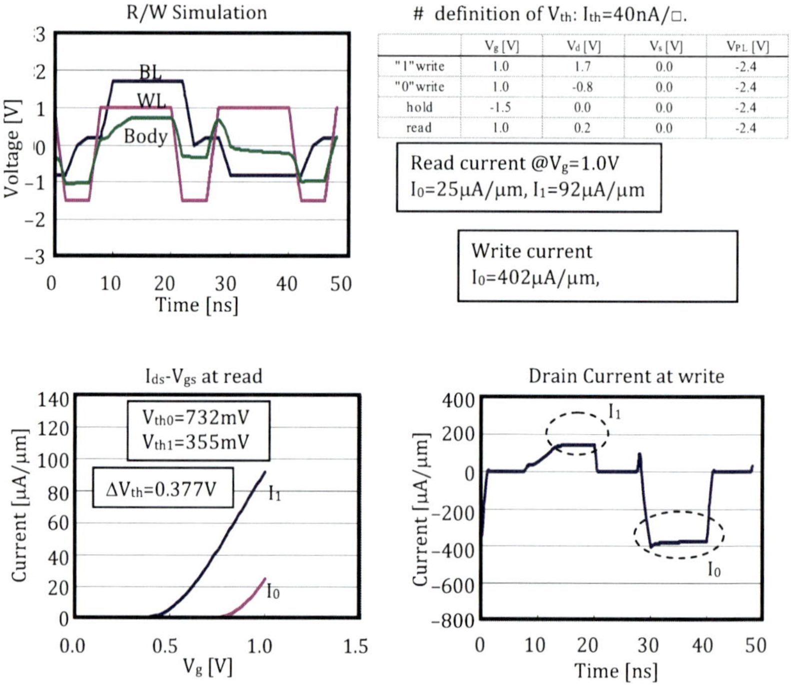

	V_g [V]	V_d [V]	V_s [V]	V_{PL} [V]
"1"write	1.0	1.7	0.0	-2.4
"0"write	1.0	-0.8	0.0	-2.4
hold	-1.5	0.0	0.0	-2.4
read	1.0	0.2	0.0	-2.4

Figure 7.4 The write and read characteristics of the cell which is simulated for a twin-cell-based high performance embedded FBRAM.

cell for the actual channel width $W = 0.06$ μm) and 402 μA/μm ($I_0 = 24.12$ μA/cell for the actual channel width $W = 0.06$ μm). The BL and the WL capacitances are calculated by performing 3D TCAD device simulations

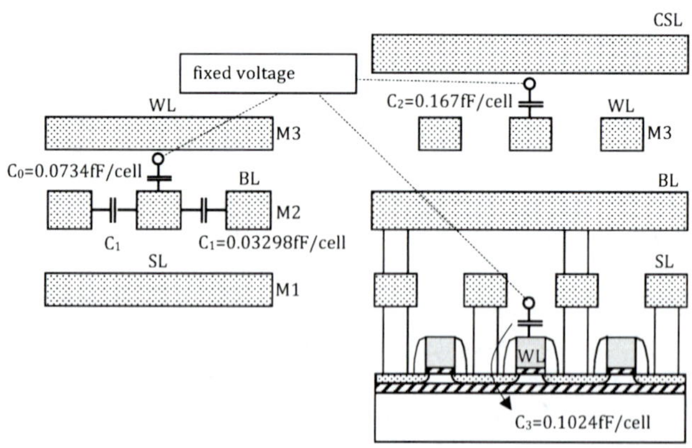

Figure 7.5 The bit line and the word line capacitances calculated by 3D TCAD device simulation for the cell shown in Fig. 7.3. The capacitance C3 includes the gate capacitance of ON state for the "1" data cell.

as are shown in Fig. 7.5. These capacitances were used for the read and the write circuit simulations for predicting the access time and the random cycle time of the FBRAM.

The 144 Kbit and 288 Kbit memory sub-arrays which will be used for the performance estimations of embedded memory macros are shown in Fig. 7.6. We studied the longer WL case of 1152 columns/WL (Type I) and the shorter WL case of 576 columns/WL with the number of cells per BL kept constant (256 cells/BL). The latter case has further divided into two cases: one is the case where 72 bits are accessible among the 576 columns (Type II) and the other is the case where 144 bitsare accessible among the 576 columns (Type III). The column, here, stands for a pair of BLs that is connected to a single S/A. The structures of these three embedded FBRAM macros are illustrated in Figs. 7.7–7.9. The S/A which will be used in the performance simulations is shown in Fig. 7.10. This S/A is basically the same as what was discussed in Fig. 6.4 in section 6.2, though some details of the S/A in Fig. 7.10 are different from Fig. 6.4. At the both ends of the S/A, n-type MOSFETs are added which are to short circuit the BLs to V_{SL} whose gates are controlled by signal EQL_L or EQL_R. In Fig. 7.10, furthermore, a p-type MOSFET pair and an n-type MOSFET pair are implemented per a BL pair controlled by FB_L or FB_R and BFB_L or BFB_R outside the n-type transfer gate MOSFETs gated by FIT_L or FIT_R, respectively, which are to be used when latched data are written or

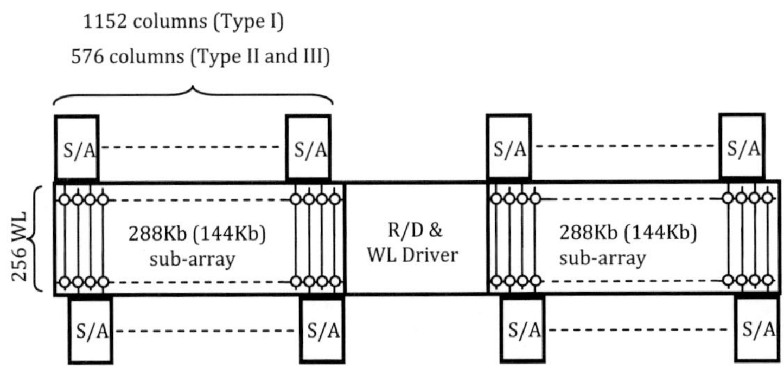

Figure 7.6 The sub-array architecture of the twin-cell-based high-performance embedded FBRAM.

restored to the cells, while the CMOS transfer gates are implemented in Fig. 6.4 for writing and restoring the latched data to c-ells instead. Another difference is that signal SAP in Fig. 6.4 is changed to a power line V_{BLH} which is the BL voltage for writing "1" in Fig. 7.10. The reason of the last change will be explained later in interpreting the simulation waveforms.

The simulation results of this S/A with WLs and BLs combined whose capacitances are assumed to be what were calculated in Fig. 7.5 are shown in Fig. 7.11. The median of the threshold voltage difference between the data "1" and the data "0" is assumed to be 0.4 V. As is shown in Fig. 7.11, signals FIT_L and FIT_R are at 0 V prior to the signal sensing operation to isolate the BLs from the sense nodes and bit lines (BL_0 and $B0_1$) and sense nodes (SN_0 and SN_1) are pre-charged at 0.8 V and 2.5 V, respectively. In actual conditions where fast sensing cycles are repeated, the sense nodes are not necessarily pre-charged to V_{BLH} = 2.5 V but to some voltage between V_{BLH} and $V_{BLH} - V_{th}$ where V_{th} is the absolute value of the threshold voltage of the p-type MOSFET constituting the p-type MOSFET flip-flop. Here, we assumed that they are pre-charged to V_{BLH} to simulate the slowest signal development condition. In this figure, BL_1 and SN_1 stand for the BL and the sense node which are connected to the "1" cell, respectively. Similarly, BL_0 and SN_0 stand for the BL and the sense node which are connected to the "0" cell, respectively. In this simulation, the source line voltage V_{SL} is fixed at 0.8 V. Therefore, if V_{SL} is taken as ground (0 V), the BL pre-charge level is 0 V and the sense node pre-charge level is 1.7 V, being regarded as the source voltage V_s and the drain voltage V_d at "1" write, respectively, i.e., V_{SL} and V_{BLH}, respectively.

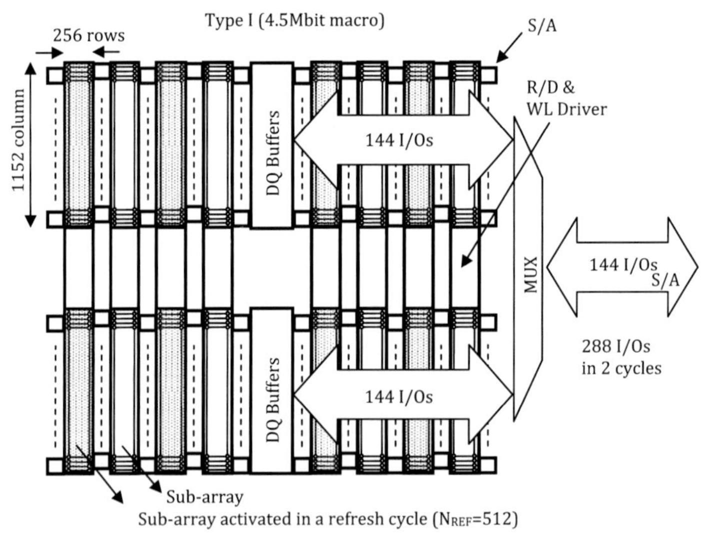

Figure 7.7 Type I structure of a 4.5 Mbit macro that is studied to estimate the performance of the twin-cell-based high-performance embedded FBRAM. Each WL runs over 1152 columns in this structure. The refresh cycle N_{REF} is 512.

The WL voltage prior to the signal sensing operation is –0.7 V, though it is not shown. This level is regarded as –1.5 V when V_{SL} is shifted to 0 V, being the same value as V_{WLL} which is shown in Fig. 7.4 as V_g at hold. The selected WL is assumed to be already raised in this simulation. When FIT and SHORT are raised high and EQL is lowered at the same time, the voltages of BLs and that of the

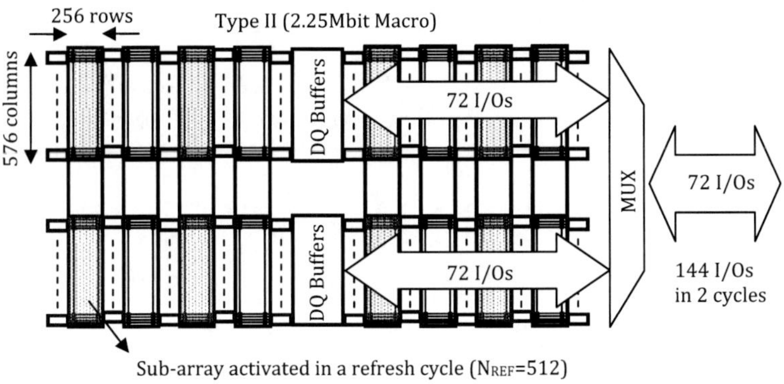

Figure 7.8 Type II structure of a 2.25 Mbit macro that is studied to estimate the performance of the twin-cell-based high-performance embedded FBRAM. Each WL runs over 576 columns in this structure. The refresh cycle N_{REF} is 512.

sense nodes are approaching and the signal is being developed gradually. According to the results, the difference in voltage between a pair of sense nodes SN_0 and SN_1 reaches 20 ~ 30 mV in about 600 ps after the pair of sense nodes are isolated, the BLs are got floating and the BLs and the sense nodes are connected. This time is defined as signal development time (SDT). The reason why the activation line of the p-type MOSFET flip-flop is connected to the constant voltage V_{BLH} instead of controlled by signal SAP as was in Fig. 6.4 is that the transition time for SAP from V_{BLL} to V_{BLH} is too long to satisfy the L2 or L3 cache random cycle specifications because the capacitance of SAP is very large when it is connected to 576 S/As (Type I) or 288 S/As (Type II and Type). In the simulation, signal SHORT isolates the sense node pair immediately after the sensing start on the contrary to the waveforms for Fig. 6.4. It is true that delaying the opening by raising SHORT makes the signal development faster than without delay as in the case of Fig. 6.5. In the particular conditions of the fluctuations in the cell and the S/A transistor characteristics, however, the optimum timing of raising SHORT high which makes the signal development the fastest was found to be at less

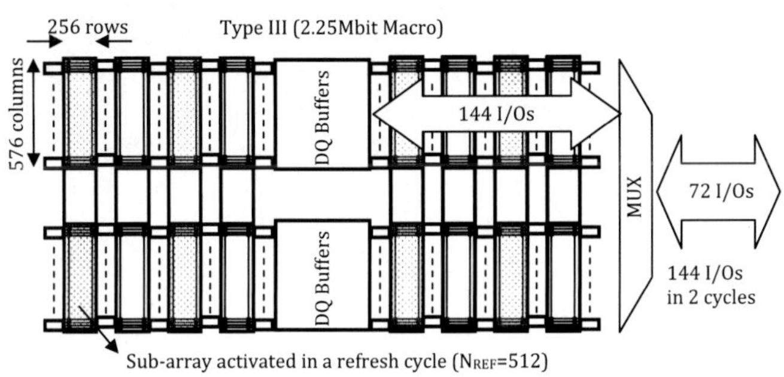

Figure 7.9 Type III structure of a 2.25 Mbit macro that is studied to estimate the performance of the twin-cell-based high-performance embedded FBRAM. Each WL runs over 576 columns in this structure as in Type II. The refresh cycle N_{REF} is 512.

than 100 ps. We, therefore, neglected the infinitesimal delay which will be very difficult to be controlled by peripheral circuits. It should be noted that the simulated waveforms in Fig. 7.11 indicate the signal development situations for the averaged cells and the averaged MOSFETs in the S/A.

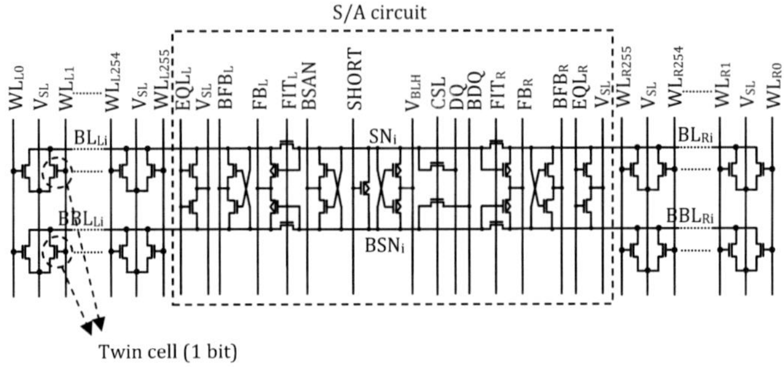

Figure 7.10 The sense amp. circuit (S/A) of *i*-th column for twin-cell-based high-performance embedded FBRAM.

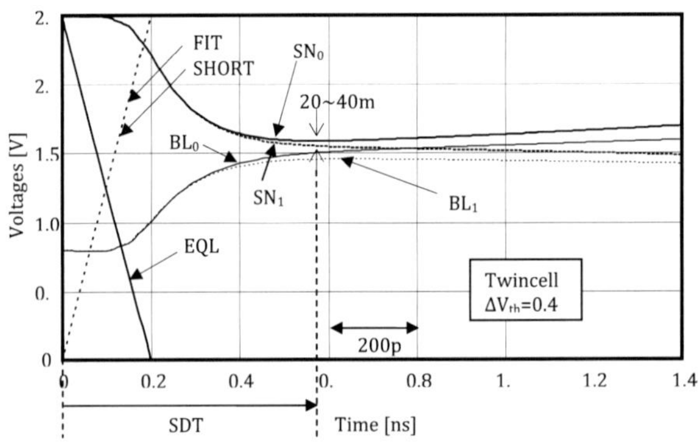

Figure 7.11 The simulated waveforms of the S/A in Fig. 7.10. The BL and WL capacitances are used which are simulated by 3D TCAD device simulations as are shown in Fig. 7.5. BL_1 (BL_0) and SN_1 (SN_0) are the bit line and the sense node which are connected to the "1" ("0") cell, respectively.

Figure 7.12 shows the results of Monte Carlo simulations for signal development of Fig. 7.11. The threshold voltages of the cells are assumed to fluctuate according to normal distributions. The difference in the averaged threshold voltages between the "0" cell and the "1" cell are assumed to be ΔV_{th} = 0.4 V. The standard deviations of the threshold voltage distributions of the data "0" cell and that of the data "1" cell are assumed to be the same as $\sigma_{V_{th}}$ = 35 mV. The threshold voltages of MOSFETs in the S/A are also assumed to fluctuate according to normal distributions. The difference in the threshold voltages of pair transistors in the S/A are assumed to be zero and the standard deviations are assumed to be 5 mV. The algorithm by which we generate random numbers to perform the Monte Carlo simulations is illustrated in Fig. 7.13. At first we pick up a S/A and the threshold voltages for the pair transistors of the p-type MOSFET flip-flop and for the pair transistors of the n-type MOSFET transfer gates are chosen randomly from the normal

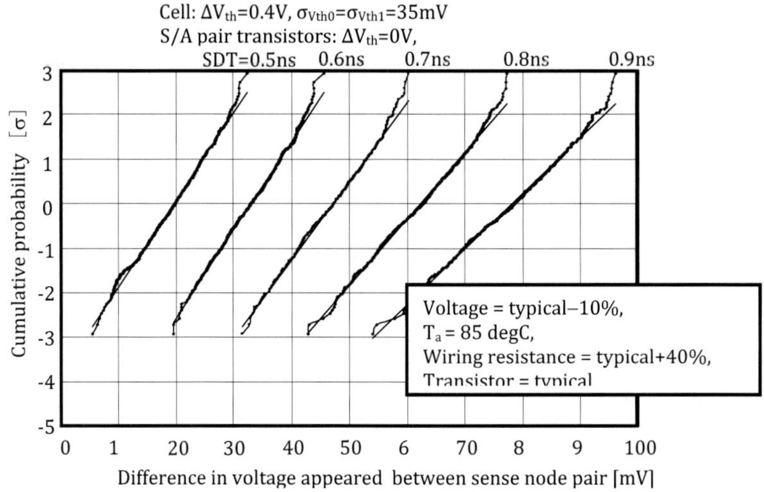

Figure 7.12 The results of Monte Carlo simulations for signal development in the twin cell scheme for high-performance embedded memories.

distribution with the same average and the 5 mV standard deviation. Next, we assign a twin cell to the S/A, the threshold voltages of the "1" cell and the "0" cell being chosen randomly from the normal distribution with $\Delta V_{th} = 0.4$ V and $\sigma_{V_{th}} = 35$ mV. By using the S/A and the twin cell, we perform a SPICE simulation to obtain waveforms similar to those in Fig. 7.11. Then we repeat the twin cell assignment *M* times and perform M simulations by using the same S/A. Next, we pick up another S/A and the pair transistors threshold

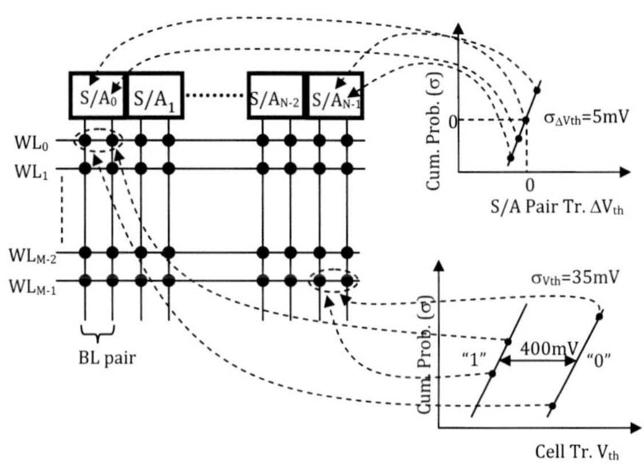

Figure 7.13 The algorithm of the Monte Carlo simulations for signal development in the twin cell scheme for high-performance embedded memories.

voltages are chosen randomly from the normal distribution mentioned above. And we chose a twin cell randomly from the normal distribution to the S/A in the same way as for the first S/A and perform SPICE simulation to obtain the waveforms. We pick up N S/As in the same way. This algorithm allows us to perform totally $M \times N$ SPICE simulations and to have $M \times N$ sets of waveforms, each set is similar to those in Fig. 7.11. The results in Fig. 7.12 is the summary of the sets of waveforms where the differences in voltage appeared between sense node pair are arranged in ascending order and represented in standard deviation of the cumulative probability for SDT = 0.5 nS, 0.6 ns, 0.7 ns, 0.8 ns, and 0.9 ns. Since it is impossible to perform the simulations for M = 256 and N = 1152 corresponding to the actual sub-array of Fig. 7.6 due to the constraint of job execution time, we performed $M \times N$ = 1000 times in which the ratio M/N is kept the same as 256/1152. The operating voltages were lowered 10% down the typical values and the wiring resistances were increased 40% up the typical values. The threshold voltages of the peripheral transistors were typical and the temperature was 85°C. Figure 7.12 teaches that the SDT which generates 20 ~ 40 mV for the worst cell with -4.5σ is found to be between 0.7 ns and 0.8 ns. To make it illustrated more clearly, Fig. 7.12 is redrawn to the cumulative probability as a function of the SDT for three differences

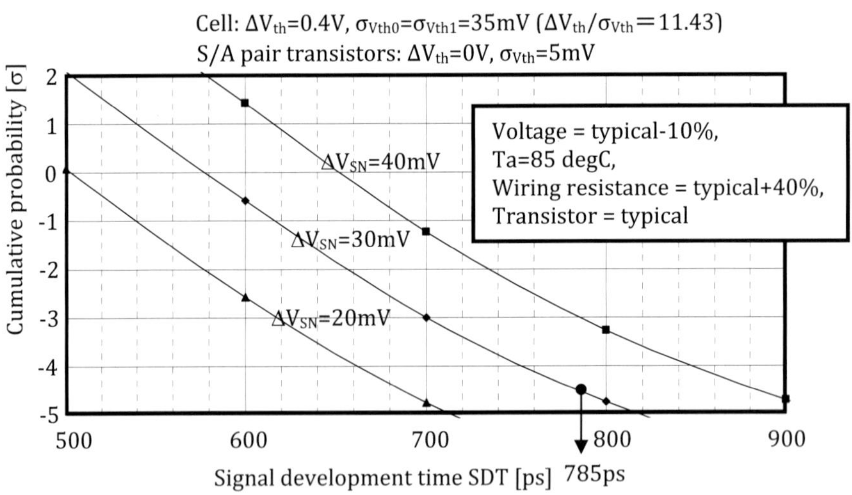

Figure 7.14 The results of Monte Carlo simulations for signal development in the twin cell scheme for high-performance embedded memories. The parameter and the variable are interchanged from Fig. 7.12, though the data used are the same. DV_{SN} is the difference in voltage appeared on a pair of sense nodes.

in voltage appeared on the pair of sense nodes of ΔV_{SN} = 40 mV, 30 mV, and 20 mV to obtain Fig. 7.14. This representation shows that the SDT which is necessary to develop 30 mV signal on the sense node pair for the worst cell of -4.5σ is 785 ps.

Generally, 30 mV can be regarded as enough signal to latch the data correctly, because it might be larger than the threshold voltage off set in the n-type MOSFET pair of the flip-flop which is to amplify and latch the signal. Therefore, the n-type flip-flop can be activated to amplify and latch the signal in 785 ps at the earliest without bit failures due to insufficient signal development in the sub-array.

Figure 7.15 shows the total cycle simulation results for the Type II macro. Although the n-type MOSFET dynamic flip-flop is activated by lowering /SAN in 350 ps after the WL rising (SDT = 350 ps) for reasons of expediency, the difference 435 ps has only to be added to the random access time t_{RAC} and the random read cycle time t_{RC} that are obtained in the simulation results shown in Fig. 7.15 to estimate the minimum access time and the read cycle time corresponding to SDT = 785 ps shown in Fig. 7.14. As a result, it is shown that the random access time t_{RAC} = 2.0 ns and the random read cycle time t_{RC} = 3.5 ns are guaranteed as the worst conditions. Since the parameters of peripheral transistors were assumed to be typical values, however, the random access time and the random read cycle

Table 7.1 Macro size estimation for Type II b in Fig. 7. 16.

X direction	Unit size [μm]	# in an array	# in a macro	Total size [μm]	Ratio
Cell	0.28	576	4	645.12	67.7%
Redundancy cell	0.28	8	4	8.92	0.9%
Stitch	3.964	4	4	63.424	6.7%
Array edge	2.236	2	4	17.888	1.9%
R/D + WLD × 2	108.5	1	2	217.0	22.8%
Total				952.325	
Cell	0.24	256	8	491.52	46.7%
Sense amp.	47.0	1	10	470.0	44.7%
Array edge	1.33	2	4	10.64	1.0%

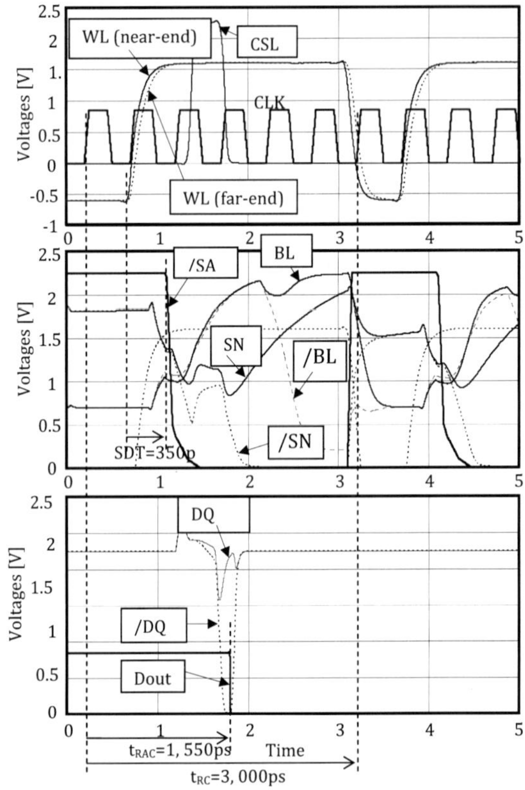

Figure 7.15 The simulation results of the total cycle of high-performance embedded FBRAM in 2 GHz clock for Type II b.

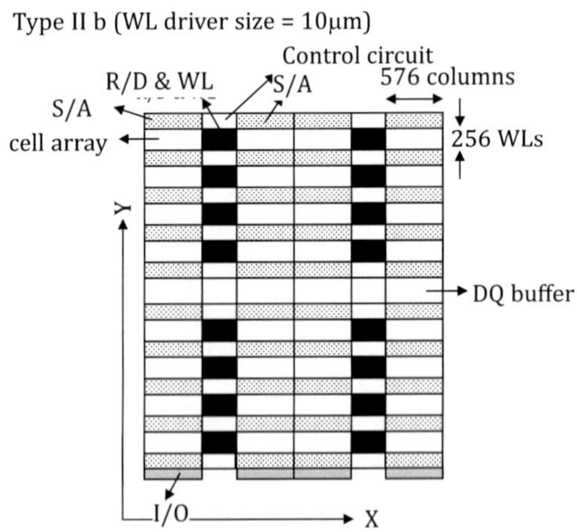

Figure 7.16 4.5 Mbit Type II b macro image whose size is estimated in Table 7.1.

time might be larger than the above values if the threshold voltages of the transistors shift to higher values than the typical ones in production. FBC is a quasi non-destructive read-out cell that does not necessary to be restored fully in read cycles. However, it is necessary for a several number of holes to be supplemented in each cycle to replenish holes lost by charge pumping phenomenon. In the above simulations, the cycle time is determined so as to wait for the BL to reach V_{BLH} = 2.25 V for the hole replenishment in prior to the BL pre-charge. Figures 7.11, 7.12, 7.14, and 7.15 are for the width of the WL driver 10 μm. This macro is named as Type IIb.

Figure 7.16 illustrates 4.5 Mbit Type IIb macro. I/O includes interface circuits and their controllers. Table 7.1 shows the detailed size of each component for the both directions. DQ buffers are the secondary sense amplifiers which transfer data amplified by S/As to I/O and the write circuits which input data from I/O to S/As. The chip size is estimated to be 1.002 mm^2 with the cell efficiency 31.6%. This macro size estimation doesn't take the redundancies and the pump circuits which generate high voltage V_{BLH} = 2.5 V from V_{CC} = 1.8 V and low voltage V_{WLL} = −0.7 V from ground into account. Figure 7.17 shows the relationship between the random read cycle and the 4.5 Mbit macro size including the redundancies and the pump circuits for Type I, Type II, and Type III. Type Ia, Type IIa, and Type IIIa are for the 4 μm narrow WL drivers, while Type Ib, Type IIb, and Type IIb are for the 10 μm wide WL drivers. As is shown in this graph, there is a trade-off between the macro size and the electrical performance. An appropriate array architecture, therefore, needs to be chosen in designing the FBRAM according to the application.

7.4 SINGLE-CELL-BASED HIGH-DENSITY MEMORY

The S/A of the single cell scheme was discussed in section 6.3 in detail. In this section, the signal development characteristics of the single cell scheme as well as the characteristics of writing data to the cell will be analyzed.

Figure 7.18 shows a part of the circuits in Fig. 6.7 that is related to the signal development operations. We simulate only two adjacent columns and

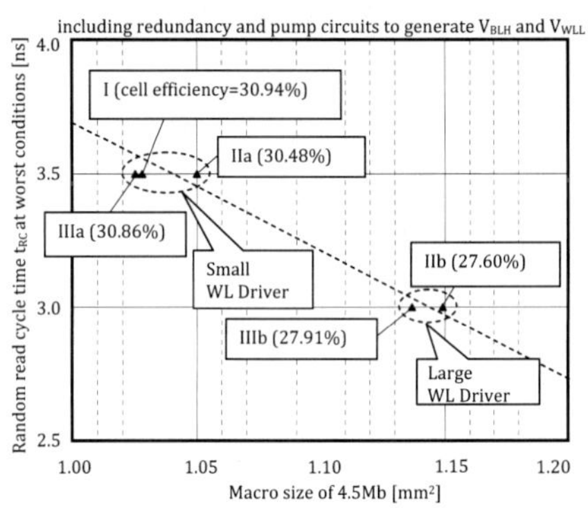

Figure 7.17 The relationship between the 4.5 Mbit macro sizes and their random read cycle time at the worst conditions.

consider the one-pair averaging of dummy cells. We performed Monte Carlo simulations where the threshold voltages of the MOSFTEs circled by dotted lines are fluctuate according to the normal distribution with the standard deviation 5 mV. The technology node we consider in this study is 45 nm with the gate length of the cell is $L_g = 75$ nm. The read currents of the memory cells circled by solid lines are assumed to fluctuate according to the normal distribution whose median and standard deviation are shown in Fig. 7.19. The SNR with respect to cell read current $(\mathrm{SNR})_{I\mathrm{cell}} = \Delta I_{\mathrm{cell}}/\sqrt{\sigma_{I_{\mathrm{cell}\,1}} \cdot \sigma_{I_{\mathrm{cell}\,0}}}$ is 21.2 at V_{BLR} = 0.45 V and V_{WLHR} = 1.15 V. The feature that $\sigma_{I_{\mathrm{cell}\,0}}$ is larger than $\sigma_{I_{\mathrm{cell}}}$ reflects the trend in the hardware measurements of the cell array at L_g = 150 nm. The BL and the WL capacitances are used that were calculated by 3D TCAD device simulations shown in Fig. 7.5.

Figure 7.20 shows the simulated waveforms of the signal development for the S/A combined with the cells in Fig. 7.18 for the typical characteristics of the MOSFETs and the cells.

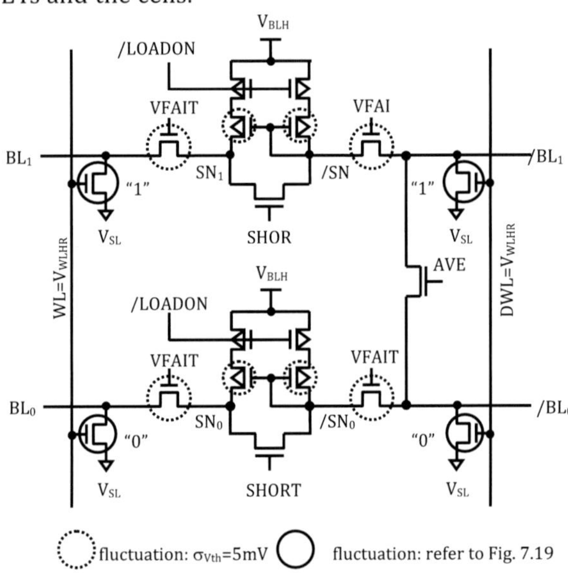

Figure 7.18 Sense amp. circuits and memory cells which are used for the Monte Carlo simulations to calculate the signal development time (SDT) of the single cell scheme.

The number of cells per BL is 1024. The S/A in Fig. 7.18 includes the n-type MOSFETs gated by signal SHORT that short-circuit pairs of sense nodes. The MOSFETs remain turned on for a while after the /LOADON lowering to develop signal. The reason is to keep the feedback gain low to avoid an amplification failure due to the sense amp. pair transistor threshold voltage mismatch and is the same as that was explained in section 6.2. The signal SHORT is lowered to isolate the pair of sense nodes for increasing the positive feedback gain after a signal is appeared on the pair of sense nodes that is larger than the possible mismatch of the pair transistors in the S/A. Figure 7.20 indicates that the signal of about 100 mV is developed on the pair of sense nodes in about 5 ns after the pair of sense nodes and the pair of BLs are equalized by raising signal VFAIT.

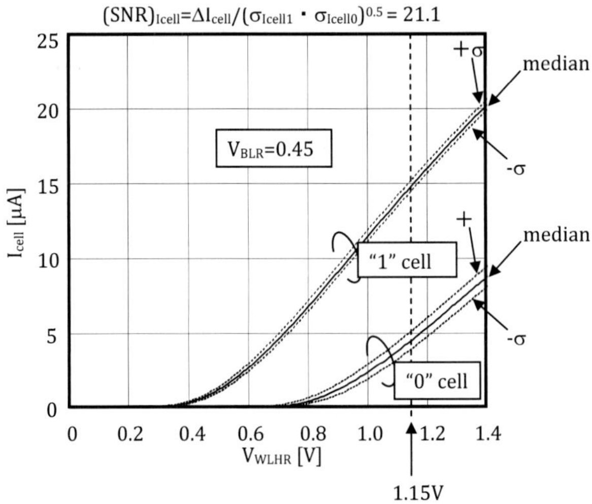

Figure 7.19 The cell read current characteristics used for the Monte Carlo circuit simulation. It is assumed to fluctuate according to the normal distribution whose median and standard deviation are shown in this graph. The signal to nose ratio with respect to cell read current is 21.2.

The results of the Monte Carlo simulation are shown in Fig. 7.21. The results show that the differences in voltage appeared on a pair of sense nodes for both the data "1" and the data "0" are 30 mV at SDT = 5.0 ns for the worst cell of -4.5σ in the case where 1024 cells are connected to a BL. This difference 30 mV can be regarded as the minimum voltage the S/A can sense the data correctly even if the characteristics of the pair transistors fluctuate.

Next, let us evaluate the write characteristics of FBC. Figure 7.22 is the results of the time which is necessary to write the data "1" to the data "0" measured for 64 kbit cells at 25°C with the "0" write BL level V_{BLL} changed as a parameter. The graph is the cumulative probability in standard deviation for the 64 kbit cells arranged in ascending order in the "0" write time measurements. The write times were measured as pass/fail judgment in a memory tester on the data which were written and sensed by S/A. The minimum write time which gives rise to the pass judgment is defined as the write time of the cell. Even the worst cell can be written to "0" in 3 ~ 4 ns after the start of writing provided the BL low level to write "0" V_{BLL} is −2.1V or lower. Figure 7.23 is the results of the time which is necessary to write the data "0" to the data "1" measured for 64 kbit cells at 25°C with the "1" write BL level V_{BLH} changed as a parameter. Even the worst cell can be written to "1" in 1ns after the start of writing provided the BL high level to write "1" V_{BLH} is 2.6 V or higher. Of course, the write time can be made shorter by increasing the absolute value of the BL write voltages.

As were simulated and measured here, the cell array read and write operations are found to complete within 5 ns by properly setting the BL voltages during write. Since the performance is comparable with that of the conventional 1T-1C DRAM, it is possible that the FBRAM can satisfy the timing specifications of the standard DRAM. The cell size is also equivalent to that of the 1T-1C DRAM or smaller. FBRAM, therefore, can be a candidate of the future 1T-1C DRAM or of the future high-density oriented embedded memories, provided that the retention time is long enough or some FBC-specific refresh which increases the refresh efficiency drastically is realized.

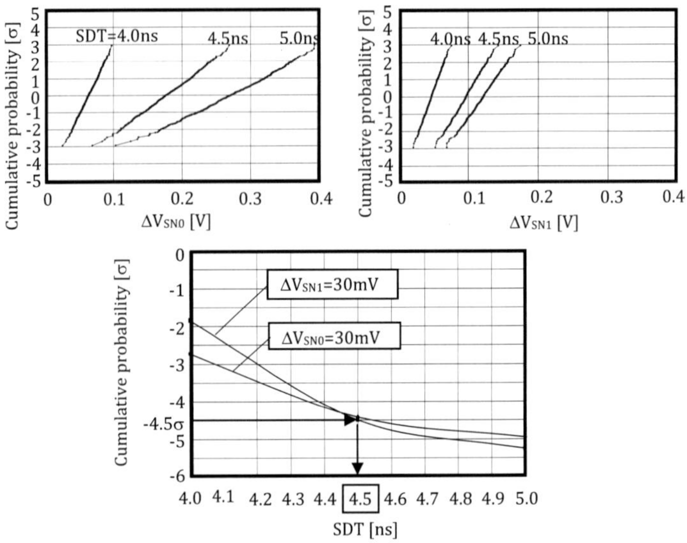

Figure 7.21 The results of Monte Carlo simulations for signal development in the single cell scheme for high-density memories. The number of cells connected to a BL is 1024.

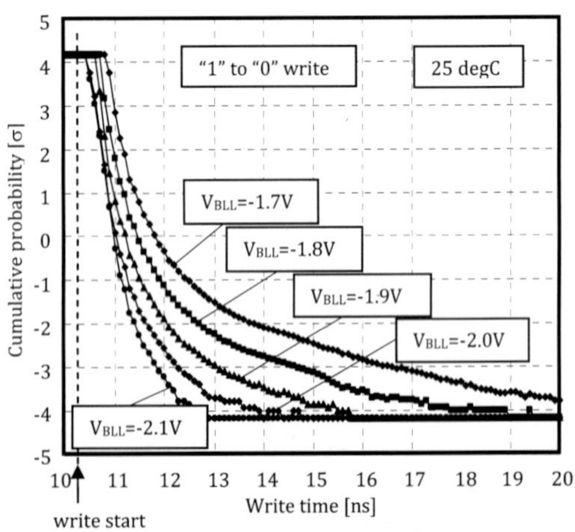

Figure 7.22 The results of "0" write speed measured for 64 kbit.

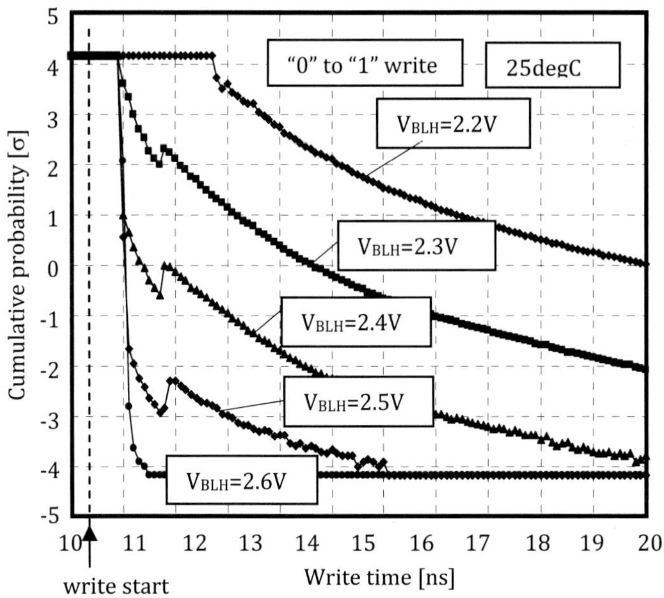

Figure 7.23 The results of "1" write speed measured for 64 kbit.

7.5 AUTONOMOUS REFRESH[17,18]

7.5.1 Concept and Impact

As was discussed in section 4.3, the retention time of FBC is shorter than that of the conventional 1T-1C DRAM cell by 2 ~ 3 orders of magnitudes, indicating that the leakage current of the floating body of FBC is at the same level as that of the DRAM capacitor, because the body capacitance even at L_g = 150 nm is about 0.1~ 0.2 fF, being smaller than that of the 1T-1C DRAM (20 ~ 30 fF) by more than 2 orders of magnitude. Therefore, the present-day FBRAM cannot be applied to a stand-alone memory, i.e. it is difficult for the present-day FBRAM to replace the conventional 1T-1C DRAM which must satisfy the retention time specifications of 64 ms.[7] The retention time of the worst cell (−4.5 ~ −5.0σ) in the memory chip must be about 100 ms or longer at 85°C with some margin included for the specifications of 64 ms to be satisfied as a product. If it is not the case, the data retention current and the refresh busy rate will increase up to an unusable level. Although the improvement of the retention time of FBC by more than 2 orders of magnitude seems difficult to be achieved, the works for enhancing the retention time of FBC is certainly necessary and important. However, another approach to decrease the data retention power and the refresh busy rate is also necessary to be investigated in parallel with the work of improving the retention time. In this section, we discuss a novel refresh method of FBC that makes it possible for many cells along a BL to be refreshed simultaneously, leading to a drastic increase of the refresh efficiency of FBRAM.

This refresh is named autonomous refresh[17,18] and is specific to FBC.[d] The conventional 1T-1C DRAM cannot perform the autonomous refresh. In the conventional refresh for the 1T-1C DRAM, the data stored in a cell is amplified by a S/A and restored to the cell prior to being degraded so much as not to be amplified by the S/A correctly. Therefore, the conventional refresh is performed in a row-by-row manner, consuming a large power in charging and discharging large BL capacitances in each cycle. In the autonomous refresh, on the other hand, the FBC's feature is exploited that it can recover from its degraded data states to its fresh binary states by itself without being amplified by a S/A. It was originally proposed by P. Fazan *et al.* in 2003 and called block refresh.[19] They showed that this feature of FBC can be used to refresh all cells in an array simultaneously to decrease the refresh power and the refresh busy rate. In this section, let us discuss the physical mechanism lying behind the autonomous refresh or the block refresh and the necessary conditions for realizing the autonomous refresh in a high-density memory. And, as an important application of the autonomous

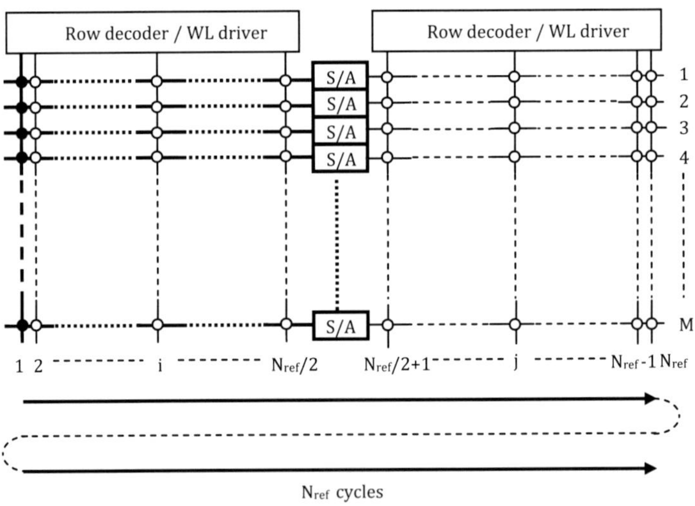

Figure 7.24 The conventional refresh method for the 1T-1C DRAM or the FBRAM. One WL is activated with *M* BLs charged in a refresh cycle to refresh *M* cells along the WL by using M S/As, requiring N_{ref} cycles to refresh all cells in the chip.

refresh, we will explain a theoretical possibility that FBC can be an SRAM cell which retains data statically without any periodical operation of BL and

[d] It is interesting that the historical 4T DRAM cell in Fig. 1.2a can also perform the autonomous refresh as was touched on briefly in section 1.1.

WL for retaining data when a double gate transistor is supposed to be used as FBC.

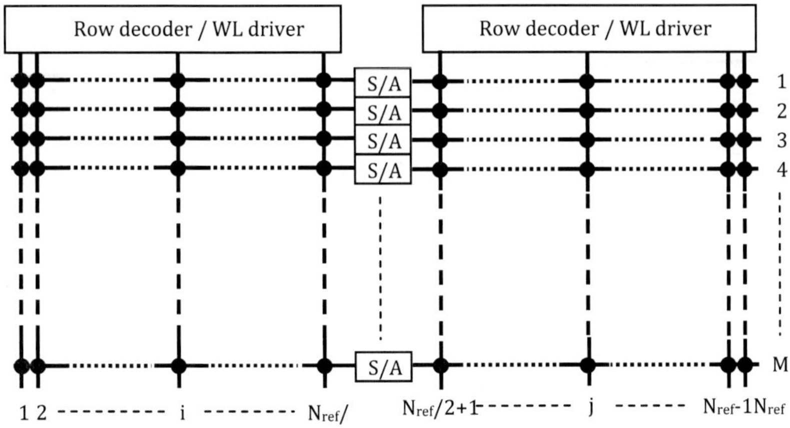

Figure 7.25 The autonomous refresh method for the FBRAM. N_{ref} WLs are activated with 2 M BLs charged in a refresh cycle to refresh all cells without S/A operations.

Now, let us calculate the refresh busy rate and the data retention current when the conventional refresh operation is applied to the FBRAM. Figure 7.24 illustrates the method of the conventional refresh applied to the FBRAM where a WL is activated and the cells along the WL are amplified by the S/As and restored to the cells in a refresh cycle and all cells are refreshed in a row-by-row manner after N_{ref} refresh cycles. If the shortest retention time among those of all cells in the memory is denoted by t_{ref}, all cells can retain their data properly when the refresh operation is performed by activating WLs consecutively in the period t_{ref}/N_{ref} or shorter. When the minimum cycle time which is necessary to perform the conventional refresh operation is denoted by $t_{RC\_conv.}$, the busy rate of the conventional refresh $RBR_{conv.}$ which is defined by the ratio of the refresh cycle time divided by the refresh period is expressed by

$$RBR_{conv.} = \frac{t_{RC\_conv.}}{t_{ref}/N_{ref}} = \frac{t_{RC\_conv.} \times N_{ref}}{t_{ref}}. \quad (7.6)$$

If we denote the capacitances of a WL and a BL as C_{WL} and C_{BL}, respectively, the DC current which the FBC consumes during the refresh as $i_{DC\_conv.}$, the duty ratio of the time for data restore divided by the refresh cycle time $t_{RC\_conv.}$ as λ and the peripheral circuit capacitance which is necessary to be charged and discharged in one conventional refresh cycle as $C_{peri\_conv.}$, the averaged current which is consumed in the conventional refresh for the FBRAM $I_{ref\_conv.}$ will be calculated as

$$
\begin{aligned}
I_{\text{ref\_conv.}} &= I_{\text{BL\_conv.}} + I_{\text{WL\_conv.}} + I_{\text{DC\_conv.}} + I_{\text{peri\_conv.}} \\
&= \frac{M \times C_{\text{BL}} \times N_{\text{ref}} \times \Delta V_{\text{BL}}}{t_{\text{ref}}} + \frac{C_{\text{WL}} \times N_{\text{ref}} \times \Delta V_{\text{WL}}}{t_{\text{ref}}} \\
&\quad + \frac{\lambda \times t_{\text{RC\_conv.}} \times M \times N_{\text{ref}}}{t_{\text{ref}}} i_{\text{DC\_conv.}} \\
&\quad + \frac{C_{\text{peri\_conv.}} \times N_{\text{ref}} \times \Delta V_{\text{peri\_conv.}}}{t_{\text{ref}}},
\end{aligned}
$$

where M is the number of BLs per WL, ΔV_{BL} and ΔV_{WL} are the voltage amplitudes of BL and WL, respectively. The first term $I_{\text{BL\_conv.}}$ in Eq. 7.7 is the current which is consumed by charging and discharging BLs, the second term $I_{\text{WL\_conv.}}$ is the current which is consumed by charging and discharging WLs, the third term $I_{\text{DC\_conv.}}$ is the DC current which flows during restoring data to FBCs, and the fourth term $I_{\text{periL\_conv.}}$ is the current which is consumed by charging and discharging the capacitances of peripheral circuits in the refresh operation. All the four currents are those averaged over refresh cycles. And $\Delta V_{\text{peri\_conv.}}$ is the amplitude of voltage at which the peripheral circuits operate in the conventional refresh cycle. In the autonomous refresh, on the other hand, all cells can be refreshed in a single cycle at once as is shown in Fig. 7.25, because the cells can be refreshed without being read by the associated S/As.[e] If we denote the cycle time which is necessary in an autonomous refresh as $t_{\text{RC\_AR}}$, the refresh busy rate of the autonomous refresh RBR_{AR} is calculated as

$$
\text{RBR}_{\text{AR}} = \frac{t_{\text{RC\_AR}}}{t_{\text{ref}}}. \tag{7.8}
$$

And the averaged current in the autonomous refresh $I_{\text{ref\_AR}}$ will be expressed as

$$
\begin{aligned}
I_{\text{ref\_AR}} &= I_{\text{BL\_AR}} + I_{\text{WL\_AR}} + I_{\text{DC\_AR}} + I_{\text{peri\_AR}} \\
&= \frac{\xi \times M \times C_{\text{BL}} \times \Delta V_{\text{BL\_AR}}}{t_{\text{ref}}} + \frac{C_{\text{WL}} \times N_{\text{ref}} \times \Delta V_{\text{WL\_AR}}}{t_{\text{ref}}} \\
&\quad + \frac{\lambda_{\text{AR}} \times t_{\text{RC\_AR}} \times M \times N_{\text{ref}}}{t_{\text{ref}}} i_{\text{DC\_AR}} + \frac{C_{\text{peri\_AR}} \times \Delta V_{\text{peri\_AR}}}{t_{\text{ref}}},
\end{aligned} \tag{7.9}
$$

[e] When a current peak is too large to operate the autonomous refresh stably, the refresh can be applied to a sub-group of the total cell arrays at a time by separating the memory into N groups. The cell arrays in each group can be autonomous-refreshed at a different timing in the period t_{ref}. In this case, though the refresh busy rate will increase from the value in Eq. 7.8 to $N \times t_{\text{RC+AC}}/t_{\text{ref}}$, the data retention current given in Eq. 7.9 does not change.

where ΔV_{BL_AR} and Δ_{VL_AR} are the voltage amplitudes of BL and WL in the autonomous refresh cycle, respectively. And i_{DC_AR} is the DC current which flows in an FBC during the autonomous refresh, λ_{AR} is the duty ratio of the time of data restore divided by the refresh cycle time t_{RC_AR}, and C_{peri_AR} and ΔV_{peri_AR} are the peripheral circuit capacitance which is necessary to be charged and discharged in one autonomous refresh cycle and the amplitude of voltage at which the peripheral circuits operate in the autonomous refresh cycle. And ξ stands for the total number of BLs divided by the number of BLs which are charged and discharged in one conventional refresh cycle or the partition number of BL. In the case of Fig. 7.25, $\xi = 2M/M = 2$.

When Eq. 7.8 is compared with Eq. 7.6, the refresh busy rate for the autonomous reh is found to be smaller than that for the conventional refresh by the factor

$$\alpha = \frac{1}{N_{ref}} \times \frac{t_{RC\_AR}}{t_{RC\_conv.}}. \tag{7.10}$$

If we assume that $N_{ref} = 1024$, $t_{RC_conv.} = 30$ns, and $t_{RC\_AR} = 100$ ns, the busy rate reduction factor is calculated to be $\alpha = 0.003255$. The improvement more than two orders of magnitude is expected, because the refreshing all cells in the array directly contributes to the reduction of the period that is occupied by refresh.

Table 7.2 compares the components of the data retention current in the conventional refresh with those in the autonomous refresh that are expressed in Eqs. 7.7 and 7.9, respectively. Two components I_{WL} and I_{DC} are basically expressed in the same formulae for the both refreshes. The components I_{BL} and I_{peri} of the autonomous refresh, however, are reduced from those of the

Table 7.2 Comparison of the data retention current components of the autonomous refresh with those of the conventional refresh

Components	Conventional	Autonomous
I_{BL}	$\frac{M \times C_{BL} \times N_{ref} \times \Delta V_{BL}}{t_{ref}}$	$\frac{\xi \times M \times C_{BL} \times \Delta V_{BL\_AR}}{t_{ref}}$
I_{WL}	$\frac{C_{WL} \times N_{ref} \times \Delta V_{WL}}{t_{ref}}$	$\frac{C_{WL} \times N_{ref} \times \Delta V_{WL\_AR}}{t_{ref}}$
I_{DC}	$\frac{\lambda \times t_{RC\_conv.} \times M \times N_{ref}}{t_{ref}} i_{DC\_conv.}$	$\frac{\lambda_{AR} \times t_{RC\_AR} \times M \times N_{ref}}{t_{ref}} i_{DC\_AR}$
I_{peri}	$\frac{C_{peri\_conv.} \times N_{ref} \times \Delta V_{peri\_conv.}}{t_{ref}}$	$\frac{C_{peri\_AR} \times \Delta V_{peri\_AR}}{t_{ref}}$

refresh by the factors of ξ/N_{ref} and $1/N_{ref}$, respectively, putting aside the difference in the voltage amplitudes and the peripheral circuit capacitances. As is shown in this table, the biggest benefit of the autonomous refresh lies in the fact that the BL charging and discharging current is drastically reduced by refreshing all cells along a BL simultaneously from the conventional refresh where only one cell is refreshed by the BL charging and discharging. Although the expressions are similar, the component I_{DC} of the autonomous refresh is much smaller than that of the conventional refresh, because the cell current which flows during the autonomous refresh i_{DC_AR} is smaller than the cell current which flows during the conventional refresh $i_{DC\_conv.R}$ by more than two orders of magnitude. Table 7.3 compares the values of all components for the both refresh in the case of 1 Gbit FBRAM with the configuration that is shown in Fig. 7.26. Each cell array consists of 16 Mbit cells arranged on 1 K WLs and 16 K BLs. The chip consists of 64 arrays. In this configuration, C_{BL} = 128 fF, ξ = 16, N_{ref} = 8 K, and M = 128 K. Other parameters that were assumed are t_{ref} = 4 ms, ΔV_{BL} = 1.2 V, ΔV_{BL_AR} = 1.0 V, C_{WL} = 31.8 pF (for 128 K cells), $\Delta V_{WL} = \Delta V_{WL\_AR}$ = 2.0 V, $t_{RC\_conv.}$ = 50 ns, t_{RC_AR} = 80 ns, λ = 0.2, λ_{AR} = 0.55, $i_{CD\_conv.}$ = 60 μA, i_{DC_AR} = 0.18 μA, $C_{peri\_conv.} = C_{peri\_AR}$ = 0.6 nF and $\Delta V_{peri\_conv.} = \Delta V_{peri\_AR}$ = = 1.5 V. As is shown in Table 7.3, the components of the autonomous refresh other than that for the

Table 7.3 Example of data retention currents for the both refreshes in the case of 1 Gbit FBRAM with the configuration shown in Fig. 7.26. The data retention time is assumed to be t_{ref} = 4 ms for the both refreshes.

Components	Conventional	Autonomous
I_{BL}	41.11 mA	0.07 mA
I_{WL}	0.008 mA	0.008 mA
I_{DC}	161 mA	0.12 mA
I_{peri}	0.12 mA	1.4×10^{-5} mA
I_{ref}	202.3 mA	0.196mA

WL charging and discharging are reduced from those of the conventional refresh by 3 ~ 4 orders of magnitude. The total data retention current of the autonomous refresh is 3 orders of magnitude smaller than that of the conventional refresh.

Figure 7.27 compares the data retention current of the 1 Gbit FBRAM having autonomous refresh function with that of the conventional 1T-1C DRAM from the 64 kbit generation through the 4 Gbit generation. The retention times at each generation for the both RAMs which were used

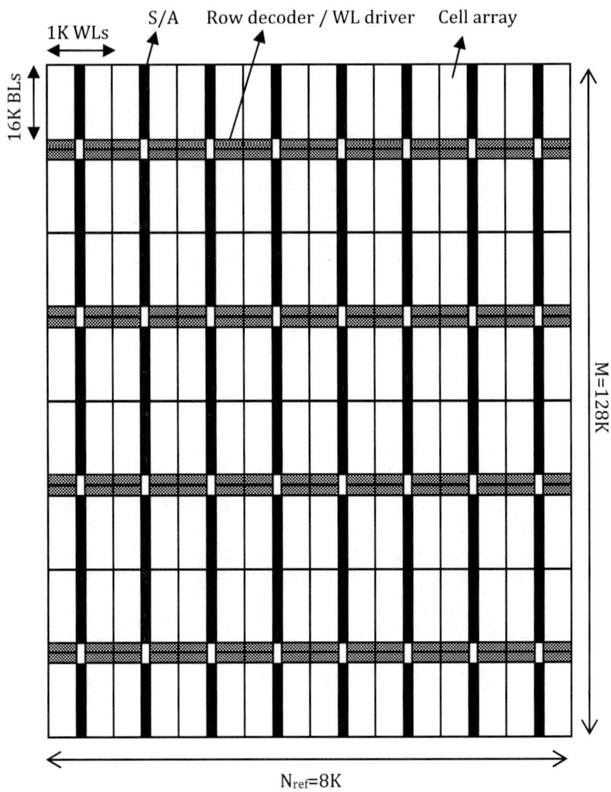

Figure 7.26 The 1 Gbit FBRAM chip architecture that is assumed in the calculation of the data retention currents.

in the calculations are shown in Table 7.4. The retention time trend in the 1T-1C DRAM was taken from that of *International Technology Roadmap for Semiconductors* (ITRS).[7] According to the trend, the retention time of the 1T-1C DRAM was increasing by the factor 2 until 64 Mbit to keep the refresh interval equal to 15.6 μs. However, the retention time cannot be increased more, being saturated at 64 ms for the generations ahead. This makes the data retention current of the 1T-1C DRAM increase exponentially from 64 Mbit as is shown in Fig. 7.27. The current increases to more than 1 mA for 1 Gbit and beyond, making it difficult to produce a low-power version to be used mainly for mobile devices. We assumed that the retention time of the FBRAM also saturates at a certain point of generation and that it increases by the factor 2 until 1 Gbit to be saturated at 4ms for the generations ahead. As is shown in Fig. 7.27, the retention current of the FBRAM with the autonomous refresh stays at a level below 1 mA even for 4 Gbit, making the autonomous refresh an attractive technology for the FBC to be used in high-density oriented stand-alone memories as well as embedded memories.

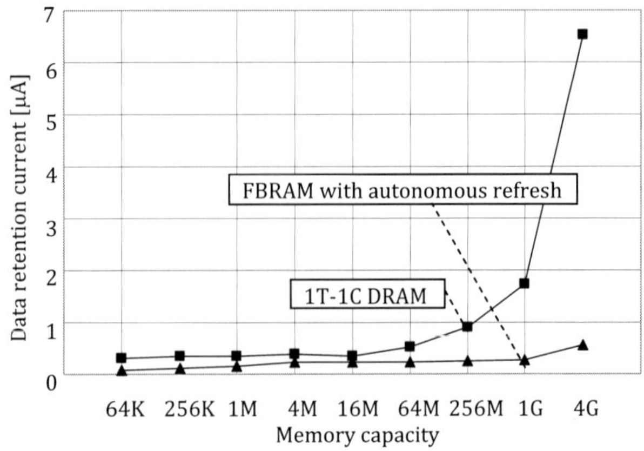

Figure 7.27 The comparison of the data retention current trend of the 1T-1CDRAM and that of the FBRAM with the autonomous refresh. The retention times of the both RAMs in the generations shown in Table 7.4 were used in the calculations.

Table 7.4 The retention times which were used in the calculations of the data retention currents in Fig. 7.27. The values for the 1T-1C DRAM are taken from the ITRS trend.[7]

	64 K	256 K	1 M	4 M	16 M	64 M	256 M	1 G	4 G
DRAM	2 ms	4 ms	8 ms	16 ms	32 ms	64 ms	64 ms	64 ms	64 ms
FBRAM	32 μs	64 μs	128 μs	256 μs	512 μs	1 ms	2 ms	4 ms	4 ms

7.5.2 Basic Principle

Now, let us take a look at the physics lying behind the autonomous refresh of FBC. When some voltage pulses are applied to the electrodes of a MOSFET whose body is electrically floating, various current components to and from the body can be observed. Now we consider a periodic application of a set of voltage pulses to FBC's electrodes. And net current input and output averaged over the period is denoted by I_{in} and I_{out}, respectively. Figure 7.28

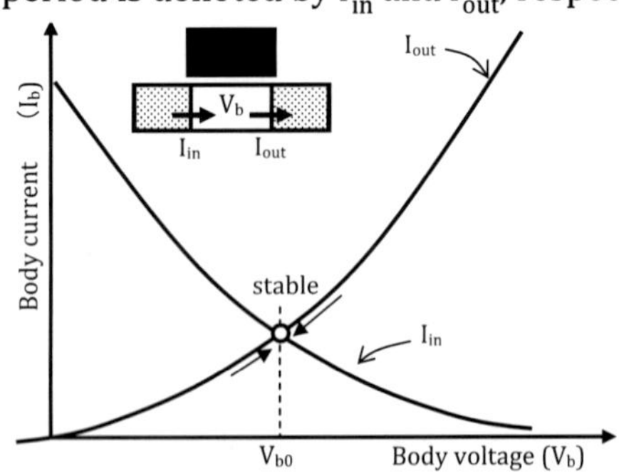

Figure 7.28 Only one stable stationary point is created when both the input current and the output current to and from the floating body are voltage-driven normal ones. All states converge to the equilibrium state after an enough long time, generating no memory functionality.

shows I_{in} and I_{out} curves as functions of body voltage V_b where both currents are driven by difference in voltage between the body and other electrodes. In this case, I_{in} and I_{out} become a monotonically decreasing function and a monotonically increasing function of V_b, respectively. Therefore, there is only one intersection between the curves which gives rise to a stationary point because

$$\frac{dV_b}{dt} = \frac{I_{in} - I_{out}}{C_b} = 0, \tag{7.11}$$

at the intersection where C_b represents the total capacitance of the body. And this stationary point is stable in the sense that any infinitesimal deviation from the state will force it to return to the stationary point. Therefore, the FBC's body voltage has only one equilibrium state if it is exposed to periodic voltage pulses which generate only voltage-driven normal currents.

However, if I_{in} has a dominant component that is an increasing function of V_b or I_{out} has a dominant component that is a decreasing function of V_b, the situation will be totally different. Figure 7.29 shows the case where I_{in} is an increasing function of V_b and its derivative dI_{in}/dV_b is larger than that of I_{out} in a certain region. This current anomaly brings about three intersections between the two curves. We have one unstable stationary point in the center which is accompanied by two stable stationary ones on both sides. Since the impact ionization current increases as the body voltage increases through the body effect, the non-voltage-driven current with a positive feedback

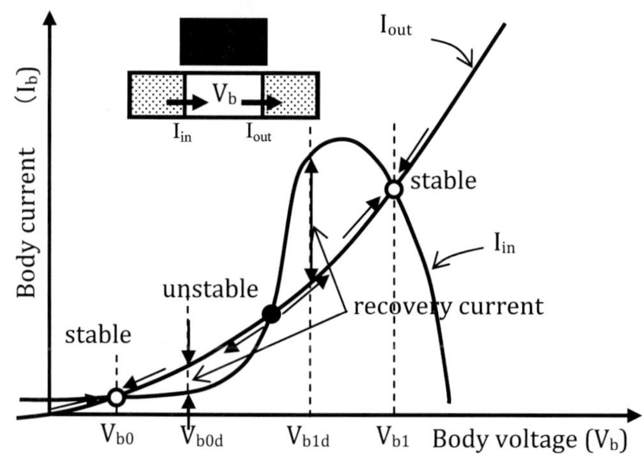

Figure 7.29 An unstable stationary point is created that is accompanied by two stable stationary points on both sides, when the input current to the floating body is an increasing function of the body voltage V_b with its derivative with respect to V_b larger than that of the output current in a certain region of V_b.

gain can be one of the candidates which exhibit the current anomaly If the binary states of "1" and "0" correspond to the two stable points, the degradation of the states during retaining data will be recovered autonomously without being read and restored by a sense amplifier. This is the mechanism of the autonomous refresh.

Figure 7.30 exemplifies a set of voltage pulses applied to FBCs in the autonomous refresh using impact ionization current as I_{in} and charge pumping current as I_{out}. There are two phases. One is the impact ionization phase with a duration $t_{i/i}$ and the other is the charge pumping phase with a duration t_{cp}. This set is repeated in a period t_{ref}. The WL voltage V_{WL_cp} in the charge pumping phase is necessary to be larger than the threshold voltage of the "0" cell V_{th0} to make the I_{out} curve as independent of V_b as possible by generating enough number of channel electrons even for the data "0" cells which are to be recombined with the holes in the body at the end of the charge pumping phase.

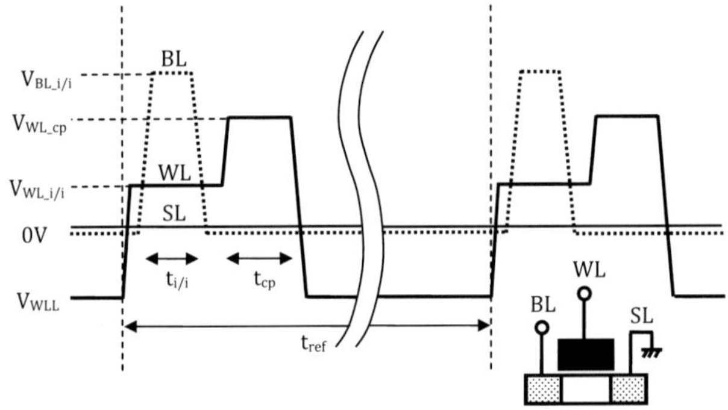

Figure 7.30 An example of a set of voltage pulses applied to FBCs in the autonomous refresh using charge pumping, where $t_{i/i}$, t_{CP}, and t_{ref} stand for the duration of an impact ionization phase, the duration of a charge pumping phase, and the period of the autonomous refresh, respectively.

7.5.3 Experiments

To predict the optimum operating points of the autonomous refresh, we performed TCAD simulations for the body current of the FBC with both the gate length L_g and the channel width W 150 nm, the gate oxide thickness t_{ox} 6 nm, the silicon thickness t_{Si} 55 nm , and the BOX thickness t_{Box} 25 nm. The current input I_{in} which is expressed by $I_{in} = T_{i/i} + I_{pn}$ is shown in Fig. 7.31, where $I_{i/i}$ is the simulated impact ionization current which is averaged by the duty ratio of $t_{i/i}/t_{ref}$ = 180 ns/100 μs and I_{pn} is also the simulated p-n diode current (positive for the reverse current and negative for the forward current) which is also averaged by the duty ratio. The BL voltage in the impact ionization phase $V_{BL\_i/i}$ is 2.4 V and the WL voltage

$V_{WL\_i/i}$ is changed from 0.2 V to 0.6 V. The reason why I_{in} curves decrease drastically into negative values at higher V_b is that the p-n diode forward current I_{pn} to GND dominates I_{in} in the region. The output current I_{out} which consists of charge pumping current I_{cp} is also drawn in Fig. 7.31. Since both the channel length and the width are 150 nm, the number of interface states N_{it} in a state-of-the-art FBC is calculated to be 2.85,[10] ncorresponding to the interface state density D_{it} 1.3×10^{10} cm^{-2}. Therefore, $I_{out} = I_{cp} = qN_{it}/t_{ref} = 4.6 \times 10^{-15}$ A, if every electron which is trapped by the state is recombined with a hole in the body. Although I_{cp} is assumed to be constant of V_b, it is expected to decrease due to the decreased channel electron density for very low V_b, because the threshold voltage of the FBC increases there.

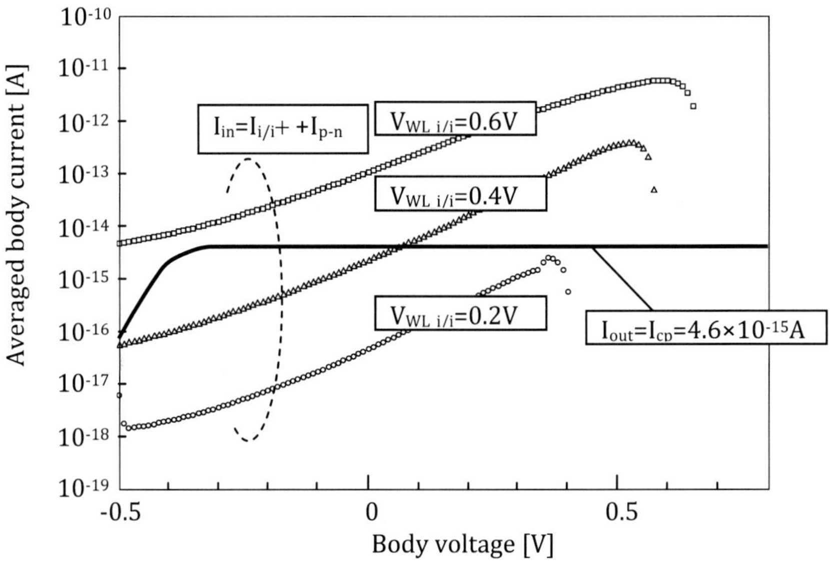

Figure 7.31 An example of a set of voltage pulses applied FBCs in the autonomous refresh using charge pumping, where $t_{i/i}$, t_{CP}, and t_{ref} stand for the duration of an impact ionization phase, the duration of a charge pumping phase, and the period of the autonomous refresh respectively.

This graph predicts that the autonomous refresh will be realized at 0.2 V < $V_{WL\_i/i}$ < 0.6 V with $V_{BL\_i/i}$ = 2.4 V for this FBC.

Experiments were performed by applying the waveforms in Fig. 7.30 many times to a single FBC with the same device parameters as those in the TCAD simulations.

The conditions are as follows: $V_{BL\_i/i}$ = 2.4 V, $0 \leq V_{WL\_i/i} \leq 0.8$ V, V_{WL_cp} = 1.5 V, V_{WLL} = −1.5 V, $t_{i/i} = t_{cp}$ = 180 ns, and t_{ref} = 100 μs. Figure 7.32a is the results for the cell read current observed after each refresh count for the cell which was written "1" initially. For $0 \leq V_{WL\_i/i} \leq 0.2$ V, the cell read current decreases from the "1" level to the "0" level after several tens counts, because the impact ionization current is smaller than the charge pumping one. For 0.3 V $\leq V_{WL\_i/i} \leq$ 0.8 V, however, the "1" read current is retained for more than 10^4 counts. This is achieved because the impact ionization current balances with the charge pumping one. Figure 7.32b the results for the cell read current observed after each refresh count for the cell which was written "0" initially. For 0.6 V $\leq V_{WL\_i/i} \leq$ 0.8 V, the cell read current increases from the "0" level to the "1" level after several thousand counts at the largest, because the impact ionization current is larger than the charge pumping one. For 0 V $\leq V_{WL\_i/i}$ $\leq$ 0.5 V, on the other hand, the cell read current remains at the "0" level.

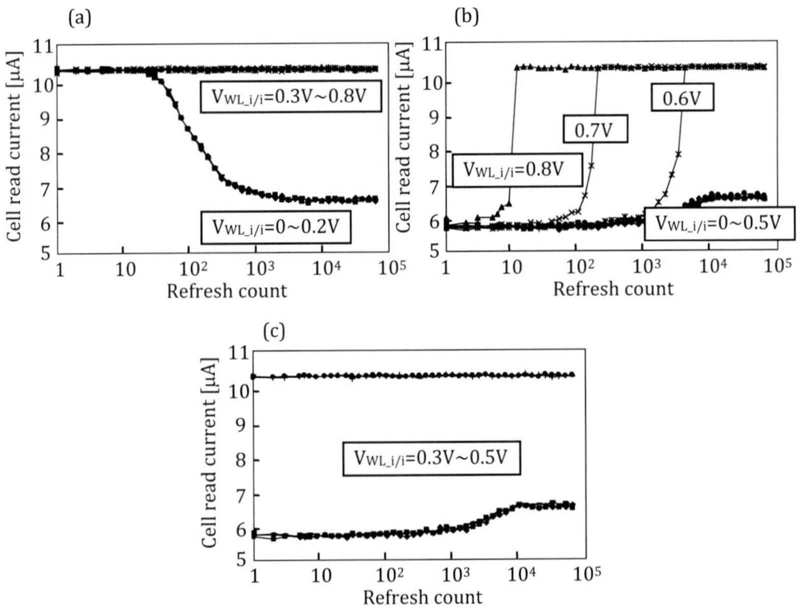

Figure 7.32 Experimental results for the autonomous refresh of Fig. 7.30 applied to a single cell. Charge pumping conditions remain unchanged, while the WL voltage at the impact ionization phase is changed from 0 V to 0.8 V with the BL voltage at the phase fixed at 2.4 V. (a) Cell read current is monitored after each autonomous refresh cycle for the cell that has the data "0" written initially. (b) Cell read current is monitored after each autonomous refresh cycle for the cell that has the data "1" written initially. (c) Both data are retained stably by the autonomous refresh for the WL voltage at the impact ionization phase 0.3 V ~ 0.5 V. All read currents are measured with the gate and the drain voltages at 1.5 V and 0.2 V, respectively.

The slight increase and the saturation of the cell read current at about 10^4 counts is due to the transient phenomenon to reach the stable stationary point V_{b0} illustrated in Fig. 7.29, because the "0" body voltage happened to be written initially to lower than the stationary point V_{b0} in this experiment. By combining the two results, we have Fig. 7.32c where both "1" data and "0" data are retained stably for 0.3 V ≤ $V_{WL\_i/i}$ ≤ 0.5 V. These results coincide with the previous predictions made by TCAD simulation and simple calculation.

The similar experiments were performed for the same FBC as in Fig. 7.32 by changing the BL voltage from 1.8 V to 2.7 V and we obtain a shmoo plot of the autonomous refresh in Fig. 7.33, where P, 0, and 1 denote pass, "1"→"0" fail and "0"→"1" fail, respectively. The balance between the impact ionization current and the charge pumping one well explains this shmoo plot.

Table 7.5 summarizes a typical set of voltages at write, hold, and read which are to be applied to FBC of L_g = 150 nm generation along with those at impact ionization phase and charge pumping phase in autonomous refresh cycles. Since the body of the FBC is fully depleted, a negative voltage (−3 V) is necessary to be constantly applied.

The key to a successful autonomous refresh is whether or not the data of all cells in a large cell array except for some replaced by redundancies are retained by a predetermined set of voltage pulses applied to the WLs and the BLs. Figure 7.34 shows the experimental results of the autonomous refresh for 32 cells in a cell array with the same device parameters as those in Fig. 7.32. Figure 7.34a,b show the cell read currents after 10^5 autonomous refresh for the data "1" cells and for the data "0" cells, respectively. The refresh period was changed as 10 μs, , and 1 ms. The initial cell read currents of the data "0" $I_{cell\ 0}$ and of the data "1" $I_{cell\ 1}$ for each cell are indicated at the left axis and the right axis in the both graphs for references, respectively. The voltage conditions are as follows: $V_{BL\_i/i}$ = 2.4 V, V_{WL_cp} = 1.5 V, and V_{WLL} = −1.5 V.

$V_{WL\_i/i}$ [V] \ $V_{BL\_i/i}$ [V]	1.8	1.9	2.0	2.1	2.2	2.3	2.4	2.5	2.6	2.7
0.2	0	0	0	0	0	0	0	0	0	0
0.3	0	0	0	0	0	P	P	P	P	P
0.4	0	0	0	P	P	P	P	P	P	P
0.5	0	0	0	P	P	P	P	P	P	1
0.6	1	1	1	1	1	1	1	1	1	1

Figure 7.33 Shmoo plot of the autonomous refresh using impact ionization and charge pumping measured for the same cell as in Fig. 7.32, where 0, 1, and P stand for "1" to "0" fail, "0" to "1" fail, and pass, respectively.

Table 7.5 Operational conditions for FBC with gate length = 150 nm, gate oxide thickness = 6 nm, silicon thickness = 55 nm, and BOX thickness = 25 nm

Operations	WL	BL	SL	Substrate
"1" write	1.5 V	2.4 V	0 V	-3.0 V
"0" write	1.5 V	-1.5 V	0 V	-3.0 V
Hold	-2.0 V	0 V	0 V	-3.0 V
Read	1.2 V	0.2 V	0 V	-3.0 V
Impact ionization phase of autonomous refresh	0.4 V	2.4 V	0 V	-3.0 V
Charge pumping phase of autonomous refresh	1.5 V	0 V	0 V	-3.0 V

The temperature is 25°C. In Fig. 7.34, the ratio $t_{i/i}/t_{ref}$ is kept constant to make the averaged input current I_{in} by the impact ionization constant: $t_{i/i}$ = 100 ns for t_{ref} = 10 μs, $t_{i/i}$ = 10 μs for t_{ref} = 100 μs, and $t_{i/i}$ = 10 μs for t_{ref} = 1 ms. Therefore, the changing of the value of t_{ref} in the experiments corresponds to moving the curve I_{out} up and down with the curve I_{in} unchanged in Fig. 7.29. The situations are illustrated in Fig. 7.35. When t_{ref} = 10 μs, I_{out} due to charge pumping becomes the largest, letting the "0" state have a large stability margin and the "1" state have a small stability margin. As is observed in Fig. 7.34a, about half of the 32 data "1" cells fail to the data "0," while the pass rate is high for the data "0" cells as is shown in Fig. 7.34b with two cells labeled A and B likely to fail. When t_{ref} = 100 μs, the stability margin of the "1" data cells increases as is shown in Fig. 7.35, making many of the "1" data cells pass as is shown in Fig. 7.34a. When t_{ref} = 1 ms, the stability margin of the data "0" cells becomes very small and some cells tend to fail as is shown in Fig. 7.34b. Figure 7.36 shows the experimental results of the charge pumping for the 32 cells in Fig. 7.34. The data "1" is written to the 32 cells initially and then the gate voltage of the cells are being pumped between V_{WL_cp} and V_{WLL}. The cells are read after each pumping cycle to monitor the number of holes lost in each cycle. The cell labeled A and B in Fig. 7.34 were found to be the two cells that have the least number of interface states as is shown in Fig. 7.36. These experiments suggest that there are several cells with very few number of interface states among many cells in a large cell array and that these cells make the total operational margin decrease for the autonomous refresh using the charge pumping for the current output.

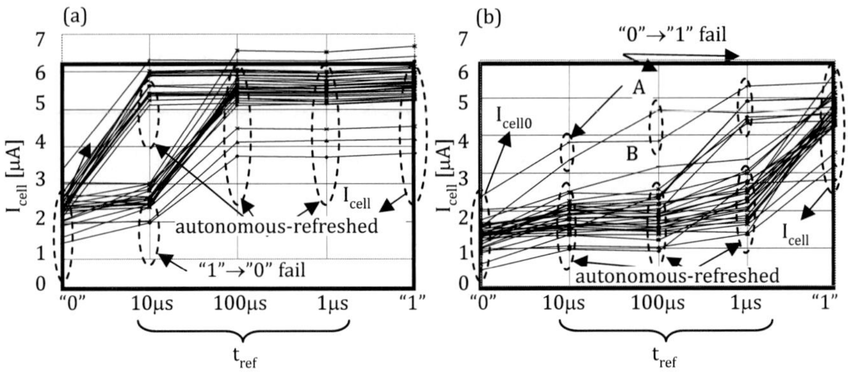

Figure 7.34 The cell read currents after 10^5 autonomous refresh cycles applied to 32 cells with (a) the data "1" written initially and (b) the data "0" written initially. The initial "0" data cell currents I_{cell0} and the "1" data cell currents I_{cell1} for each cell are indicated at the left and the right y-axes, respectively.

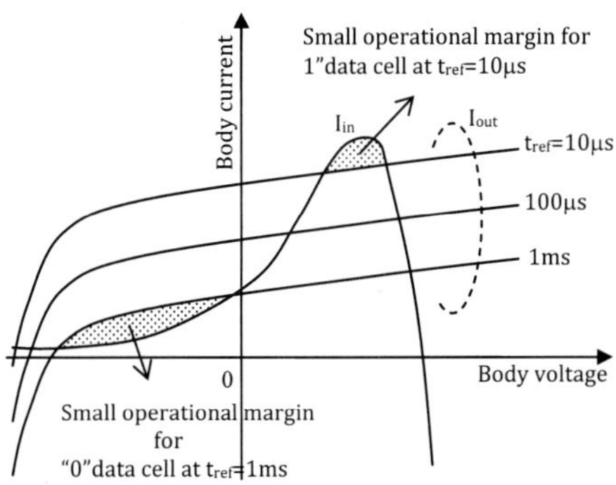

Figure 7.35 The change in the relation between I_{in} and I_{out} when the value of t_{ref} is changed with the ratio $t_{i/i}/t_{ref}$ kept constant.

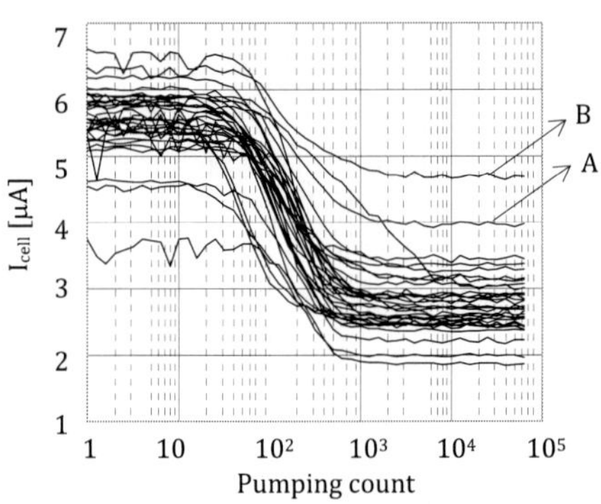

Figure 7.36 The characteristics of charge pumping for 32 cells in Fig. 7.34. The cells labeled A and B in Fig. 7.34 were found to be the two cells that have the least number of interface states at the channels among the 32 cells.

7.5.4 Necessary Conditions of Autonomous Refresh Using Charge Pumping for High-Density Memories

Now, let us investigate necessary conditions for the autonomous refresh using charge pumping as current output to be performed successfully for high-density memories. Figure 7.37 illustrates current input Iin to the floating body and current output Iout from the body both averaged in

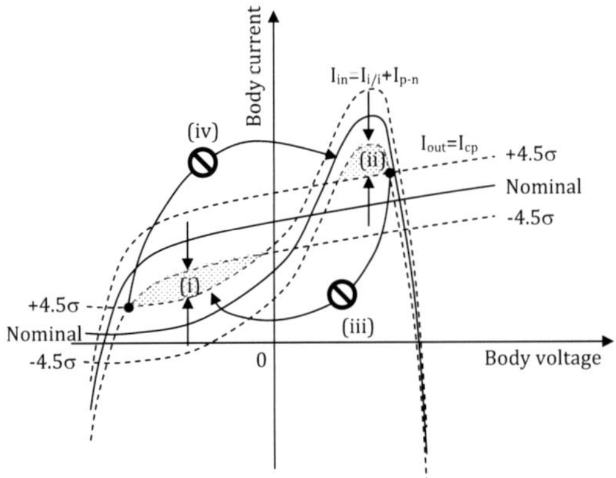

Figure 7.37 The necessary conditions of realizing the autonomous refresh using charge pumping as I_{out} for high-density memories. The nominal, +4.5σ and −4.5σ indicate the median and the statistical fluctuations for I_{in} and I_{out} within a cell array, respectively.

the autonomous refresh cycles as a function of the body voltage V_b with statistical variations among many cells within a large cell array taken into account. The label "Nominal" stands for the median value and "±4.5σ" stand for the statistical variations. The following four conditions are necessary to hold for the autonomous refresh to be applied successfully: (i) A "wing" of the butterfly curves must exist for the "0" state, (ii) a "wing" of the butterfly curves must exist for the "1" state, (iii) the "1" state must not go beyond the unstable stationary point in a single charge pumping cycle, and (iv) the "0" state must not go beyond the unstable stationary point in a single impact ionization cycle. The conditions (iii) and (iv) are necessary, because the impact ionization current and the charge pumping current do not flow at the same time but alternately as shown in Fig. 7.30. However, the condition (iv) is satisfied automatically if the condition (iii) holds, because the charge pumping current for the data "1" cell is larger than the impact ionization current for the data "0" cell, as is indicated in Fig. 7.37. Therefore, the necessary conditions for the autonomous refresh using charge pumping to be successfully applied are (i), (ii), and (iii).

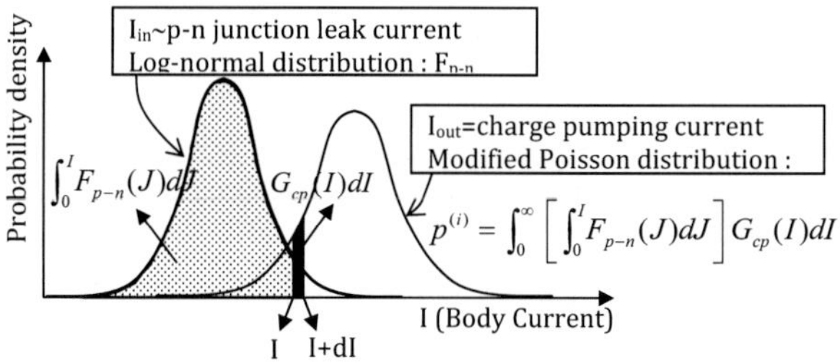

Figure 7.38 The graph indicating the method to calculate the probability $p^{(i)}$ for cells to satisfy the necessary condition (i) or, in other words, the probability $p^{(i)}$ for a "0" cell to have I_{out} larger than I_{in}.

At first, let us derive the criteria from the condition (i). The probability $P^{(i)}$ for a "0" cell to have $I_{out} = I_{cp}$ larger than $I_{in} \cong I_{p-n}$ is calculated, as is shown in Fig. 7.38, to be

$$p^{(i)} = \int_0^\infty \left[\int_0^I F_{p-n}(J)dJ\right] G_{cp}(I)dI, \tag{7.12}$$

where $F_{p-n}(I)$ is a log-normal distribution in a cell array for the p-n junction leak current and $G_{cp}(I)$ is a modified Poisson distribution in the cell array for the charge pumping current both for the data "0" cell, i.e.,

$$F_{p-n}(I) = \frac{1}{\sqrt{2\pi}\cdot\ln(10)\cdot\sigma_{\log(I_{p-n})}\cdot I}\exp\left\{-\frac{[\log(I)-\log(I_{p-n0})]^2}{2\cdot\sigma_{\log(I_{p-n})}{}^2}\right\}, \tag{7.13}$$

$$G_{cp}(I) = \left(\frac{1}{k}\right)\frac{e^{-N_{it}}\cdot N_{it}{}^{\text{Int}[I\cdot t_{ref}/(q\cdot k)]}}{(\text{Int}[I\cdot t_{ref}/(q\cdot k)])!}\frac{t_{ref}}{q}, \tag{7.14}$$

where I is current, I_{p-n0} and $\sigma_{\text{logt}(I_{p-n})}$ are, respectively, the average and the standard deviation in logarithmic scale of the p-n junction leak current, N_{it} is the averaged number of interface states in a cell, t_{ref} is the refresh period defined in Fig. 7.30, q is the absolute value of the electron's charge, and k is a reduction factor of the modified Poisson distribution function. The modified Poisson distribution $Q(n; \lambda, k)$ with average λ and reduction factor k is expressed by

$$Q(n;\lambda,k) = \left(\frac{1}{k}\right)\frac{e^{-\lambda}\cdot\lambda^{\text{Int}[n/k]}}{(\text{Int}[n/k])!}, \tag{7.15}$$

where $I_{nt}[x]$ stands for the largest integer not greater than x. This distribution is reduced from the Poisson distribution

$$P(n;\lambda) = \frac{e^{-\lambda}\cdot\lambda^n}{n!}, \tag{7.16}$$

by k as is shown in Fig. 7.39. The modified Poisson distribution $Q(n;\,24,\,1/4)$ has the same average value as the Poisson distribution $P(n;\,6)$, while the standard deviation of the former distribution is smaller than that of the latter one. And it was found that the measured distribution of the number

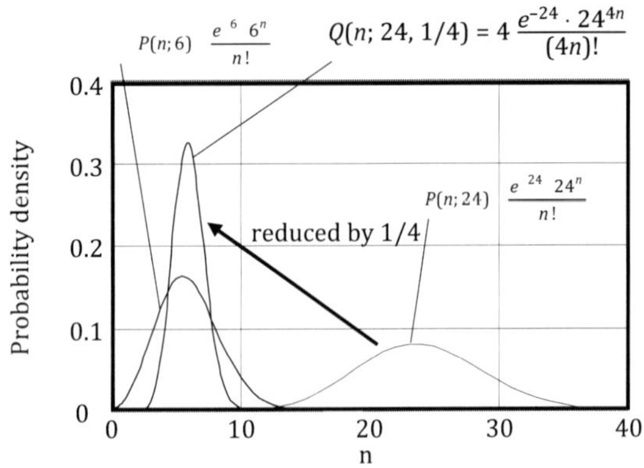

Figure 7.39 The relationship between a Poisson distribution and a modified Poisson distribution.

of holes lost by gate pumping of a MOSFET follows a modified Poisson distribution rather than a Poisson distribution as is shown in Fig. 7.40. The reason why the number of holes lost by a WL pumping obeys a modified Poisson distribution function can be explained as follows: Although the number of interface states obeys a Poisson distribution function $P(n;\,\lambda)$, all the interface states do not trap electrons and all trapped electrons at the interface states do not recombine with the holes in the floating body. If the efficiency at which the electrons are trapped at the interface states is written as η_{trap} and the efficiency at which the holes are recombined with the electrons trapped at the interface states is written as $\eta_{\text{recombine}}$, the number of holes lost by a WL p umping will obey the modified Poisson distribution function $Q(n;\,\lambda,\,k)$ with the reduction factor k expressed as And

$$k = \eta_{\text{trap}} \times \eta_{\text{recombine}}. \tag{7.17}$$

η_{trap} will depend on the density of the channel electrons in a given period of channel inversion which increases as the gate voltage $V_{\text{WL\_cp}}$ increases and $\eta_{\text{recombine}}$ will depend on the density of the body holes in a given period of channel accumulation which increases as the gate voltage V_{WLL} decreases. Here, let us assume that $I_{\text{p - n0}} = 5.0 \times 10^{-17}$A and $\sigma_{\log(I_{\text{p- n}})} = 0.33$ based on hard ware measurement results for the FBC with $L_{\text{g}} = 150$ nm. And k is assumed to be 1/4 based on parameter fitting results of the modified Poisson distribution to charge pumping experiments for the FBC with $V_{\text{WLL}} = -1.5$ V. And t_{ref} is set to be 1 ms. By calculating $p^{(i)}$ in Eq. 7.12 under these assumptions and setting the failure rate $f^{(i)} = 1 - p^{(i)} \leq 3$ppm which is the condition that failed cell rate over population is less than -4.5σ, we have the condition

$$N_{it} \geq 45. \tag{7.18}$$

This means that there will be a significant minority of cells where holes input into the floating body through the p-n junction leak cannot be got rid of by its weak charge pumping current unless the average of the Poisson distribution for the number of the interface states in cells is larger than a certain number. If the p-n junction leak current per cell decreases as the cell size shrinks, the criteria (7.18) can be relaxed for the future generations. However, we assume that the p-n junction leak current is constant even if the cell shrinks for estimating the worst case.

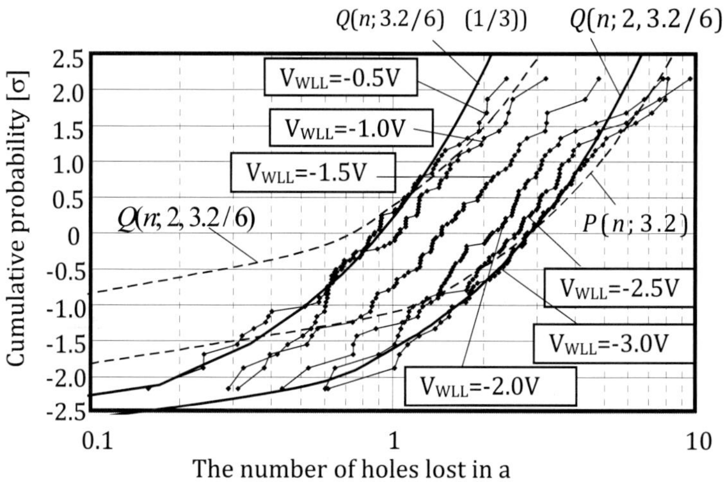

Figure 7.40 The modified Poisson distributions that are fitted to the measured distributions of number of holes lost by a charge pumping cycle for various conditions of WL low level.

As was discussed in Fig. 6.24 of section 6.6, the averaged number of interface states per cell with L_g = 150 nm is 2.85. However, since the measurements in Fig. 6.24 were performed at V_{WLL} = −2.0 V, the number ΔN_{hole} = 2.85 which gives the peak frequency should be regarded as not N_{it} but $N_{it}/3$ as can be understood in Fig. 7.39 and also in the difference in peak between the measured data and the Poisson curve in Fig. 6.24. Then $N_{it} = 2.85 \times 3 = 8.55$ for state-of-the-art FBC. Anyway, the condition (7.18) is not satisfied by state-of-the-art FBC, explaining the situation that some "0" data cells remain failed even for t_{ref} = 10 μs in Fig. 7.34b. It is worth noting that the condition (7.18) corresponds to the interface state density $D_{it} \geq 1.67 \times 10^{11}$ cm^{-2} for the FBC with L_g = 150 nm and W = 180 nm.

Next, let us derive the criteria from the condition (ii). The probability $p^{(i)}$ for a "1" cell to have $I_{out} = I_{cp}$ smaller than $I_{in} = I_{i/i}$ is

$$p^{(ii)} = \int_0^{\infty} \left[\int_0^{I} G_{cp}(J)dJ \right] H_{i/i}(I)dI, \qquad (7.19)$$

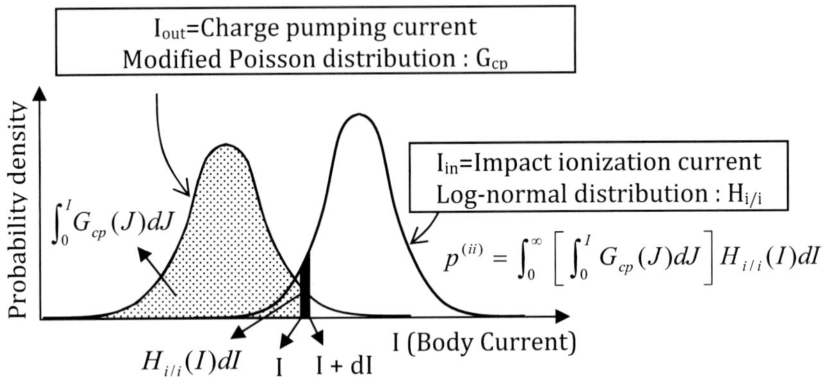

Figure 7.41 The graph indicating the method to calculate the probability $p^{(ii)}$ for cells to satisfy the necessary condition (ii) or, in other words, the probability $p^{(ii)}$ for a "1" cell to have I_{out} smaller than I_{in}.

calculated, as is shown in Fig. 7.41 to be where we assume that the charge pumping current distribution for the "1" data cells is the same as that of the data "0" cells and that the impact ionization current distribution in the cell array for the data "1" cells is where $I_{i/i0}$ and $\sigma_{\log(I_{i/i})}$ are the average and the

$$H_{i/i}(I) = \frac{1}{\sqrt{2\pi} \cdot \ln(10) \cdot \sigma_{\log(I_{i/i})} \cdot I} \exp\left\{ -\frac{[\log(I) - \log(I_{i/i0})]^2}{2 \cdot \sigma_{\log(I_{i/i})}^{\ 2}} \right\}, \qquad (7.20)$$

standard deviation in logarithmic scale of the impact ionization current. By substituting Eq. 7.14 with $N_{it} = 45$, $t_{ref} = 1$ ms, and $k = 1/4$ and Eq. 7.20 into Eq. 7.19 and setting the failure rate $f^{(ii)} = 1 - p^{(ii)} \leq 3$ppm which is the condition that failed cell rate over population is less than -4.5σ, we obtain Fig. 7.42 which is a 2D margin map for condition (ii). In this graph, N_{hole_in} means the averaged number of holes input into the body per cycle defined by $N_{hole\_in} = I_{i/i0} \cdot = I_{i/i}\ t_{i/i}/q$, and $\sigma_{\log(N_{hole\_in})}$ is its standard deviation in logarithmic scale, which is equal to $\sigma_{\log(I_{i/i})}$. The two parameters should be within the pass area for condition (ii) to hold. This margin map regulates the extent to which the impact ionization current fluctuates for cells in a high-density memory.

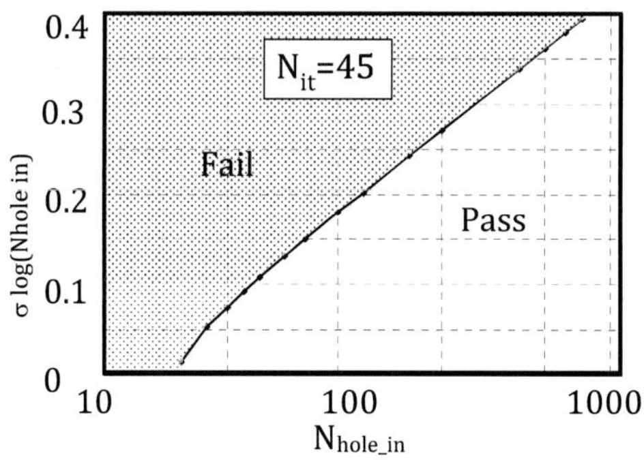

Figure 7.42 Margin map represented in the plane of the averaged number of holes input into the body in an impact ionization cycle $N_{\text{hole\_in}}$ and its standard deviation in logarithmic scale $\sigma_{\log(N_{\text{hole\_in}})}$ derived from the necessary condition (ii) of the autonomous refresh.

Then, let us derive the criteria from the condition (iii). The probability $p^{(\text{iii})}$ for a "1" cell not to lose its hole charge larger than a critical value Q_C is calculated, as is shown in Fig. 7.43, to be

$$p^{(\text{iii})} = \int_{-\infty}^{\infty}\left[\int_{Q+Q_C}^{\infty} f(R)dR\right]g(Q)dQ, \tag{7.21}$$

the distribution function in the cell array of the charge of holes accumulated in the "1" cell's body $f(Q)$ and that of the charge of holes lost in a charge pumping cycle $g(Q)$ are expressed as

$$f(Q) = \frac{1}{\sqrt{2\pi}\cdot\sigma_{Q_{\text{hole1}}}}\exp\left\{\frac{-[Q - Q_{\text{hole1}}]^2}{2\cdot\sigma_{Q_{\text{hole1}}}{}^2}\right\}, \tag{7.22}$$

$$g(Q) = \left(\frac{1}{k}\right)\frac{e^{-N_{\text{it}}}\cdot N_{\text{it}}{}^{\text{Int}[Q/(q\cdot k)]}}{(\text{Int}\,[Q/(q\cdot k)])!}\frac{1}{q} \tag{7.23}$$

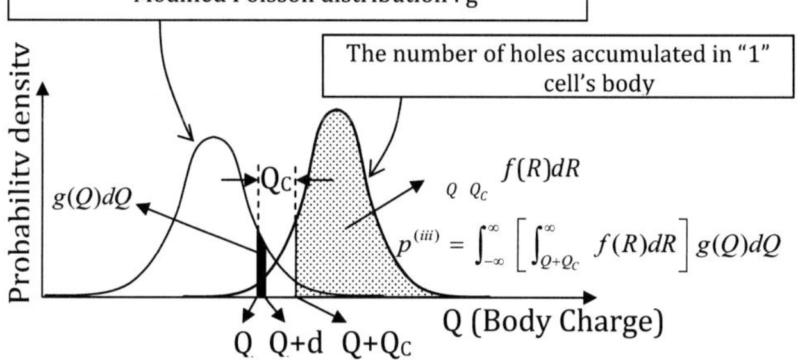

Figure 7.43 The graph indicating the method to calculate the probability $p^{(\text{iii})}$ for cells to satisfy the necessary condition (iii) or, in other words, the probability $p^{(\text{iii})}$ for a "1" cell not to lose its hole charge larger than Q_C.

where Q is charge, To check the condition (ii), we need to measure and Q_{hole1} and σ_{hole1} are the average and the standard deviation of the hole charge in "1" cell bodies. By substituting Eqs. 7.22 and 7.23 with $N_{it} = 45$, $t_{ref} = 1\text{ms}$, and $k = 1/4$ into Eq. 7.21 and setting the failure rate $f^{(iii)} = 1 - p^{(iii)} \leq 3\text{ppm}$ which is the condition that failed cell rate over population is less than -4.5σ, we obtain Fig. 7.44 which is a 2D margin map for condition (iii). Here, we assumed $Q_C = Q_{\text{hole1}}/2$ and that "0" cell has no holes. The parameters $N_{\text{hole 1}} = Q_{\text{hole1}}/q$ and $\sigma_{N_{\text{hole1}}} = \sigma_{\text{hole1}}/q$ should be within the pass area for condition (iii) to hold. This margin map regulates the extent to which the number of holes accumulated in "1" cell bodies fluctuates for cells in a high-density memory.

Until now, the necessary conditions for the autonomous refresh using impact ionization as I_{in} and charge pumping as I_{out} have been derived. It is already known that the regulation on the averaged number of interface states in a cell N_{it}, Eq. 7.18, is not satisfied by state-of-the-art FBCs. Therefore, it is needed to increase the interface state density to $\cong 10^{11}\text{cm}^{-2}$, for example, by adding some process steps and/or modifying the existing process. However, other two conditions which are represented in Figs. 7.42 and 7.44 are also needed to be checked whether or not state-of-the-art FBC satisfies them.

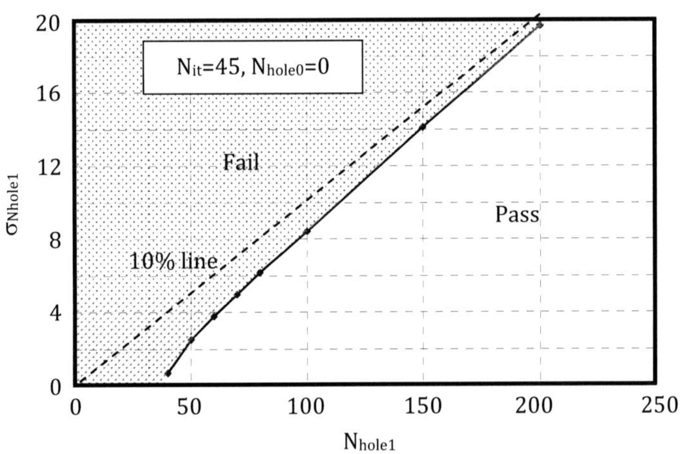

Figure 7.44 Margin map represented in the plane of the averaged number of holes accumulated in "1" cell bodies N_{hole1} and its standard deviation σN_{hole1} derived from the necessary condition (iii) of the autonomous refresh.

To chect the condition (ii), we need to measure impact ionization current $I_{i/i}$ for many cells in a cell array. Direct measurements of $I_{i/i}$ is impossible, since FBC in a cell array has no body contact. Therefore, we devise a novel method to measure $I_{i/i}$ indirectly. Figure 7.45a shows a "0" to "1" write characteristics of the FBC with $L_g = 150\text{nm}$. $I_{i/i}$ is proportional to to dI_{cell}/dt_W where I_{cell} is cell read current and t_W is "1" write duration:

$$I_{\mathrm{i/i}} = \frac{C_{\mathrm{b}} \cdot C_{\mathrm{ox}}}{C_{\mathrm{Si}} \cdot g_{\mathrm{m}}} \frac{dI_{\mathrm{cell}}}{dt_{\mathrm{W}}} \tag{7.24}$$

where C_b, C_{ox}, C_{Si}, and g_m are body capacitance, gate oxide capacitance, silicon capacitance per unit area, and transconductance of the FBC, respectively. To assess the condition (ii), we need to evaluate the peak value $(dI/_{\mathrm{cell}}/dt_{\mathrm{W}})_{\mathrm{peak}}$ as shown in Fig. 7.45b. Figure 7.46 shows the $(dI_{\mathrm{cell}}/dt_{\mathrm{W}})_{\mathrm{peak}}$ distribution in a cell array for the FBC with L_g = 150 nm at $V_{\mathrm{WL\_i/i}}$ = 1.5 V and $V_{\mathrm{BL\_i/i}}$ = 2.4.

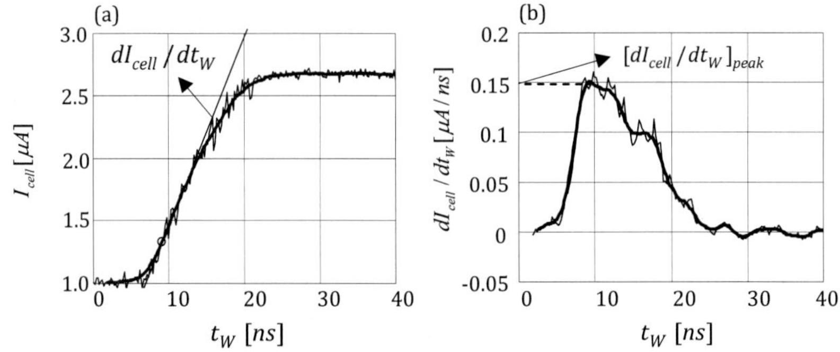

Figure 7.45 (a) Cell read current measured as a function of time during which data "1" continues to be written to a cell that was initially set to "0." (b) Derivative of the cell current with respect to the write duration time. The experiment was performed for the FBC with L_g = 150 nm and the BL and for the WL voltages for "1" write with $V_{\mathrm{BL\_i/I}}$ = 2.4 V and $V_{\mathrm{WL\_i/I}}$ = 1.5 V, respectively. The read operation is performed with the gate voltage at 1.1 V.

The standard deviation of $(dI_{\mathrm{cell}}/dt_{\mathrm{W}})_{\mathrm{peak}}$ in logarithmic scale is estimated to be1/9. The WL voltage at the impact ionization phase $V_{\mathrm{WL\_i/i}}$ for the autonomous refresh is around 0.5 V. However, our measurement explained above has very poor accuracy around $V_{\mathrm{WL\_i/i}}$ = 0.5 V. So we extrapolate the standard deviation from larger $V_{\mathrm{WL\_i/i}}$ to 0.5 V as shown in Fig. 7.47. The standard deviation at $V_{\mathrm{WL\_i/i}}$ = 0.5 V will be less than 0.1. The standard deviation of $N_{\mathrm{hole\_in}}$ at peak in logarithmic scale is equal to that of $(I_{\mathrm{i/i}})_{\mathrm{peak}}$ in logarithmic scale which must be equal or smaller than that of $(dI_{\mathrm{cell}}/dt_{\mathrm{W}})_{\mathrm{peak}}$ in logarithmic scale:

$$\sigma_{\log([N_{\mathrm{hole\_in}}]_{\mathrm{peak}})} = \sigma_{\log([I_{\mathrm{i/i}}]_{\mathrm{peak}})} \leq \sigma_{\log([dI_{\mathrm{cell}}/dt_{\mathrm{W}}]_{\mathrm{peak}})} < 0.1. \tag{7.25}$$

Therefore, $\sigma_{\log}(N_{\mathrm{hole\_in}})_{\mathrm{peak}}$ will be less than 0.1. On the other hand, the peak current of $I_{\mathrm{i/i}}$ at $V_{\mathrm{WL\_i/i}}$ = 0.5 V and $V_{\mathrm{BL\_i/i}}$ = 2.4 V is calculated to be 19 nA in TCAD. Therefore, $N_{\mathrm{hole\_in}}$ which is the averaged number of holes input to the body in an impact ionization cycle is calculated to be 60 at the peak for $t_{\mathrm{i/i}}$ = 5 ns. This point (60, 0.1), stays within the pass area of Fig. 7.42. This result shows that the condition (ii) is satisfied if the condition (i) is fulfilled.

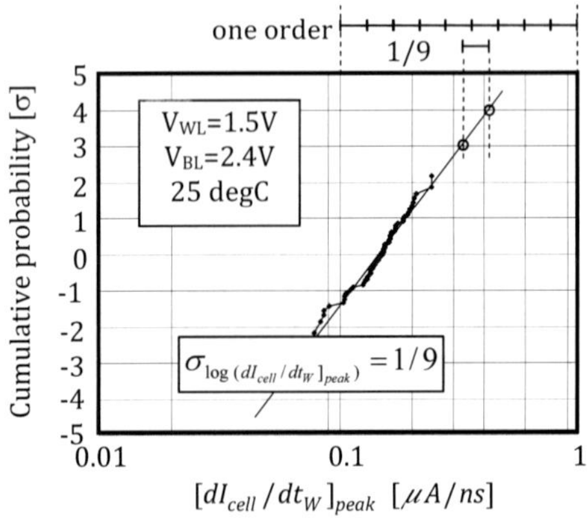

Figure 7.46 Cumulative distribution of the measured peak derivative of the cell read current with respect to the write duration time for multiple cells in a cell array. The data were taken for the FBCs in a cell array with L_g = 150 nm and the BL and WL voltages for "1" write with $V_{BL\_i/1}$ = 2.4 V and $V_{WL\_i/1}$ = 1.5 V, respectively.

Now we check condition (iii). Figure 7.48 shows TCAD Monte Carlo simulation results of hole number fluctuation of FBC with L_g = 150 nm.[14] The fluctuation originates from dopant atom's fluctuations. We take fluctuations of both donor in source and drain and acceptor in body into account. This graph indicates that $N_{hole1} = \langle N_1 \rangle - \langle N_0 \rangle = 500$ and $\sigma_{Nhole1} = 20$, where $\langle N_1 \rangle$ and $\langle N_0 \rangle$ stand for the averaged number of holes accumulated in

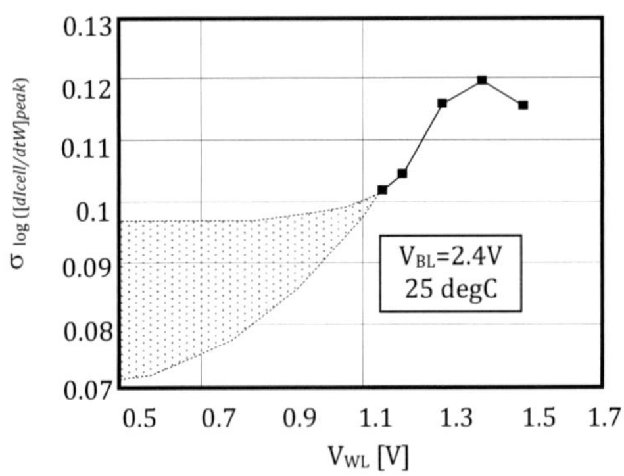

Figure 7.47 Prediction of the standard deviation of the distribution of the peak derivative of the cell read current with respect to the write duration time made by extrapolation from larger word line voltages to the smaller word line voltage (0.5 V) used in actual autonomous refresh operations. The bit line voltage is fixed at 2.4 V.

a cell of data "1" and "0," respectively. This point (500, 20) is also within the pass area of Fig. 7.44. This result shows that the condition (iii) is also satisfied if the condition (i) is fulfilled.

As a result, high-density memories using state-of-the-art FBC is shown to be autonomous-refreshed by introducing some additional process which increases the number of interface states. Other conditions, namely those for the impact ionization current fluctuation and the hole number fluctuation for the "1" cell, are shown to be satisfied by state-of-the-art FBC.

7.5.5 Autonomous Refresh of Direct Tunneling Current: Single-Transistor SRAM Cell

Another candidate for I_{out} is gate oxide direct tunneling current. For this autonomous refresh, the gate to generate I_{in} by impact ionization current $I_{i/i}$ and that to draw I_{out} by gate oxide direct tunneling current I_{DT} need to be different as shown schematically in Fig. 7.49, because the voltages of the gate for $I_{i/i}$ and for I_{DT} are opposite in sign with respect to V_{SL}, making the body potential shift too much in the transition between the two operating points and making the I_b - V_b butterfly curves too slim or with no wings. Therefore, some double gate MOSFET structures such as a planar double gate MOSFET or an independent gate Fin FET etc. are mandatory for this type of autonomous refresh.

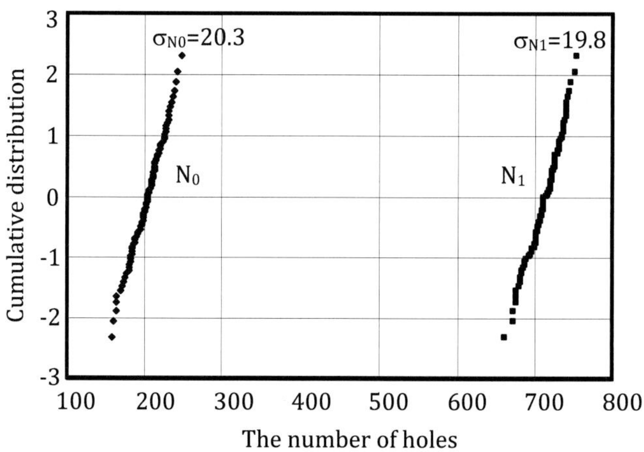

Figure 7.48 TCAD Monte Carlo simulation results for the fluctuation of the number of holes accumulated in a cell's body. The FBC size is $L_g = W = 150$ nm.

Figure 7.50 is TCAD results which simulate $I_{in} = I_{i/i} + I_{p-n} + I_{GIDL}$ and $I_{out} = I_{DT}$ for the FBC with W/L_g = 70 nm/75 nm, $t_{fox} = t_{box}$ = 2.2 nm, and t_{Si} = 21 nm. The back gate voltage is constant at −1.8 V for the direct tunneling. The impact ionization cycle with $V_{BL\_i/i}$ = 1.5 V and $V_{WL\_i/i}$ = 1.5 V has a duration $t_{i/i}$ = 200 ns in a period t_{ref} = 100 μs as is shown in Fig. 7.51. It is shown that an unstable stationary point is successfully generated between two stable stationary points.

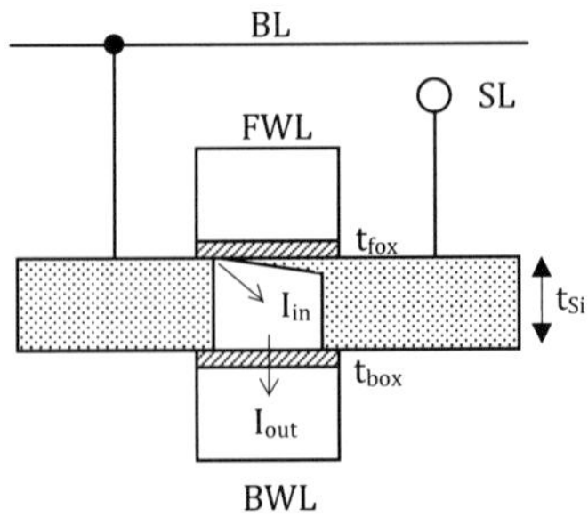

Figure 7.49. Example of the MOSFET structure with two independent gates coupled to a floating body that can perform the autonomous refresh using impact ionization and the gate direct tunneling one.

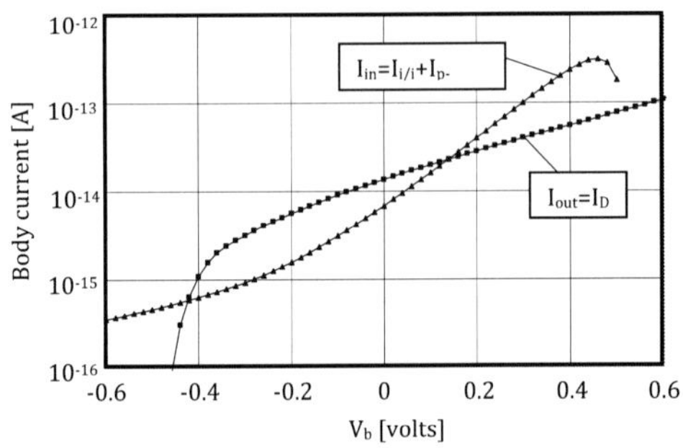

Figure 7.50 TCAD simulation results for impact ionization current I_{in}, p-n junction current $I_{p\text{-}n}$, and GIDL one I_{GIDL}, which are included in $I_{in} = I_{i/I} + I_{p\text{-}n} + I_{GIDL}$, and that for gate oxide direct tunneling current I_{DT}, which is included in I_{out} for the double-gate FBC with W/L_g = 70 nm/75 nm, $t_{fox} = t_{box}$ = 2.2 nm, and t_{Si} = 21 nm (for the definition of the parameters, refer to Fig. 7. 49). The back gate voltage is constantly fixed at −1.8 V for the direct tunneling. Impact ionization occurs during 200 ns in the 100 μs period with both the BL and the WL voltages being 1.5 V (refer to Fig. 7.51 for the waveforms).

One merit of the autonomous refresh using gate direct tunneling over that using charge pumping is that both I_{DT} and $I_{i/i}$ can be flowing in static conditions, i.e., with the voltages of the front and the back WLs and the BL constant during retaining data. Although dynamic pumping of the WL is indispensable for the autonomous refresh using charge pumping phenomenon, the pulsed impact ionization current which is assumed in the simulation of Fig. 7.50 can be diluted over the entire period of data retention by reducing the BL voltage and/or WL voltage, making constant voltage application for impact ionization during the data retention as is shown in Fig. 7.52 where the gate direct tunneling current is also flowing constantly to BWL to which a constant negative voltage is applied. Therefore, SRAM can be achieved by using FBC. This single-transistor SRAM cell is theoretically expected to have the simplest structure ever reported as far as authors know.[20,21] The data retention current in this refresh scheme is determined only by the cell DC current on the contrary to the autonomous refresh using charge pumping current where the AC current due to charging the WL and the BL capacitances must be added to the cell's DC current. We can roughly estimate from Fig. 7.50 the maximum data retention current of 1 Gbit SRAM using this refresh as $1.0 \times 10^9 \times 1.0 \times 10^{-13}$ A = 100 μA which is quite low for a high-speed RAM whose access time can be in nanosecond range.[22] It is also beneficial that the gate direct tunneling current is more controllable than the charge pumping current whose origin comes from the interface states which is essentially out of control. In-depth analysis as well as detailed experiments will be expected to be made for realization of the autonomous refresh using gate oxide direct tunneling. The autonomous refresh is a technology that mitigates the retention

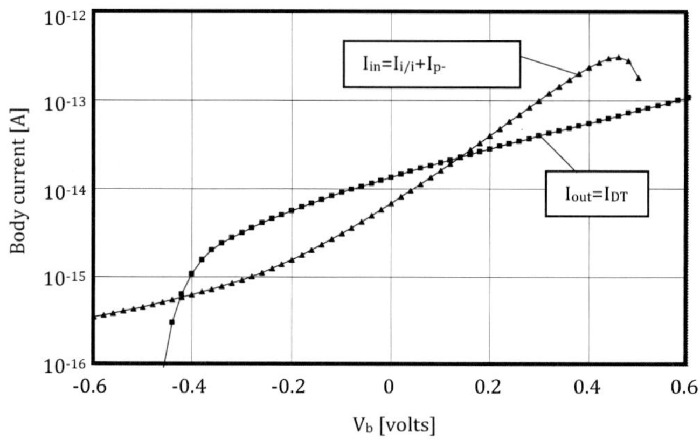

Figure 7.51 A set of voltage pulses that is applied to FBC with double-gate structure to obtain the TCAD simulation in Fig. 7.50 for the autonomous refresh using gate direct tunneling current.

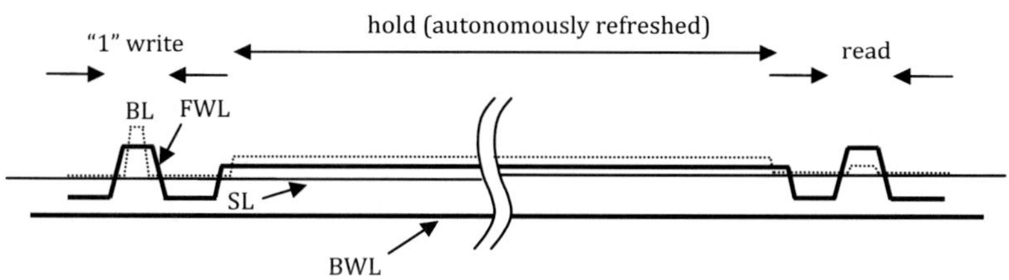

Figure 7.52 Example of a set of voltage pulses applied to FBC with double-gate structure for the autonomous refresh using gate direct tunneling current along with a normal "1" write cycle and a read cycle. This operation can hold data without dynamic voltage application, making the FBC an SRAM cell.

time specifications of FBC to meet the refresh power and the busy rate of stand-alone DRAMs and it can make full use of its advantages when combined with other leakage current reduction works on FBC.[23]

References

1. M. Horstmann, D. Greenlaw, Th. Feudel, A. Wei, K. Frohberg, G. Burbach, M. Gerhardt, M. Lenski, R. Stephan, K. Wieczorek, M. Schaller, J. Hohage, H. Ruelke, J. Klais, P. Huebler, S. Luning, R. van Bentum, G. Grasshoff, C. Schwan, J. Cheek, J. Buller, S. Krishnan, M. Raab and N. Kepler, "Sub-50nm gate length SOI transistor development for high performance microprocessor," *Mater. Sci. Eng.*, B 114–115, pp. 3–8, 2004.
2. J. Warnock, D. Wendel, T. Aipperspach, E. Behnen, R. A. Cordes, S. H. Dhong, K. Hirairi, H. Murakami, S. Onishi, D. C. Pham, J. Pille, S. D. Posluszny, O. Takahashi and H. Wen, "Circuit design techniques for a first-generation Cell broadband engine processor," *IEEE J. Solid-State Circuits*, vol. 41, no. 8, pp. 1692–1706, Aug. 2006.
3. R. Ranica, A. Villaret, P. Malinge, P. Mazoyer, D. Lenoble, P. Candelier, F. Jacquet, P. Masson, R. Bouchakour, R. Fournel, J. P. Schoellkopf and T. Skotnicki, "A one transistor cell on bulk substrate (1T-Bulk) for low-cost and high density eDRAM," *2004 Symposium on VLSI Technology Digest of Technical Papers*, pp. 128–129, June 2004.
4. T. Yamada, K. Takahashi, H. Oyamatsu, H. Nagano, T. Sato, I. Mizushima, S. Nitta, T. Hojo, K. Kokubun, K. Yasumoto, Y. Matsubara, T. Yoshida, S. Yamada, Y. Tsunashima, Y. Saito, S. Nadahara, Y. Katsumata, M. Yoshimi and H. Ishiuchi, "An embedded DRAM technology on SOI/Bulk Hybrid structure formed with SEG process for high-end SOC application," *2002 Symposium on VLSI Technology Digest of Technical Papers*, pp. 112–113, June 2002.
5. K. H. Yeo, C. W. Oh, S.-M. Kim, M.-S. Kim, C.-S. Lee, S.-Y. Lee, M. Li, H.-J. Cho, E.-J. Yoon, S.-H. Kim, J.-D. Chow, D.-W. Kim, D. Park and K. Kim, "80 nm 512 M DRAM with emhanced data retention time using partially-insulated cell array transistor (PiCAT)," *2004 Symposium on VLSI Technology Digest of Technical Papers*, pp. 30–31, June 2004.
6. C. H. Stapper, "The effects of wafer to wafer defect density variations on integrated circuit defect and fault distributions," *IBM J. Res. Develop.*, vol. 29, no. 1, pp. 87–97, 1985.
7. *International Technology Roadmap for Semiconductors (ITRS)*, 2007 [Online]. Available: http://www.itrs.net.
8. T. Ohsawa, K. Fujita, K. Hatsuda, T. Higashi, M. Morikado, Y. Minami, T. Shino, H. Nakajima, K. Inoh, T. Hamamoto and S. Watanabe, "An 18.5 ns 128 Mb SOI DRAM with a floating body cell," *ISSCC Dig. Tech. Papers*, pp. 458–459, Feb. 2005.
9. T. Ohsawa, K. Fujita, K. Hatsuda, T. Higashi, T. Shino, Y. Minami, H. Nakajima, M. Morikado, K. Inoh, T. Hamamoto, S. Watanabe, S. Fujii and T. Furuyama, "Design of a 128 Mb SOI DRAM using the floating body cell (FBC)," *IEEE J. Solid-State Circuits*, vol. 41, no. 1, pp. 135–145, Jan. 2006.

10. K. Fujita, T. Ohsawa, R. Fukuda, F. Matsuoka, T. Higashi, T. Shino and Y. Watanabe, "Array architecture of floating body cell (FBC) with quasi-shielded open bit line scheme for sub-40 nm node," *IEEE Int. SOI Conf.*, pp. 31–32, Oct. 2008.

11. T. Ohsawa, K. Fujita, T. Higashi, Y. Iwata, T. Kajiyama, Y. Asao and K. Sunouchi, "Memorydesign using one-transistor gain cell on SOI," *ISSCC Dig. Tech. Papers*, pp. 152–153, Feb. 2002.

12. T. Ohsawa, K. Fujita, T. Higashi, Y. Iwata, T. Kajiyama, Y. Asao and K. Sunouchi, "Memory design using a one-transistor gain cell on SOI," *IEEE J. Solid-State Circuits,* vol. 37, no. 11, pp. 1510–1522, Nov. 2002.

13. T. Shino, T. Higashi, N. Kusunoki, K. Fujita, T. Ohsawa, N. Aoki, H. Tanimoto, Y. Minami, T. Yamada, M. Morikado, H. Nakajima, K. Inoh, T. Hamamoto and A. Nitayama, "Fully-depleted FBC (Floating body cell) with enlarged signal window and excellent logic process compatibility," *IEDM Tech. Dig.*, pp. 281–284, Dec. 2004.

14. H. Furuhashi, T. Shino, T. Ohsawa, F. Matsuoka, T. Higashi, Y. Minami, H. Nakajima, K. Fujita, R. Fukuda, T. Hamamoto and A. Nitayama, "Scaling scenario of floating body cell (FBC) suppressing Vth variation due to random dopant fluctuation," *IEEE Int. SOI Conf.*, pp. 33–34, Oct. 2008.

15. T. Hamamoto, Y. Minami, T. Shino, N. Kusunoki, H. Nakajima, M. Morikado, T. Yamada, K. Inoh, A. Sakamoto, T. Higashi, K. Fujita, K. Hatsuda, T. Ohsawa and A. Nitayama, "A floating-body cell fully compatible with 90 nm CMOS technology node for a 128 Mb SOI DRAM and its scalability," *IEEE Trans. Electron Devices,* vol. 54, no. 3, pp. 563–571, March 2007.

16. T. Ohsawa, T. Higashi, K. Fujita, K. Hatsuda, N. Ikumi, T. Shino, H. Nakajima, Y. Minami, N. Kusunoki, A. Sakamoto, J. Nishimura, T. Hamamoto and S. Fujii, "A 128 Mb floating body RAM (FBRAM) on SOI with multi-averaging scheme of dummy cell," *2006 Symposium on VLSI Circuits Digest of Technical Papers*, pp. 224–225, June 2006.

17. T. Ohsawa, R. Fukuda, T. Higashi, K. Fujita, F. Matsuoka, T. Shino, H. Furuhashi, Y. Minami, H. Nakajima, T. Hamamoto, Y. Watanabe, A. Nitayama and T. Furuyam, "Autonomous refresh of floating body cell (FBC)," *IEDM Tech. Dig.*, pp. 801–804, Dec. 2008.

18. T. Ohsawa, R. Fukuda, T. Higashi, K. Fujita, F. Matsuoka, T. Shino, H. Furuhashi, Y. Minami, H. Nakajima, T. Hamamoto, Y. Watanabe, A. Nitayama and T. Furuyama, "Autonomous refresh of floating body cell due to current anomaly of impact ionization," *IEEE Trans. Electron Devices*, vol. 56, no. 10, pp. 2302–2311, Oct. 2009 (in press).

19. P. Fazan, S. Okhonin and M. Nagoga, "A new block refresh concept for SOI floating body memories," *IEEE Int. SOI Conf.*, pp. 15–16, Sept. 2003.

20. K. Sakui, T. Hasegawa, T. Fuse, S. Watanabe, K. Ohuchi and F. Masuoka, "A new static memory cell based on the reverse base current effect of bipolar transistor," *IEEE Trans. Electron Devices*, vol. 36, no. 6, pp. 1215–1217, June 1989.

21. R. Roy, F. Nemati, K. Young, B. Bateman, R. Chopra, S.-O. Jung, C. Show and H.-I. Cho, "Thyristor-based volatile memory in nano-scale CMOS," *ISSCC Dig. Tech. Papers*, pp. 2612–2613, Feb. 2006.

22. K. Hatsuda, K. Fujita and T. Ohsawa, "A 333MHz random cycle DRAM using the floating body cell," *Proc. IEEE 2005 Custom Integrated Circuit Conf.*, pp. 259–262, Sept. 2005.

23. T.-S. Jang, J.-S. Kim, S.-M. Hwang, Y.-H. Oh, K.-M. Rho, S.-J. Chung, S.-O. Chung, J.-G. Oh, S. Bhardwaj, J. Kwon, D. Kim, M. Nagoga, Y.-T. Kim, S.-Y. Cha, S.-C. Moon, S.-W. Chung, S.-J. Hong and S.-W. Park, "Highly scalable Z-RAM with remarkably long data retention for DRAM application," *2009 Symposium on VLSI Technology Digest of Technical Papers*, pp. 234–235, June 2009.

Chapter 8

Floating Body Cell Development Activities and Future Directions

The past results of research and development for floating body cell are reviewed and categorized to foresee right directions in which the future works should be focused on to make floating body RAM a real product with attractive features.

8.1 PAST ACTIVITIES

Before we forecast future studies and developments that are necessary in making floating body cell (FBC) phased into real memory products, let us examine FBC patents issued or applied and FBC papers published in academic journals or presented at international conferences for them to be categorized. This may underscore important activities that will highlight future directions. There have appeared about 100 publications regarding FBC in the form of papers printed in journals or presented in major conferences and of issued patents[a] in the nine years since the original studies appeared that directly triggered the present-day activities of FBC research and development that are continuing till now, putting aside several independent pioneering works reviewed in section 1.3. They are categorized in Fig. 8.1 which illustrates the flows of the FBC technologies that have been in public domain originated from the cell concept proposal in 2001 to various directions until 2009.

From 2001 to 2002, the idea of FBC was proposed in the two organizations independently.[1–5] In February 2002, a paper on 512 kbit memory using FBC was presented at the *International Solid-State Circuits Conference* (*ISSCC*),[6] followed by a paper publication in November 2002.[7] The evaluation results of the memory were and *Symposium on VLSI Circuits*[9] both

[a] Here, only patents regarding the basic ideas of FBC are referred to. It is certain that more than 100 patents have been issued worldwide till now and that much more are pending.

Floating Body Cell: A Novel Capacitor-less DRAM Cell
By Takashi Ohsawa and Takeshi Hamamoto

www.panstanford.com

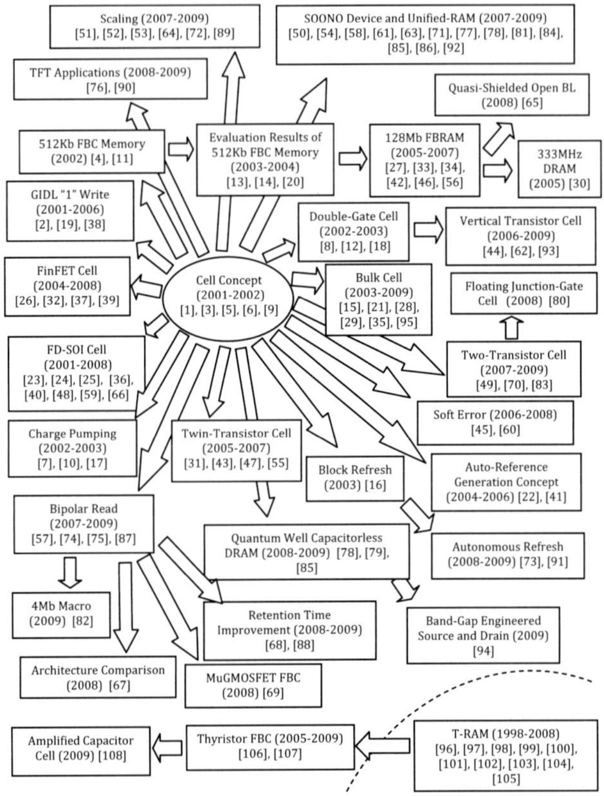

Figure 8.1 Technology flows of FBC from 2001 through 2009.

in June 2003, and *Symposium on VLSI Technology* in June 2004.[10] These initial works proposed the cell basic concept and its operational methods combined with verifications confirmed by simulations and experiments.

Almost at the same time, FBC with double gate structure was proposed.[11–13] The data "1" are purged to the data "0" by applying a positive voltage pulse to the back-gate of the cell.

It is worth noting that a paper about the effect of the charge pumping phenomenon on the FBC characteristics was published at the very early stage of FBC development in May 2002.[14] This paper along with related ones[15,16] made a big impact on the sense amplifier circuit (S/A) design of floating body RAM (FBRAM).[17,18]

The FBC built on bulk silicon substrate instead of on SOI substrate was developed from 2003 through 2005.[19–22] The p-type cell's body is made floating by isolating it from the p-type bulk silicon by n-type well. The 8 Mb memory was designed and fabricated with the measurement results presented in June 2005, at *Symposium on VLSI Circuits*.[23] Though the cell size of the FBC built on bulk silicon substrate becomes large because of the cell-to-cell isolation along the bit line (BL) direction, a paper was presented at *International Conference on Solid State Devices and Materials* (*SSDM*) in October 2009, which proposed a $6F^2$ size bulk FBC whose source and drain are isolated from the bulk by silicon dioxide layers.[24] The source and the drain are made of poly-silicon and the p-n junction area between the body and the source/drain is controlled to be small, leading to the increased number of holes accumulated for the data "1" cell.

An FBC-specific refresh method (Block Refresh) was disclosed at *IEEE International SOI Conference* in October 2003.[25] The paper claimed that the method can refresh all cells in an array simultaneously to improve the refresh efficiency drastically. This work prompted the investigation on physics behind the block refresh and a general concept of the autonomous refresh was reached, suggesting a possibility of operating FBC as a static random access memory (SRAM) cell by using the gate direct tunneling current for extracting holes from the body.[26,27]

The proposal and development of FBC built on a fully depleted SOI (FD-SOI) have been made since 2002.[28–35] This type of FBC is one of the possible structures for future scalable cells.

At *International Electron Devices Meeting* (*IEDM*) in December 2003, the "1" write method by the band-to-band tunneling or the Gate-Induced Drain Leakage (GIDL) was proposed,[36] followed by a paper publication in April 2006.[37] The idea was originally described in a US patent filed in December 2001.[38] The method provides a low power write, because the cell does not flow the channel current on the contrary to the impact ionization method.

At *IEDM* in December 2004, Fin FET FBC was proposed.[39] The Fin FET FBC has been investigated till recently.[40–42] This 3D cell structure is expected to extend the generation of FBC to the technology nodes that are thought difficult to be reached by the simple planar FBC.

Another 3D transistor, the vertical transistor, was also applied to FBC. At *SSDM* in September 2006, a paper was presented on an FBC using surrounding gate transistor (SGT).[43] The cell array was fabricated with partially depleted silicon pillars on bulk silicon substrate. A smaller vertical transistor FBC was disclosed in a letter paper in June 2008, which uses a fully depleted silicon pillar with independently controlled double gate, one of which is negatively biased to make a potential well for holding holes.[44] An interesting idea on a cell array and its operation for the vertical double gate transistor FBC was proposed with some simulation results at *SSDM* in October 2009, that can achieve $4F^2$ cell size with simple fabrication process.[45] The operation is little bit complicated to make each cell selectively written and read. This architecture is worth being verified by hardware measurements for an actual cell array. The vertical transistor FBC is attractive especially because it can be built on bulk silicon substrate.

As an idea of circuit to distinguish cell read data using an automatically calibrated reference current, a method of automatic reference current generation for sensing the cell signal was reported in June 2004, at *Symposium on VLSI Circuits*,[46] followed by a paper publication in June 2006.[47]

At *ISSCC* in February 2005, a design of 128 Mb FBRAM was presented.[17] This design was based on the finding of the effect of the charge pumping phenomenon on the data "1" state degradation by word line (WL) pumping.[14–16] A very accurate reference current for sensing the signal is generated by averaging many pairs of dummy cells with data "1" cell and the data "0" explained in section 6.4. The design detail was also published in a journal article.[18] A various evaluation results on this FBRAM were constantly published for a long time.[48–51] High speed RAM design based on the twin cell scheme was presented at *IEEE Custom Integrated Circuits Conference* (*CICC*) in September 2005, showing the potential of 333 MHz random cycle operation in simulation.[52] On the other hand, at *IEEE International SOI Conference* in October 2008, an open BL array architecture of FBC which is highly immune against inter BL coupling noise was presented that has no penalty on the access time and the cycle time degradation explained in section 6.6.[53]

Though it is not a single-transistor cell, an idea of twin-transistor RAM (TTRAM) cell was proposed at *CICC* in September 2005.[54] This cell consists of an FBC plus a select MOS transistor to avoid the BL disturb. The extended method of writing to TTRAM was presented at *Symposium on VLSI Circuits* in June 2006[55] and also published in journals.[56,57] Another idea of two-transistor FBC was presented at *IEEE International SOI Conference* in October 2007,[58] followed by a paper publication in June 2008.[59] This is an FBC whose floating body is connected to another MOSFET's gate. This MOSFET forms a sensing transistor. Recently, the idea of the two-transistor FBC in which a gated diode instead of the floating body of a MOSFET is connected to the sensing transistor to reduce the cell size was proposed.[60] A letter paper was published on a novel idea of the FBC whose floating body plays as a floating gate of a read transistor.[61]

A soft error rate of FBC was measured in an accelerated alpha particle test[62] and was also estimated by using a simulator under the assumption of neutron exposure with the spectrum at sea level in New York City.[63] The former measurements exhibited about 1000 FITs/Mbit, while the latter estimated less than 10 FITs/Mbit. Further investigation is required for assessing soft error rates for scaled FBCs in various structures.

An analysis of FBC scalability was presented at *IEEE International SOI Conference* in October 2008,[64] which was made by using TCAD simulation with signal uncertainty considered based on random dopant fluctuations. The scaling potential of the cell is one of the most intensely investigated themes in FBC.[65–69]

A new method of reading signal from FBC was disclosed at *IEDM* in December 2007, which utilizes current of a parasitic bipolar transistor built in the MOSFET structure instead of channel current of the MOSFET.[70] It claimed that a large signal between the data "1" and the "0" is obtained and that there is no need to apply a negative back-gate bias even for FD-SOI FBC because holes are accumulated under the front-gate that is not positively biased. The last feature is attractive from a scalability point of view, because a $4F^2$ size FBC is achievable with the structure of the SGT or the multiple-gate MOSFET.[71] This work was followed by two papers presented at *IEDM* in December 2008, which confirm the usefulness of this read method.[72,73] There was also a letter paper published that analyzed the parasitic bipolar effect by using device simulation.[74] It focused on the method of "1" write by using the effect and found a constraint in timing offset between BL reset and WL reset for successful write. It is worth noting that FBC with this read method provides a perfect nondestructive read-out cell. A paper presented at *IEEE International SOI Conference* in October 2008 concluded that the read using the parasitic bipolar is effective in large signal window for a scaled Fin FET FBC over the MOS channel current read.[75]

An FBC can be implemented into a non-volatile memory cell having a charge trapping layer in the gate insulator. One is a silicon-on-ONO device[76,77] and the other is a unified-RAM.[78–88] The former consists of an FBC built on ONO layer. The latter is an FBC with a gate dielectric having a charge trapping layer within it. Although there remains much room to discuss, they may widen the application of FBC.

A paper on an FBC using polycrystalline silicon thin-film transistor (poly-Si TFT) was published in November 2008.[89] The FBC is aimed for the application of flat panel display and has trenched body to improve the signal ΔV_{th} as well as the retention time. Another paper on an FBC using poly-Si TFT was published which is supposed to be used in 3D layer stackable high-density memories.[90]

It is interesting that several studies appear on an FBC which stores holes in a quantum-well with hetero structures such as $Si/Si_{(1-x)}Ge_x/Si$ instead of in a floating body isolated by silicon dioxide (SOI) or buried n-well.[83,91] The paper cited before[86] is also in the same direction in the sense that the cell utilizes a quantum-well by the hetero structure of $Si/Si_{1-y}C_y/Si$ to store holes. An idea of using SiC for the source and the drain of FBC was presented at *SSDM* in October 2009, which can make a deeper potential well than the conventional FBC, enhancing the signal window.[92] This transistor structure has also been studied for an efficient stressor to enhance the drain current in SOI MOSFET devices.

At *ISSCC* in February 2009, a paper was presented describing a 4 Mbit memory macro using FBC which is designed to be implemented into microprocessors as cache.[93] Its measured 2 ns access time and 4 ns random cycle time with a wide operating window, drawing an attention whether embedded memories are being realized at last by using FBC. For the stand-alone memory application, a paper was published at *Symposium on VLSI Technology* in June 2009, which reported an improved retention time characteristics.[94] A paper on a model of leakage current of FBC was presented at *IEEE International SOI Conference* in October 2008.[95]

Prior to the onset of the FBC memory investigation, a memory cell using a thyristor (T-RAM) was proposed and has been studied.[96–105] The T-RAM cell is an SRAM cell without the necessity of being refreshed with data stored in binary stable states generated by the negative differential resistance of a thyristor. At *IEDM* in December 2005, a paper was presented which proposed a new operational mode of the T-RAM which stores data dynamically with the power shut down during retaining data to reduce the data retention power.[106] This cell can be regarded as one category of FBC variations. A vertical thyristor cell was also proposed to make the cell size $4F^2$.[107] An idea was presented at *IEDM* in December 2009, on two-transistor cell, one of which acts as a bipolar device to use the collector current enhanced by the base current for discharging the BL capacitor.[108] This cell is also very close to FBC in its basic principle of storing majority carries in the floating body of a MOSFET.

Figure 8.2 shows the number of papers published or presented and key US patents filed in each year from 2001 to 2009 on FBC. The number is steadily increasing year by year, reaching more than 10 papers per year in 2007.

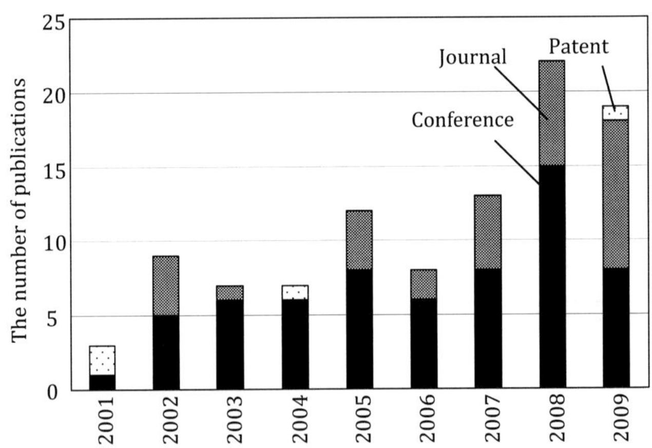

Figure 8.2 The number of the papers published or presented and key US patents issued or applied in each year on FBC.

8.2 TECHNOLOGY VARIATIONS

Table 8.1 shows merits and demerits of the FBC transistor structures along with their suitable read operational modes, i.e., the MOS channel current read mode, the bipolar current read mode, and the thyristor current read mode.

8.2.1 2D Cells

There are five 2D cell structures: bulk transistor, SOI transistor, band-gap engineered transistor, double gate transistor on SOI, and thyristor on SOI. The original bulk transistor cell is just a simple n-type MOFFET whose body is made floating by using deep n-well as is shown in Fig. 8.3. Therefore, each cell needs to be isolated from adjacent ones by shallow trench isolation, making the cell size $10 \sim 12F^2$.[19–23] On the other hand, the bulk cell with quasi-SOI structure which is shown in Fig. 8.4 can shrink the cell to $6F^2$ size by using deep isolation of the source and the drain from the bulk silicon which reaches the deep n-well at the cost of using a complex fabrication process.[24]

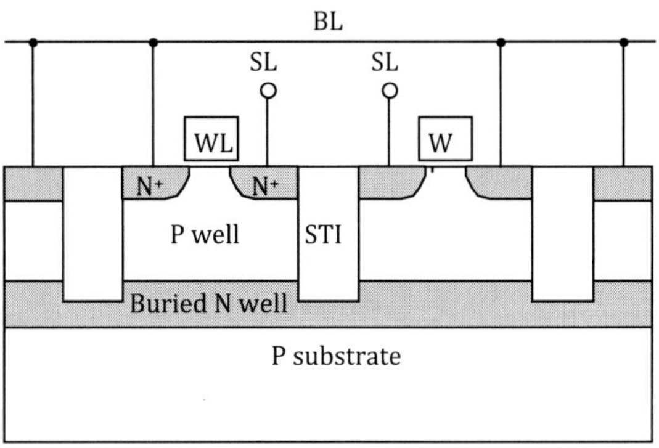

Figure 8.3 FBC on bulk silicon substrate.

The partially depleted silicon-on-insulator (PD-SOI) transistor cell, one of the oldest FBCs which is illustrated in Fig. 8.5, is a standard with respect to the figure of merit. The demerit of the PD-SOI transistor cell comes from the fact that the impurity doping concentration in the body which is inevitably high for scaled cells leads to a large threshold voltage fluctuation, degrading the signal-to-noise ratio (SNR). To avoid this degradation,

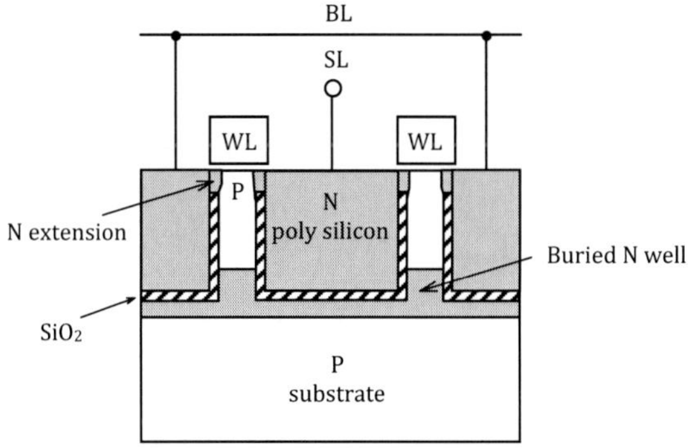

Figure 8.4 Quasi-SOI FBC on bulk silicon substrate.

Table 8.1 Comparison among FBC transistor structures with suitable read modes

Dim.	Transistor structures		Read mode	Merits	Demerits
2D	Bulk transistor	Conventional	MOS	Low SC#	• Large cell size • Large RDF%
		Quasi-SOI		Low SC	• Large RDF • Complex integration
	SOI transistor	PD-SOI	MOS	Standard	Large RDF
		FD-SOI	MOS	Large SNR$	Thin SB& required
	Double gate transistor on SOI		MOS	• Large SNR • Relaxed BL disturb	• Difficult GA* • Thin SB required
	Band-gap engineered transistor	Quantum-well on bulk Si.	MOS	• Low SC • Large SNR	• Complex integration
		Wide-gap S/D on SOI		Large SNR	Complex integration
	Thyristor on SOI		Thyristor	Large SNR	Large cell size
3D	Fin transistor on SOI	Independent double gate	MOS	• Large SNR • Relaxed BL disturb	• Thin SB required • Large cell size
		Π gate	Bipolar	Large SNR	Read BL disturb
	Vertical transistor on bulk Si.	Independent double gate	MOS	• Low SC • Large SNR • Relaxed BL disturb • $4F^2$	• Complex array control • Read current drivability
		Surrounding gate	Bipolar	• Low SC • $4F^2$	• Read BL disturb • Complex integration
	Vertical thyristor on bulk Si.		Thyristor	• Large SNR • Low SC • $4F^2$	Complex integration

Substrate cost %: Random dopant fluctuation
$ Signal-to-noise ratio &: Silicon body
* Gate alignment

the SOI substrate should be thin and lightly or non-doped, leading to the FD-SOI transistor cell as is shown in Fig. 8.6. The challenge will be a very thin silicon film thickness for scaled cells. One of the most serious problems for an ultra-thin SOI body is its difficulty of accumulating enough number of holes due to the quantum confinement of carriers.[66] The double gate transistor cell illustrated in Fig. 8.7 has a merit of eliminating the BL "0" disturb by positively biasing the back-gate to eject holes instead of negatively biasing the associated BL (see 2.6.2.6 in section 2.6). However, the 2D (planar) double gate transistor is very two scaled gates aligned. The thin silicon film is also required in this option. Though they are subject to their own complex fabrication processes, the bad-gap engineered transistor cells have possibilities to eliminate or mitigate some of the demerits owned by the other 2D transistor cells. The FBC having a quantum-well on bulk silicon substrate can store holes within the quantum-well introduced by a hetero

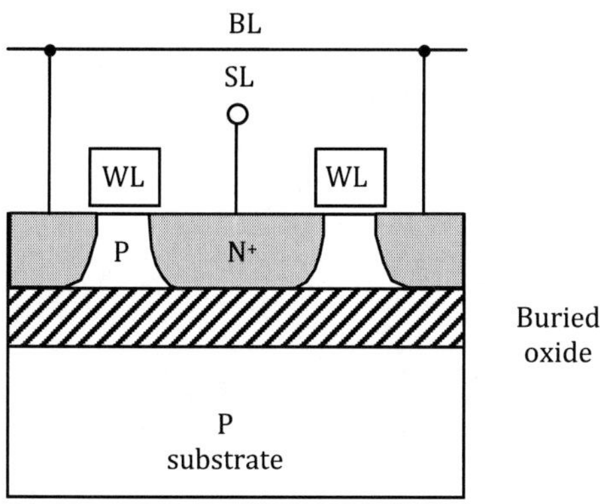

Figure 8.5 FBC on PD-SOI substrate.

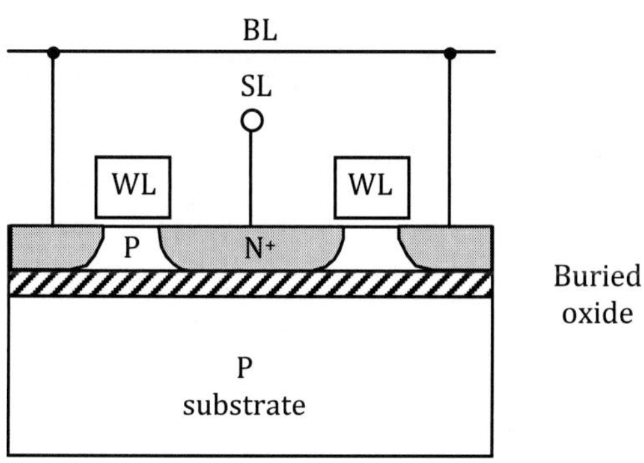

Figure 8.6 FBC on FD-SOI substrate.

structure such as $Si/Si_{(1-x)}Ge_x/Si$ and $Si_{1-y}C_y/Si$. By controlling the position of the quantum-well from the silicon channel, the signal of the cell can be enhanced without thinning the silicon film. This type of FBC is depicted in Fig. 8.8. Though further investigation for the options is still needed, this idea

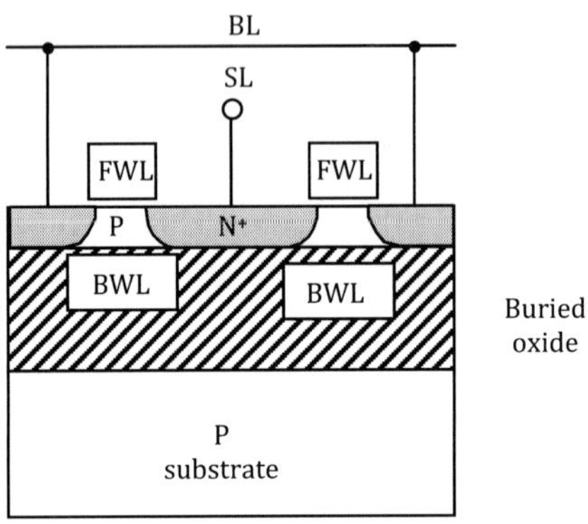

Figure 8.7 Double gate FBC.

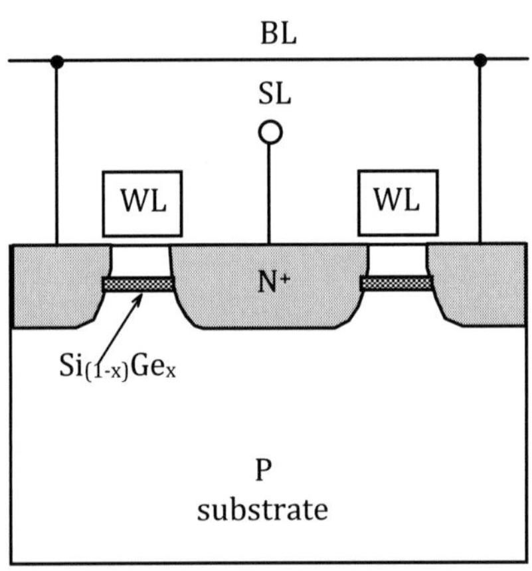

Figure 8.8 Quantum-well FBC on bulk silicon substrate.

may be used for the future scaled FBCs. The wide-gap source/drain transistor cell on SOI illustrated in Fig. 8.9 is another interesting idea for storing more holes in the body than the conventional FBC. The characteristics may mitigate the trend of thinning the body and the quantum confinement effect. Though the 2D thyristor cell shown in Fig. 8.10 has a large SNR, its cell size becomes larger. The scaled thyristor cell will be made full use of in its 3D version as will be explained later.

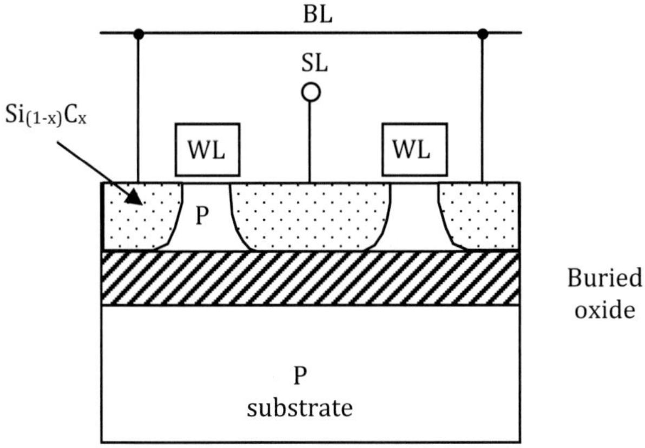

Figure 8.9 Wide-gap source and drain FBC on SOI.

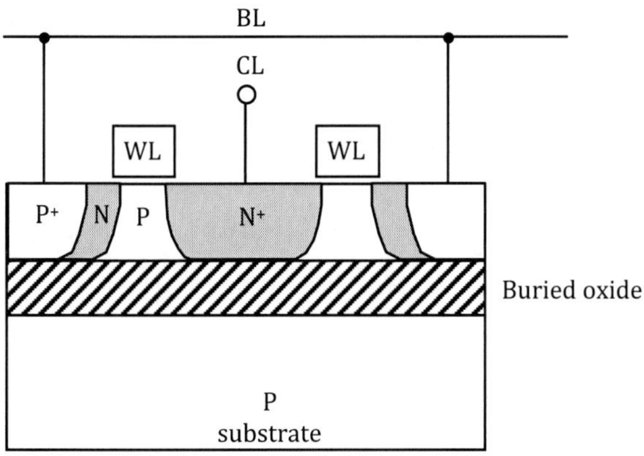

Figure 8.10 Thyristor FBC on SOI.

8.2.2 3D Cells

Many drawbacks in the 2D cells can be totally eliminated or partially mitigated by making the direction of the cell current flow perpendicular to the silicon surface, converting them to one type of 3D cells using a vertical transistor or a vertical thyristor. Another type of 3D cells, fin transistor cells, has their current direction still parallel to the silicon surface, though their currents are not limited within the silicon surface. Therefore, the fin transistor cells has basically the similar performance as their 2D counterparts except for the body volume increase which has a benefit of reducing the SNR degradation due to random dopant fluctuation.

The independent double gate fin transistor cell on SOI is the 3D version of the double gate transistor cell on SOI. A plan view and the cross-sectional views along BL and SL for an example of the independent double gate fin transistor cell are illustrated in Fig. 8.11. The cell size will be larger than the 2D counterpart. However, the difficulty of the alignment of the two gates is eliminated because of its gate electrode orientation which is perpendicular to the silicon surface. The Π gate transistor cell on SOI which is shown in Fig. 8.12 can be operated with the bipolar current read mode in scaled cells. The cell size will be the same as the standard 2D cell ($6F^2$). One concern is high voltage application for triggering the parasitic bipolar action in read and "1" write cycles, leading to BL disturb. Independent double gate vertical transistor cell on bulk silicon has many merits. The recently proposed cell which is shown in Fig. 8.13 has a small cell size ($4F^2$) to be fabricated with a simple process.[45] One of concerns will be its complex array operation to avoid WL disturb in read and write cycles. Since the channel length in this cell is independent of its footprint, it can use a transistor several generations behind with the cutting-edge cell size. However, the decrease of the cell read current derivatively will be another concern, because W/L_g reduces as cell size shrinks with the height of the pillar ($\sim L_g$) kept constant. The surrounding gate vertical transistor cell on bulk silicon with $4F^2$ cell size whose structure is depicted in Fig. 8.14 be read by the bipolar current read mode, the BL disturb being a concern as is usual for this mode. Since the fabrication process will be different from the standard ones, there is a necessity of developing and optimizing the suitable process. To reduce the BL voltages in read and write cycles, the vertical thyristor cell illustrated in Fig. 8.15 is worth being studied because the triggering voltage of a thyristor is much lower than that of the parasitic bipolar transistor with its base floating. This cell also requires some complex fabrication process.

8.3 FUTURE DIRECTIONS

From the above review of the past activities on FBC technology development, it is clear that the simple 2D transistor cells will hit walls in scaling the cell sizes. They are mainly large random dopant fluctuation leading to reduction of SNR and ultra-thin silicon body leading to reduction of the number of accumulated holes due to the quantum confinement of carriers. In this respect, the following five types will be the candidates for the future scaled FBC: (1) the quantum-well transistor cell on bulk silicon, (2) the wide-gap source and drain transistor on SOI, (3) the independent double gate vertical transistor cell on bulk silicon, (4) the surrounding gate vertical transistor cell on bulk silicon, and (5) the vertical thyristor cell on bulk silicon. They have just started to be studied in recent years, none of them having been fully developed yet.

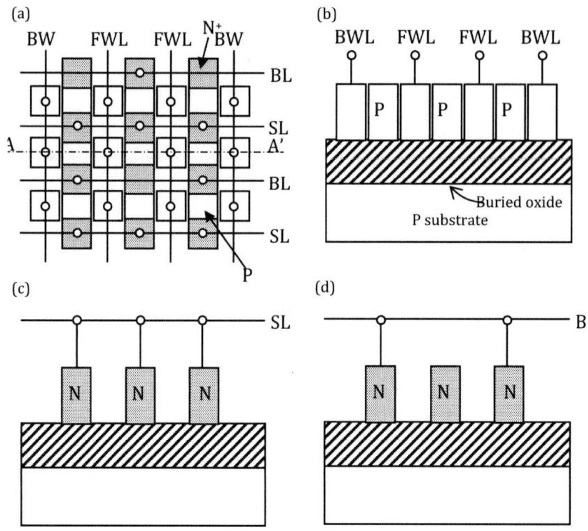

Figure 8.11 Independent double gate fin FBC on SOI: (a) plan view, (b) cross-sectional view along A–A′, (c) along SL, and (d) along BL.

The cell with only one demerit which can be overcome by the maturity of process is the vertical thyristor cell on bulk silicon. Though the p-type region in this thyristor is coupled by a gate electrode as a capacitor, the overlap between the gate and the two n-type diffusion layers besides the p-type

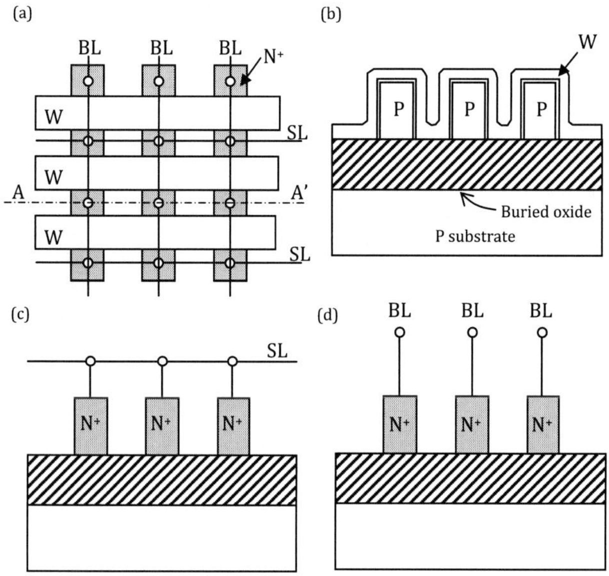

Figure 8.12 Π gate FBC on SOI: (a) plan view, (b) cross-sectional view along WL, (c) along SL, and (d) along A–A′.

region can be eliminated with some margin of distance, because there is no need for the structure to operate as a MOSFET. Therefore, the GIDL can be reduced drastically from that of other cells, leading to a relatively long retention time. And since the thyristor's *I–V* curve has two stably stationary points with a proper load line, the cell can be refreshed without being read by the S/A, making it possible for the cell to be refreshed autonomously just like what was explained in section 7.5. This further reduces the data retention current and the refresh busy rate. However, improving the retention time characteristics is unchangeably one of the biggest challenges in realizing FBC memories.

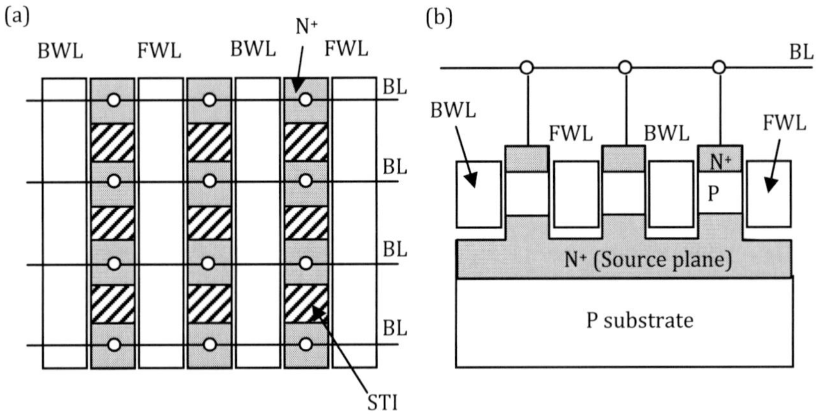

Figure 8.13 Independent double gate vertical transistor FBC on bulk silicon substrate: (a) plan view, (b) cross-sectional view along BL.

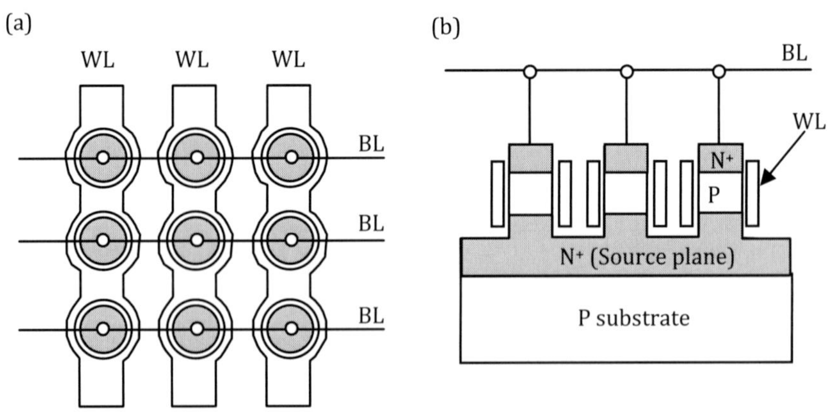

Figure 8.14 Surrounding gate transistor FBC on bulk silicon substrate: (a) plan view, (b) cross-sectional view along BL.

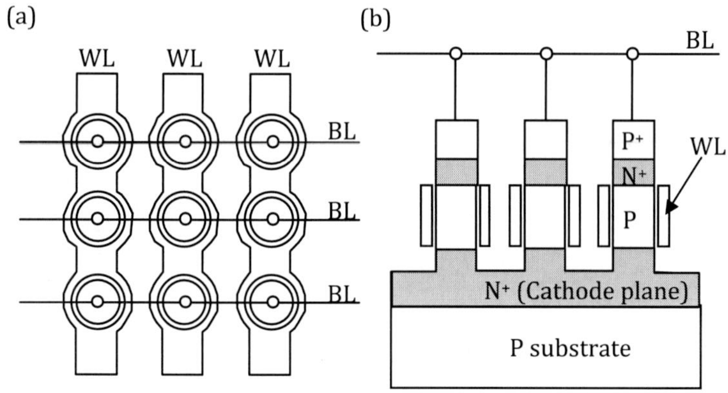

Figure 8.15 Vertical thyristor FBC on bulk silicon substrate: (a) plan view, (b) cross-sectional view along BL.

References

1. T. Ohsawa, "Semiconductor memory device with floating storage bulk region and method of manufacturing the same," *U.S. Patent 6,621,725*, Sep. 2003 (filed in July 2001).
2. S. Okhonin, M. Nagoga, J. M. Sallese and P. Fazan, "SOI capacitor-less 1T-DRAM concept," *IEEE Int. SOI Conf.*, pp. 153–154, Oct. 2001.
3. S. Okhonin, M. Nagoga, J. M. Sallese and P. Fazan, "A capacitor-less 1T-DRAM cell," *IEEE Electron Device Lett.*, vol. 23, no. 2, pp. 85–87, Feb. 2002.
4. P. Fazan, S. Okhonin, M. Nagoga and J.-M. Sallese, "A simple 1-transistor capacitor-less memory cell for high performance embedded DRAMs," *Proc. IEEE 2002 Custom Integrated Circuit Conf.*, pp. 99–102, May 2002.
5. P. Fazan, S. Okhonin, M. Nagoga, J. M. Sallese, L. Portmann, R. Ferrant, M. Kayal, M. Pastre, M. Blagojevic, A. Borschberg and M. Declercq, "Capacitor-less 1-transistor DRAM," *IEEE Int. SOI Conf.*, pp. 10–13, Oct. 2002.
6. T. Ohsawa, K. Fujita, T. Higashi, Y. Iwata, T. Kajiyama, Y. Asao and K. Sunouchi, "Memory design using one-transistor gain cell on SOI," *ISSCC Dig. Tech. Papers*, pp. 152–153, Feb. 2002.
7. T. Ohsawa, K. Fujita, T. Higashi, Y. Iwata, T. Kajiyama, Y.Asao and K. Sunouchi, "Memory design using a one-transistor gain cell on SOI," *IEEE J. Solid-State Circuits*, vol. 37, no. 11, pp. 1510-1522, Nov. 2002.
8. K. Inoh, T. Shino, H. Yamada, H. Nakajima, Y. Minami, T. Yamada, T. Ohsawa, T. Higashi, K. Fujita, T. Ikehashi, T. Kajiyama, Y. Fukuzumi, T. Hamamoto and H. Ishiuchi, "FBC (Floating Body Cell) for embedded DRAM on SOI," *2003 Symposium on VLSI Technology Digest of Tech. Papers*, pp. 63–64, June 2003.
9. T. Ohsawa, T. Higashi, K. Fujita, T. Ikehashi, T. Kajiyama, Y. Fukuzumi, T. Shino, H. Yamada, H. Nakajima, Y. Minami, T. Yamada, K. Inoh and T. Hamamoto,

"A memory using one-transistor gain cell on SOI (FBC) with performance suitable for embedded DRAM's," *2003 Symposium on VLSI Circuits Digest of Tech. Papers*, pp. 93–96, June 2003.

10. T. Shino, T. Higashi, K. Fujita, T. Ohsawa, Y. Minami, T. Yamada, M. Morikado, H. Nakajima, K. Inoh, T. Hamamoto and A. Nitayama, "Highly scalable FBC (Floating Body Cell) with 25nm BOX structure for embedded DRAM applications," *2004 Symposium on VLSI Technology Digest of Technical Papers*, pp. 132–133, June 2004.
11. C. Kuo, T.-J. King and C. Hu, "A capacitorless double-gate DRAM cell," *IEEE Electron Device Lett.*, vol. 23, no. 6, pp. 345–347, June 2002.
12. C. Kuo, T.-J. King and C. Hu, "A capacitorless double-gate DRAM cell design for high density applications," *IEDM Tech. Dig.*, pp. 843–846, Dec. 2002.
13. C. Kuo, T.-J. King and C. Hu, "A capacitorless double-gate DRAM technology for sub-100-nm gate embedded and stand-alone memory applications." *IEEE Trans. Electron Devices*, vol. 50, no. 12, pp. 2408–2416, Dec. 2003.
14. S. Okhonin, M. Nagoga and P. Fazan, "Principles of transient charge pumping on partially depleted SOI MOSFETs," *IEEE Electron Device Lett.*, vol. 23, no. 5, pp. 279–281, May 2002.
15. S. Okhonin, M. Nagoga and P. Fazan, "Transient charge pumping for partially and fully depleted SOI MOSFETs," *IEEE Int. SOI Conf.*, pp. 171–172, Oct. 2002.
16. S. Okhonin, M. Nagoga and P. Fazan, "Charge pumping effects in partially depleted SOI MOSFETs," *IEEE Int. SOI Conf.*, pp. 74–75, Sept. 2003.
17. T. Ohsawa, K. Fujita, K. Hatsuda, T. Higashi, M. Morikado, Y. Minami, T. Shino, H. Nakajima, K. Inoh, T. Hamamoto and S. Watanabe, "An 18.5ns 128Mb SOI DRAM with a floating body cell," *ISSCC Dig. Tech. Papers*, pp. 458–459, Feb. 2005.
18. T. Ohsawa, K. Fujita, K. Hatsuda, T. Higashi, T. Shino, Y. Minami, H. Nakajima, M. Morikado, K. Inoh, T. Hamamoto, S. Watanabe, S. Fujii and T. Furuyama, "Design of a 128-Mb SOI DRAM using the floating body cell (FBC)," *IEEE J. Solid-State Circuits*, vol. 41, no. 1, pp. 135–145, Jan. 2006.
19. A. Villaret, R. Ranica, P. Mazoyer, P. Chandelier, F. Jacquet, S. Cristoloveanu and T. Skotnicki, "Triple-well nMOSFET evaluated as a capacitor-less DRAM cell for nanoscale low-cost & high density applications," *2003 Silicon Nanoelectronics Workshop Abstracts*, pp. 40–41, June 2003.
20. R. Ranica, A. Villaret, P. Malinge, P. Mazoyer, D. Lenoble, P. Candelier, F. Jacquet, P. Masson, R. Bouchakour, R. Fournel, J.P. Schoellkopf and T. Skotnicki, "A one transistor cell on bulk substrate (1T-Bulk) for low-cost and high density eDRAM," *2004 Symposium on VLSI Technology Digest of Technical Papers*, pp. 128–129, June 2004.
21. R. Ranica, A. Vollaret, G. Gasiot, P. Mazoyer, P. Roche, P. Cadelier, F. Jacquet, P. Masson, R. Bouchakour, R. Fournel, J. P. Schoellkopf and T. Skotnicki, "Scaled 1T-bulk devices built with CMOS 90nm technology for low-cost eDRAM

applications," *2005 Symposium on VLSI Technology Digest of Technical Papers*, pp. 38–39, June 2005.

22. A.Villaret, R. Ranica, P. Milinge, P. Masson, B. Martinet, P. Mazoyer, P. Candelier and T. Skotnicki, "Further insight into the physics and modeling of floating-body capacitorless DRAMs," *IEEE Trans. Electron Devices*, vol. 52, no. 11, pp. 2447–2454, Nov. 2005.
23. P. Milinge, P. Candelier, F. Jacquet, S. Martin, R. Ranica, A. Villaret, P. Mazoyer, R. Fournel and B. Allard, "An 8Mb DRAM using a 1TBulk cell," *2005 Symposium on VLSI Circuits Digest of Technical Papers*, pp. 358–361, June 2005.
24. D. Wu, R. Huang, P. Tang, Y. Zhang and U. Wang, "Performance improvement of the capacitorless DRAM cell with quasi-SOI structure based on bulk substrate," *Ext. Abs. Int. Conf. Solid State Devices Materials*, pp. 146–147, Oct. 2009.
25. P. Fazan, S. Okhonin and M. Nagoga, "A new block refresh concept for SOI floating body memories," *IEEE Int. SOI Conf.*, pp. 74–75, Sept. 2003.
26. T. Ohsawa, R. Fukuda, T. Higashi, K. Fujita, F. Matsuoka, T. Shino, H. Furuhashi, Y. Minami, H. Nakajima, T. Hamamoto, Y. Watanabe, A. Nitayama and T. Furuyama, "Autonomous refresh of floating body cell (FBC)," *IEDM Tech. Dig.*, pp. 801–804, Dec. 2008.
27. T. Ohsawa, R. Fukuda, T. Higashi, K. Fujita, F. Matsuoka, T. Shino, H. Furuhashi, Y. Minami, H. Nakajima, T. Hamamoto, Y. Watanabe, A. Nitayama and T. Furuyama, "Autonomous refresh of floating body cell due to current anomaly of impact ionization," *IEEE Trans. Electron Devices*, vol. 56, no. 10, pp. 2302–2311, Oct. 2009.
28. T. Ohsawa, "Semiconductor memory device," *U.S. Patent 6,617,651*; Sep. 2003 (filed in Sept. 2001).
29. R. Ranica, A. Villaret, C. Fenouillet-Beranger, P. Malinge, P. Mazoyer, P. Masson, D. Delille, C. Charbuillet, P. Cadelier and T. Skotnicki, "A capacitor-less DRAM cell on 75nm gate length, 16nm thin fully depleted SOI device for high density embedded memories," *IEDM Tech. Dig.*, pp. 277–280, Dec. 2004.
30. T. Shino, T. Higashi, N. Kusunoki, K. Fujita, T. Ohsawa, N. Aoki, H. Tanimoto, Y. Minami, T. Yamada, M. Morikado, H. Nakajima, K. Inoh, T. Hamamoto and A. Nitayama, "Fully-depleted FBC (Floating body cell) with enlarged signal window and excellent logic process compatibility," *IEDM Tech. Dig.*, pp. 281–284, Dec. 2004.
31. T. Shino, T. Ohsawa, T. Higashi, K. Fujita, N. Kusunoki, Y. Minami, M. Morikado, H. Nakajima, K. Inoh, T. Hamamoto and A. Nitayama, "Operation voltage dependence of memory cell characteristics in fully depleted floating-body cell," *IEEE Trans. Electron Devices*, vol. 52, no. 10, pp. 2220–2226, Oct. 2005.
32. T. Shino, N. Kusunoki, T. Higashi, T. Ohsawa, K, Fujita, K. Hatsuda, N. Ikumi, F. Matsuoka, Y. Kajitani, R. Fukuda, Y. Watanabe, Y. Minami, A. Sakamoto, J. Nishimura, H. Nakajima, M. Morikado, K. Inoh, T. Hamamoto and A. Nitayama, "Floating body RAM technology and its scalability to 32nm node and beyond," *IEDM Tech. Dig.*, pp. 569–572, Dec. 2006.

33. J. G. Fossum, Z. Lu and V. P. Trivedi, "New insights on "capacitorless" floating-body DRAM cells," *IEEE Electron Device Lett.*, vol. 28, no. 6, pp. 513–516, June 2007.
34. I. Ban, U. E. Avci, D. L. Lencke and P. L. D. Chang, "A scaled floating body cell (FBC) memory with high-k+metal gate on thin-silicon and thin-box for 16-nm technology node and beyond," *2008 Symposium on VLSI Technology Digest of Technical Papers*, pp. 92–93, June 2008.
35. U. E. Avci, I. Ban, D. L. Kencke and P. L. D. Chang, "Floating body cell (FBC) memory for 16-nm technology with low variation on thin silicon and 10-nm BOX," *IEEE Int. SOI Conf.*, pp. 29–30, Oct. 2008.
36. E. Yoshida and T. Tanaka, "A design of a capacitorless 1T-DRAM cell using gate-induced drain leakage (GIDL) current for low-power and high-speed embedded memory," *IEDM Tech. Dig.*, pp. 913–916, Dec. 2003.
37. E. Yoshida and T. Tanaka, "A capacitorless 1T-DRAM technology using Gate-Induced Drain Leakage (GIDL) current for low-power and high-speed embedded memory," *IEEE Trans. Electron Devices*, vol. 53, no. 4, pp. 692–697, April 2006.
38. F. Horiguchi, T. Ohsawa, Y. Iwata and T. Yamada, "Semiconductor memory device," *U.S. Patent 6,548,848*, April 2003 (filed in Dec. 2001).
39. T. Tanaka, E. Yoshida and T. Miyashita, "Scalability study of a capacitorless 1T-DRAM:from single-gate PD-SOI to double-gate FinDRAM," *IEDM Tech. Dig.*, pp. 919–922, Dec. 2004..
40. E. Yoshida, T. Miyashita and T. Tanaka, "A study of high scalable DG-FinDRAM," *IEEE Electron Device Lett.*, vol. 26, no. 9, pp. 655–657, May 2005.
41. M. Nagoga, S. Okhonin, C. Bassin, P. Fazan, W. Xiong, C. R. Cleaveli, T. Schulz, K. Schruefer, M. Gostowski, P. Patruno and C. Maleville, "Retention characteristics of Zero-capacitor RAM (Z-RAM) cell based on FinFET and Tri-Gate devices," *IEEE Int. SOI Conf.*, pp. 203–204, Oct. 2005.
42. I. Ban, U. E. Avci, U. Shah, C. E. Barns, D. L. Kencke and P. Chang, "Floating body cell with independently-controlled double gates for high density memory," *IEDM Tech. Dig.*, pp. 573–576, Dec. 2006.
43. H. Jeong, Y. S. Lee, S. Kang, I. H. Park, W. Y. Choi, H. Shin, J. D. Lee and B.-G. Park, "Capacitorless DRAM cell with highly scalable surrounding gate structure," *Ext. Abs. Int. Conf. Solid State Devices Materials*, pp. 574–575, Sept. 2006.
44. M. G. Ertosun, H. Cho, P. Kapur and K. C. Saraswat, "A nanoscale vertical doudle-gate single-transistor capacitorless DRAM," *IEEE Electron Device Lett.*, vol. 29, no. 6, pp. 615–617, June 2008.
45. W. Kwon and T.-J. King Liu, "A highly scalable $4F^2$ DRAM cell utilizing a doubly gated vertical channel," *Ext. Abs. Int. Conf. Solid State Devices Materials*, pp. 142–143, Oct. 2009.
46. M. Bolagojevic, M. Pastre, M. Kayal, P. Fazan, S. Okhonin, M. Nagoga and M. Declercq, "SOI capacitor-less 1-transistor DRAM sensing scheme with automatic reference generation," *2004 Symposium on VLSI Circuits Digest of Technical Papers*, pp. 182–183, June 2004.

47. M. Blagojevic, M. Kayal, M. Pastre, L. Harik, M. J. Declercq, S. Okhonin and P. Fazan, "Capacitorless 1T DRAM sensing scheme with automatic reference generation," *IEEE J. Solid-State Circuits*, vol. 41, no. 6, pp. 1463–1470, June 2006.

48. Y. Minami, T. Shino, A. Sakamoto, T. Higashi, N. Kusunoki, K. Fujita, K. Hatsuda, T. Ohsawa, N. Aoki, H. Tanimoto, M. Morikado, H. Nakajima, K. Inoh, T. Hamamoto and A. Nitayama, "A floating body cell (FBC) fully compatible with 90nm CMOS technology (CMOSIV) for 128Mb SOI DRAM," *IEDM Tech. Dig.*, pp. 307–310, Dec. 2005.

49. T. Ohsawa, T. Higashi, K. Fujita, K. Hatsuda, N. Ikumi, T. Shino, H. Nakajima, Y. Minami, N. Kusunoki, A. Sakamoto, J. Nishimura, T. Hamamoto and S. Fujii, "A 128Mb floating body RAM (FBRAM) on SOI with multi-averaging scheme of dummy cell," *2006 Symposium on VLSI Circuits Digest of Technical Papers*, pp. 224–225, June 2006.

50. T. Hamamoto, Y. Minami, T. Shino, N. Kusunoki, H. Nakajima, M. Morikado, T. Yamada, K. Inoh, A. Sakamoto, T. Higashi, K. Fujita, K. Hatsuda, T. Ohsawa and A. Nitayama, "A floating-body cell fully compatible with 90-nm CMOS technology node for a 128-Mb SOI DRAM and its scalability," *IEEE Trans. Electron Devices*, vol. 54, no. 3, pp. 563–571, March 2007.

51. F. Matsuoka, T. Ohsawa, T. Higashi, H. Furuhashi, K. Hatsuda, K. Fujita, R. Fukuda, N. Ikumi, T. Shino, Y. Minami, H. Nakajima, T. Hamamoto, A. Nitayama and Y. Watanabe, "FBC's Potential of $6F^2$ single cell operation in multi-Gbit memories confirmed by a newly developed method for measuring signal sense margin," *IEDM Tech. Dig.*, pp. 39–42, Dec. 2007.

52. K. Hatsuda, K. Fujita and T. Ohsawa, "A 333MHz random cycle DRAM using the floating body cell," *Proc. IEEE 2005 Custom Integrated Circuit Conf.*, pp. 259–262, Sept. 2005.

53. K. Fujita, T. Ohsawa, R. Fukuda, F. Matsuoka, T. Higashi, T. Shino and Y. Watanabe, "Array architecture of floating body cell (FBC) with quasi-shielded open bit line scheme for sub-40nm node," *IEEE Int. SOI Conf.*, pp. 31–32, Oct. 2008.

54. F. Morishita, H. Noda, T. Gyohten, M. Okamoto, T. Ipposhi, S. Maegawa, K. Dosaka and K. Arimoto, "A capacitorless twin-transistor random access memory (TTRAM) on SOI," *Proc. IEEE 2005 Custom Integrated Circuit Conf.*, pp. 435–438, Sept. 2005.

55. K. Arimoto, F. Morishita, I. Hayashi, T. Gyohten, H. Noda, T. Ipposhi and K. Dosaka, "A configurable enhanced T^2RAM macro for system-level power management unified memory," *2006 Symposium on VLSI Circuits Digest of Technical Papers*, pp. 182–183, June 2006.

56. F. Morishita, I. Hayashi, T. Gyohten, H. Noda, T. Ipposhi, H. Shimano, K. Dosaka and K. Arimoto, "A configurable enhanced TTRAM macro for system-level power management unified memory," *IEEE J. Solid-State Circuits*, vol. 42, no. 4, pp. 853–861, April 2007.

57. K. Arimoto, F. Morishita, I. Hayashi, K. Dosaka, H. Shimano and T. Ipposhi, "A high-density scalable twin transistor RAM (TTRAM) with verify control for SOI platform memory IPs," *IEEE J. Solid-State Circuits*, vol. 42, no. 11, pp. 2611–2619, Nov. 2007.

58. J. G. Fossum, Z. Lu, W. Zhang, V. P. Trivedi, L. Mathew and M. Sadd, "A novel two-transistor floating-body memory cells," *IEEE Int. SOI Conf.*, pp. 105–106, Oct. 2007.

59. Z. Lu, J. G. Fossum, W. Zhang, V. P. Trivedi, L. Mathew and M. Sadd, "A novel two-transistor floating-body/gate cell for low-power nanoscale embedded DRAM," *IEEE Trans. Electron Devices*, vol. 55, no. 6, pp. 1511–1518, June 2008.

60. Z. Lu, J. G. Fossum, J.-W. Yang, H. R. Harris, V. P. Trivedi, M. Chu and S. E. Thompson, "A simplified superior floating-body/gate DRAM cell," *IEEE Electron Device Lett.*, vol. 30, no. 3, pp. 282–284, March 2009.

61. P.-F. Wang and Y. Gong, "A novel 4.5F^2 capacitorless semiconductor memory device," *IEEE Electron Device Lett.*, vol. 29, no. 12, pp. 1347–1348, Dec. 2008.

62. D. Fisch, R. Beffa and C. Bassin," Soft error performance of Z-RAM floating body memory," *IEEE Int. SOI Conf.*, pp. 111–112, Oct. 2006.

63. N. Kameshiro, T. Watanabe, T. Ishii, T. Mine, T. Sano, H. Ibe, S. Akiyama, K. Yanagisawa, T. Ipposhi, T. Iwamatsu and Y. Takahashi, "A fully logic-compatible, 3-transistor, SESO-memory cell featuring 0.1-FIT/Mb soft error, 100-MHz random cycle, and 100-ms retention," *2008 Symposium on VLSI Circuits Digest of Technical Papers*, pp. 122–123, June 2008.

64. H. Furuhashi, T. Shino, T. Ohsawa, F. Matsuoka, T. Higashi, Y. Minami, H. Nakajima, K. Fujita, R. Fukuda, T. Hamamoto and A. Nitayama, "Scaling scenario of floating body cell (FBC) suppressing Vth variation due to random dopant fluctuation," *IEEE Int. SOI Conf.*, pp. 33–34, Oct. 2008.

65. T. Shino, T. Ohsawa, T. Hamamoto and A. Nitayama, "Overview and future challenges of floating body RAM technologies," *Ext. Abs. Int. Conf. Solid State Devices Materials*, pp. 226–227, Sept. 2007.

66. N. Z. Butt and M. A. Alam, "Scaling limit of double-gate and surround-gate Z-RAM cells," *IEEE Trans. Electron Devices*, vol. 54, no. 9, pp. 2255–2262, Sept. 2007.

67. D. Fisch, "Innovative approach to drive floating body Z-RAM embedded memory to 32nm and beyond," *IEEE Int. SOI Conf.*, pp. 101–102, Oct. 2007.

68. T. Hamamoto and T. Ohsawa, "Overview and future challenges of floating body RAM (FBRAM) technology for 32nm technology node and beyond," *Proc. 38th European Solid-State Device Research Conf.*, pp. 25–29, Sept. 2008.

69. T. Hamamoto and T. Ohsawa, "Overview and future challenges of floating body RAM (FBRAM) technology for 32 nm technology node and beyond," *Solid-State Electronics*, vol. 53, pp. 676–683, July 2009.

70. S. Okhonin, M. Nagoga, E. Carman, R. Beffa and E. Faraoni, "New generation of Z-RAM," *IEDM Tech. Dig.*, pp. 925–928, Dec. 2007.

71. S. Okhonin, M. Nagoga, C.-W. Lee, J.-P. Colinge, A. Afzalian, R. Yan., N. D. Akhavan, W. Xiong, V. Sverdlov, S. Selberherr and C. Mazure, "Ultra-scaled Z-RAM cell," *IEEE Int. SOI Conf.*, pp. 157–158, Oct. 2008.

72. K.-W. Song, H. Jeong, J.-W. Lee, S. I. Hong, N.-K. Tak, Y.-T. Kim, Y. L. Choi, H. S. Joo, S. H. Kim, H. J. Song, Y. C. Oh, W.-S. Kim, Y.-T. Lee, K. Oh and C. Kim, "55nm capacitor-less 1T DRAM cell transistor with non-overlap structure," *IEDM Tech. Dig.*, pp. 797–800, Dec. 2008.

73. H. J. Bae, H. J. Song, Y. L. Choi, S. H. Kim, S. I. Hong, C. W. Oh, D.-W. Kim, G. Jin, K. S. Oh and W.-S. Lee, "Evaluation of 1T RAM using various operation modes with SOONO (Silicon-On-ONO) device," *IEDM Tech. Dig.*, pp. 805–808, Dec. 2008.

74. Z. Zhou, J. G. Fossum and Z. Lu, "Physical insights on BJT-based 1T DRAM cells," *IEEE Electron Device Lett.*, vol. 30, no. 5, pp. 565–567, May 2009.

75. N. Collaert, M. Rosmeulen, M. Rakowski, R. Rooyackers, L. Witters, A. Veloso, J. Van Houdt and M. Jurczak, "Comparison of scaled floating body RAM architecture," *IEEE Int. SOI Conf.*, pp. 35–36, Oct. 2008.

76. C. W. Oh, N. Y. Kim, H. J. Song, S. I. Hong, S. H. Kim, Y. L. Choi, H. J. Bae, D. U. Choi, Y. S. Lee, D.-W. Kim, D. Park and B.-I. Ryu, "Floating body DRAM characteristics of Silicon-On-ONO (SOONO) devices for system-on-chip (SoC) applications," *2007 Symposium on VLSI Technology Digest of Technical Papers*, pp. 168–169, June 2007.

77. E. J. Yun, H. J. Song, S. I. Hong, S. H. Kim, Y. L. Choi, H. J. Bae, N. Y. Kim, C. W. Oh, D.-W. Kim and D. Park, "Analysis of sensing margin in Silicon-On-ONO (SOONO) device for the capacitor-less RAM applications," *IEEE Int. SOI Conf.*, pp. 103–104, Oct. 2007.

78. J.-W. Han, S.-W. Ryu, C. Kim, S. Kim, M. Im, S. J. Choi, J. S. Kim, K. H. Kim, G. S. Lee, J. S. Oh, M. H. Song, Y. C. Park, J. W. Kim and Y.-K. Choi, "An unified-RAM (URAM) cell for multi-functioning capacitorless DRAM and NVM," *IEDM Tech. Dig.*, pp. 929–932, Dec. 2007.

79. J.-W. Han, S.-W. Ryu, S. Kim, C.-J. Kim, J.-H. Ahn, S.-J. Choi, K. J. Choi, B. J. Cho, J. S. Kim, K. H. Kim, G. S. Lee, J. S. Oh, M. H. Song, Y. C. Park, J. W. Kim and Y.-K., Choi, "Band offset FinFET-based URAM (Unified-RAM) built on SiC for multi-functional NVR and capacitorless 1T-DRAM," *2008 Symposium on VLSI Technology Digest of Technical Papers*, pp. 200–201, June 2008.

80. J. -W, S.-W. Ryu, C.-J. Kim, S. Kim, M. Im, S. J. Choi, J. S. Kim, K.H.Kim, G. S. Lee, J. S. Oh, M. H. Song, Y. C. Park, J. W. Kim and Y.-K. Chol, "Partially depleted SONOS FinFET for unified RAM (URAM)—Unified function for high-speed IT DRAM and nonvolatile memory," IEEE Electron Device Lett, vol. 29, no. 7, pp. 781-783, July 2008.

81. J.-W. Han, S.-W. Ryu, S. Kim, C.-J. Kim, J.-H. Ahn, S.-J. Choi, J. S. Kim, K. H. Kim, G. S. Lee, J. S. Oh, M. H. Song, Y. C. Park, J. W. Kim and Y.-K. Choi, "A bulk finFET unified-RAM (URAM) cell for multifunctioning NVM and capacitorless 1T-DRAM," *IEEE Electron Device Lett.*, vol. 29, no. 6, pp. 632–634, June 2008.

82. S.-J. Choi, J.-W. Han, S. Kim, D.-H. Kim, M.-G. Jang, J.-H. Yang, J. S. Kim, K. H. Kim, G. S. Lee, J. S. Oh, M. H. Song, Y. C. Park, J. W. Kim and Y.-K. Choi, "High speed flash memory and 1-T DRAM on dopant segregated Schottky barrier (DSSB) Fin FET SONOS device for multi-functional SoC applications," *IEDM Tech. Dig.*, pp. 223–226, Dec. 2008.

83. J.- W. Han, S. W. Ryu, S. Kim, C.-J. Kim, J.-H. Ahn, S.-J. Choi, K. J. Choi, B. J. Cho, J. S. Kim, K. H. Kim, G. S. Lee, J. S. Oh, M. H. Song, Y. C. Park, J. W. Kim and Y.-K. Choi, "Energy band engineered unified-RAM (URAM) for multi-functioning 1T-DRAM and NVM," *IEDM Tech. Dig.*, pp. 227–230, Dec. 2008.

84. J.-W. Han, S.-W. Ryu, S.-J. Choi and Y. K. Choi, "Gate-induced drain leakage (GIDL) programming method for soft-programming-free operation in unified RAM (URAM)," *IEEE Electron Device Lett.*, vol. 30, no. 2, pp. 189–191, Feb. 2009.

85. S.-W. Ryu, J.-W. Han, C.-J. Kim, S.-J. Choi, S. Kim, J.-S. Kim, K. H. Kim, J.-S. Oh, M.-H. Song, G.-S. Lee, Y. C. Park, J. W. Kim and Y.-K. Choi, "Refinement of unified random access memory," *IEEE Trans. Electron Devices*, vol. 56, no, 4, pp. 601–608, April 2009.

86. J.-W. Han, S.-W. Ryu, C.-J. Kim, S.-J. Choi, S. Kim, J.-H. Ahn, D.-H. Kim, K. J. Choi, B. J. Cho, J.-S. Kim, K. H. Kim, G.-S. Lee, J.-S. Oh, M.-H. Song, Y.-C. Park, M.-H. Song, Y. C. Park, J. W. Kim and Y.-K. Choi, "Energy-band-engineered unified-RAM (URAM) cell on buried $Si_{1-y}C_y$ substrate for multifunctioning flash memory and 1T-DRAM," *IEEE Trans. Electron Devices*, vol. 56, no. 4, pp. 641–647, April 2009.

87. J.-W. Han, C.-J. Kim, S.-J. Choi, D.-H. Kim, D.-I. Moon and Y.-K. Choi, "Gate-to-source/drain nonoverlap device for soft-program immune unified RAM (URAM)," *IEEE Electron Device Lett.*, vol. 30, no. 5, pp. 544–546, May 2009.

88. J.-W. Han, D.-I. Moon, D.-H. Kim and Y.-K. Choi, "Parasitic BJT read method for high-performance capacitorless 1T-DRAM mode in unified RAM," *IEEE Electron Device Lett.*, vol. 30, no. 10, pp. 1108–1110, Oct. 2009.

89. J.-T. Lin, K.-D. Huang and B.-T. Jheng, "Performances of a capacitorless 1T-DRAM using polycrystalline silicon thin-film transistors with trenched body," *IEEE Electron Device Lett.*, vol. 29, no. 11, pp. 1222–1225, Nov. 2008.

90. J.-W. Han, S.-W. Ryu, D.-H. Kim, C.-J. Kim, S. Kim, D.-I. Moon, S.-J. Choi and Y.-K. Choi, "Fully depleted polysilicon TFTs for capacitorless 1T-DRAM," *IEEE Electron Device Lett.*, vol. 30, no. 7, pp. 742–744, July 2009.

91. M. G. Ertosun, P. Kauper and K. C. Saraswat, "A highly scalable capacitorless double gate quantum well single transistor DRAM: 1T-QW DRAM," *IEEE Electron Device Lett.*, vol. 29, no. 12, pp. 1405–1407, Dec. 2008.

92. P. Tang, D. Wu, L. Zhang and R. Huang, "Highly scalable capacitorless DRAM cell on thin-body with band-gap engineered source and drain," *Ext. Abs. Int. Conf. Solid State Devices Materials*, pp. 144–145, Oct. 2009.

93. A. Singh, M. Ciraula, D. Weiss, J. Wuu, P. Bauser, P. de Champs, H. Daghighian, D. Fisch, P. Graber and M. Bron, "A 2ns-read-latency 4Mb embedded floating-body memory macro in 45nm SOI technology," *ISSCC Dig. Tech. Papers*, pp. 460–461, Feb. 2009.

94. T.-S. Jang, J.-S. Kim, S.-M. Hwang, Y.-H. Oh, K.-M. Rho, S.-J. Chung, S.-O. Chung, J.-G. Oh, S. Bhardwaj, J. Kwon, D. Kim, M. Nagoga, Y.-T. Kim, S.-Y. Cha, S.-C. Moon, S.-W. Chung, S.-J. Hong and S.-W. Park, "Highly scalable Z-RAM with remarkably long data retention for DRAM application," *2009 Symposium on VLSI Technology Digest of Technical Papers*, pp. 234–235, June 2009.

95. A. Nayfeh, V. Koldyaev, P. Beaud, M. Nagoga and S. Okhonin, "A leakage current model for SOI based floating body memory that includes the Poole-Frenkel effect," *IEEE Int. SOI Conf.*, pp. 75–76, Oct. 2008.

96. F. Nemati and J. D. Plummer, "A novel high density, low voltage SRAM cell with a vertical NDR device," *1998 Symposium on VLSI Technology Digest of Technical Papers*, pp. 66–67, June 1998.

97. F. Nemati and J. D. Plummer, "A novel thyrister-based SRAM cell (T-RAM) for high-speed, low-voltage, giga-scale memories," *IEDM Tech. Dig.*, pp. 283–286, Dec. 1999.

98. F. Nemati, H.-J. Cho, S. Robins, R. Gupta, M. Tarabbia, K. J. Yang, D. Hayes and V. Gopalakrishnan, "Fully planar 0.56mm^2 T-RAM cell in a 130nm SOI CMOS logic technology for high-density high-performance SRAMs," *IEDM Tech. Dig.*, pp. 273–276, Dec. 2004.

99. M. Ershov, F. Nemati, R. Gupta, V. Gopalakrishnan, R. Gooty, M. Tarabbia, K. J. Yang, S. Banna, D. Hayes, H. J. Cho and S. Robins, "Optimization of substrate doping for back-gate control in SOI T-RAM memory technology," *IEEE Int. SOI Conf.*, pp. 215–216, Oct. 2005.

100. R. Roy, F. Nemati, K. Young, B. Bateman, R. Chopra, S.-O. Jung, C. Show and H.-J. Cho, "Thyristor-based volatile memory in nano-scale CMOS," *ISSCC Dig. Tech. Papers*, pp. 632–633, Feb. 2006.

101. K. J. Yang, R. N. Gupta, S. Banna, F. Nemati, H.-J. Cho, M. Ershov, M. Tarabbia, D. Hayes and S. T. Robins, "Optimization of nanoscale thyristors on SOI for high-performance high-density memories," *IEEE Int. SOI Conf.*, pp. 113–114, Oct. 2006.

102. T. Sugizaki, M. Nakamura, M. Yanagita, M. Honda, M. Shinohara, T. Ikuta, T. Ohchi, K. Kugimiya, R. Yamamoto, S. Kanda, I. Yamamura, K. Yagami and T. Oda, "Ultra high-speed novel bulk thyristor-SRAM (BT-RAM) cell with selective epitaxy anode (SEA)," *IEDM Tech. Dig.*, pp. 157–160, Dec. 2006.

103. T. Sugizaki, M. Nakamura, M. Yanagita, M. Shinohara, T. Ikuta, T. Ohchi, K. Kugimiiya, S. Kanda, K. Yagami and T. Oda, "Low voltage/sub-ns operation bulk thyristor-SRAM (BT-RAM) cell with double selective epitaxy emitters (DEE)," *2007 Symposium on VLSI Technology Digest of Technical Papers*, pp. 170–171, June 2007.

104. T. Sugizaki, M. Nakamura, M. Yanagita, M. Shinohara, T. Ikuta, T. Ohchi, K. Kugimiya, S. Kanda, K. Yagami and T. Oda, "Extremely low-voltage and high-speed operation bulk thyristor-SRAM/DRAM (BT-RAM) cell with triple selective layers (TEL)," *IEDM Tech. Dig.*, pp. 933–936, Dec. 2007.

105. T. Sugizaki, M. Nakamura, M. Yanagita, M. Shinohara, T. Ikuta, T. Ohchi, K. Kugimiya, S. Kanda, K. Yagami and T. Oda, "35-nm gate-length and ultra low-voltage (0.45V) operation bulk thyrirtor-SRAM/DRAM (BT-RAM) cell with triple epitaxy layers (TELs)," *2008 Symposium on VLSI Technology Digest of Technical Papers*, pp. 198–199, June 2008.

106. H.-J. Cho, F. Nemati, R. Roy, R. Gupta, K. Yang, M. Ershov, S. Banna, M. Tarabbia, C. Salling, D. Hayes, A. Mittal and S. Robins, "A novel capacitor-less DRAM cell using thin capacitively-coupled thyristor (TCCT)," *IEDM Tech. Dig.*, pp. 321–324, Dec. 2005.

107. S. Slesazeck, "Integrated circuit comprising a thyristor and method of controlling a memory cell comprising a thyristor," *US Patent Application Publication* US 2009/0213648, Aug. 2009 (filed in Dec. 2008)

108. H.-J. Cho and M.-R. Lin, "Novel DRAM cell with amplified capacitor for embedded application," *IEDM Tech. Dig.*, pp. 263–266, Dec. 2009.

Index